KEVIN MORENO

UUMC NEUROPATHOLOGY SECTION

To Kevin

It's great to meet an MD. with a biochemistry back-ground. Good luck in your MRS future

[illegible signature]

8 Dec 2001

All good wishes for your future in Clinical MRS

Sincerely

[illegible signature]

Paradise 12/9/01

Magnetic Resonance Spectroscopy Diagnosis of Neurological Diseases

Magnetic Resonance Spectroscopy Diagnosis of Neurological Diseases

Else Rubæk Danielsen

Brian Ross

Huntington Medical Research Institute
Pasadena, California

MARCEL DEKKER, INC. NEW YORK • BASEL

Libarary of Congress Cataloging-in-Publication Data

Danielsen, Else Rubæk
Magnetic resonance spectroscopy diagnosis of neurological diseases /Else Rubæk Danielsen, Brian Ross
p. cm.
Includes index.
ISBN: 0-8247-0238-7
1. Nuclear magnetic resonance spectroscopy—Diagnositic use. 2. Brain—Diseases—diagnosis. I. Ross, Brian. II. Title.
[DNLM: 1. Brain Diseases—diagnosis. 2. Nuclear Magnetic Resonance. 3. Brain Chemistry. 4. Brain Diseases—pathology. WL 348 D186m 1998]
RC386.6.N83D36 1998
616.8′047548—dc21
DNLM/DLC
for Library of Congress 98-44756
CIP

This book is printed on acid-free paper.

Headquarters
Marcel Dekker, Inc.
270 Madison Avenue, New York, NY 10016
tel: 212-696-9000; fax: 212-685-4540

Eastern Hemisphere Distribution
Marcel Dekker AG
Hutgasse 4, Postfach 812, CH-4001 Basel, Switzerland
tel: 44-61-261-8482; fax: 44-61-261-8896

World Wide Web
http://www.dekker.com

The publisher offers discounts on this book when ordered in bulk quantities. For more information, write to Special Sales/Professional Marketing at the headquarters address above.

Current printing (last digit):
10 9 8 7 6 5 4 3 2

PRINTED IN THE UNITED STATES OF AMERICA

Preface

Who is this book for? Many thousands of patients around the world undergo MRS examinations of the brain. MRS is "high-tech" medicine, and there is a serious lack of clinically relevant books to which physicians and others outside the small field of magnetic resonance physics may turn for guidance. We conceived the idea of an atlas of common neuropathologies as reflected in the MRS that could fill this gap for the busy clinician/neurologist. This book is the result: some 72 clinical cases, and well over 100 spectra are presented in a standard format with only a minimum of technical information. With an eye to practical needs, the maximum of help in reading and interpreting each spectrum is provided in the associated text. Clinical spectra covering most of the common (and some not-so-common) neurological entities are shown. Reading this book, and then acquiring and reading some MRS studies of patients on their own should allow all clinicians to confidently read MRS studies as they appear in the literature and in their own Clinical practices. Finally, for the practicing neuroradiologist faced with the task of writing a report for the use of the referring doctor and for the insurance provider (thereby ensuring reimbursement for their work), one reading of this book should establish a routine and enhance the intellectual and diagnostic interest of this newly expanding branch of their specialty.

The origins of this book are to be found in a series of clinical magnetic resonance spectroscopy courses run by the staff of the Huntington Medical Research Institute's (HMRI) MRS Unit in Pasadena, California between 1995 and 1997. Case studies assembled for that tutorial-style course proved educational to all, participants and teachers alike, and provided the idea for a teaching manual that would be less ephemeral than patient files and MRIs on X-ray film. Several of the often repeated difficulties encountered during these courses have been taken into account and incorporated in a method of reading each spectrum that, we hope, minimizes errors, highlights artifacts and maximizes useable clinical information.

However, even earlier than the Huntington courses, a weekly clinical readout of brain MRS modeled on the usual medical grand round was conducted for the benefit of referring physicians (mostly, but not exclusively, neurologists or neurosurgeons), staff radiologists, surgical and medical residents of affiliated hospitals, and HMRI research fellows. This organization, which came into existence at HMRI during the 1992 Presidential campaign adopted a name unmistakably that of the soon-to-be White House cat—Socks! SoCCSS, the Southern California Clinical Spectroscopy Service, remains in existence to provide a forum for sharing clinical interpretation and advice among many MRS proponents throughout southern California and the world. A number of the instructive cases assembled in this book come from this source. This fact alone should convince the skeptic that MRS can be a rigorous diagnostic tool. The message we wish to convey (beyond the central one, that there is valuable diagnostic information in the noninvasive biochemical analysis of the brain) is that when applied in a rigorous way, proton MRS achieves greater diagnostic power than might have been expected. The bold diagnostic patterns are recognized by everyone, and even survive technical limitations of less-than-optimal MR scanners. But a new level of confidence is achieved as the neurologist sees how robustly reproducible—even across state and international boundaries—are even the finest details of the MR brain spectrum. At this point, MRS has achieved the status of a routine medical screening test, in which the known and the hitherto unknown features of the scan can be incorporated in the physician's decision tree to arrive at a new, more certain, or a more rapid diagnosis. Throughout this book there are clinical brain spectra, the initial meaning of which was not immediately clear, but which became so only after several days or weeks and after highly focused consideration of other clinical or laboratory data available in the workup of the patient. This exciting process stimulated many of us to share our approach with our colleagues in medicine. We hope this will expand the number of patients to whom MRS is offered and also will more deeply involve experts from neurological disciplines and beyond who through their better understanding of the variations of clinical neurological disease, can stretch the clinically useful interpretation of MRS beyond the somewhat limited range we offer here.

We thought hard about the format of this book and are particularly grateful for the encouragement and ideas offered by Graham Garratt and Kerry Doyle at Marcel Dekker, Inc. After consulting many colleagues, in the end we have taken a highly didactic approach. Hopefully this will not unduly offend those already expert. Our hope is that by the simple-minded approach, anyone will become comfortable reading a brain MRS; no one will be put off by the mystique of peaks or the imagined complexity of biochemistry and thereby become inhibited from discovering the diagnostic value of the scan. Instead we hope the MR spectrum will become what it actually is, a fingerprint of brain chemistry, which is to be read for its diagnostic value. Normal and abnormal will be instantly recognized by the clinician, much as the abnormal EKG, protein electrophoretic strip, and M-mode cardiac ultrasound are currently recognizable. Once this is routine, MR spectra should hold no terrors.

Else Rubæk Danielsen
Brian Ross

Acknowledgments

Thanks are due to many who helped in the production of this book: first, former and current colleagues at Huntington Medical Research Institutes, who have contributed in many ways, in particular by acquiring MR spectra, but also by providing invaluable suggestions, and assisting in general (in alphabetical order: Stefan Blüml, David Dubowitz, Thomas Ernst, Neil Farrow, Angela Geissler, Luke Haseler, Tuan Hoang, Keiko Kanamori, Roland Kreis, Jung-Hee Lee, Jennifer Mandigo, Thomas Michaelis, Rex Moats, Mary Munoz, Truda Shonk, Sandra de Silva, Jeannie Tan, John Videen). Nationally and internationally, there has been a necessary engineering effort to sustain MRS capability on clinical MR scanners over many years even when its ultimate clinical utility was in doubt. These people, mentioning some but not by any means all, showed faith in their own MR physics skills, when financial imperatives often indicated otherwise (in alphabetical order: David Gurr, Ralph Hurd, Peter Luyten, Peter Martin, James Murdoch, Thomas Raidy, Pom Salasuta, Rolf Sauter, James Tropp, Peter Webb, Wicklow); teachers of several MRS Courses (in alphabetical order: Edguardo Arcinue, Emanuel Cabanis, Russell Caldwell, David Dubowitz, Patricia Gifford, Luke Haseler; Bruce Miller, Rex Moats, Joseph Norfray, Truda Shonk (who also acquired a large number of the spectra from HMRI), Leslie Watson (who also acquired all spectra from Rancho Los Amigos); students who attended courses and taught all of us a great deal. This includes MR technologists, physicians, MR field-service engineers, and MR physicists, all of whom brought some special expertise to our handling of the basic MRS examination and quality assurance.

Ms. Maureen Brooks deserves special mention for providing reimbursement seminars, through which very few of the participants could afford to doze! This attention was rewarded in 1998 by the issuance of the first CPT code for clinical magnetic resonance spectroscopy; this, as every American reader will be aware, is an accolade that every new medical technology needs in order to be accepted into routine clinical practice. Without any note of cynicism, if MRS were not so recognized, it could hardly survive in the face of intense competition from a host of neurological diagnostic tools available to the modern physician.

Very special thanks are due to Dr. Roland Kreis, Boswell Fellow at HMRI and California Institute of Technology (1991–1993), who implemented the first quantitative localized MRS pulse sequences, obtained the first data in healthy volunteer of all ages, and initiated a vast number of the clinical studies upon which this book relies so heavily. A large number of figures (see below) are taken from the published works of Dr. Kreis.

Also deserving special thanks are Dr. Stefan Blüml, Research Fellow at HMRI and the Schulte Research Foundation; Dr. Thomas Michaelis, Boswell Fellow at HMRI and California Institute of Technology (1994–1995); and

Dr. Thomas Ernst, Research Fellow at HMRI, who all have been invaluable sources for discussion and support during the preparation of the book and in the years before, as well as having contributed to maintaining the clinical MRS program.

Faith in the future of clinical MRS has been constant and reflected in encouragement of the Executive Director (Dr. W. Opel), chairman and members of the Board of Trustees of HMRI.

Acknowledgments

Several of the figures shown were produced at HMRI by or in part by the following: Kreis: 2.11, 2.12, 3.2, 3.3, 3.4, 3.22, 3.34, 4.2, 4.3, 6.2D, 6.3, 6.4 panel (i), 6.5B; Blüml: 3.24, 3.6, 6.6, 6.7; Ernst: 3.27, 3.31, 4.2, 4.3, 6.1, 6.2, 6.3, 6.6, 6.5B; Haseler: 3.24, 6.6, 6.7; Mandigo: 2.3, 3.2, 3.9, 4.1G; Michaelis: 6.5A; Moats: 6.1; Shonk: 6.1, 6.2.

We are grateful for permission to reproduce the following published figures (other acknowledgements are provided in the text):

Academic Press (Magn. Reson. Med.): 2.12
Williams & Wilkins (Magn. Reson. Med.): 3.36, 6.4(i)
Society of Magnetic Resonance in Medicine: 3.1
John Wiley & Sons (NMR in Biomed.): 3.2, 3.14, 3.18
Eur. J. Radiol.: 3.4, 6.2D
RSNA: 3.4, 6.1 (modified), 6.2A, 6.8
(JCAT): 3.5, 3.8
Based on Agarwal: 3.7 and 3.17
J Biochem. (modified): 3.10
Stockler et al. : 3.15
Moffett et al.: 3.19
Pediatrics: 3.24, 6.7
Saunders (or modified from): 3.31, 6.2C, 6.4(ii–iv)
Keusler et al.: 3.32
Istituto Superiore di Sanita, Rome, Italy, 1994 (Eurospin Annual): 3.33
Am. J. Neuroradiol.: 6.1B
Lancet: 6.2B
Raven Press (MR Quarterly): 6.3A, 6.5A
Cambridge University Press: 6.3B, 6.5B,C
J. Clin. Invest.: 6.3C,D
modified from JMRI: 6.6.

Contents

Preface *iii*
Acknowledgments *v*

CHAPTER 1: INTRODUCTION TO MAGNETIC RESONANCE SPECTROSCOPY 1
I. A Brief History of Clinical Magnetic Resonance Spectroscopy 1
II. Motivation 2
A. A Working Knowledge of Brain Chemistry in Medicine 2
B. Clinical Indications for MRS 2
C. How to Read and Interpret the MR Spectrum 2
III. A More Detailed Look at MRS: The Single Volume of Interest 3
IV. Philosophy of Single Volume of Interest MRS 3
V. Method of Single VOI Adopted in This Book 4
VI. Three Simple Rules for Understanding an MR Spectrum of the Brain 4
A. Rule One: Spectra Are Read in Terms of Peak Ratios 4
B. Rule Two: Uniform Clinical Methods Yield Identical Spectra 4
C. Rule Three: MRS Is a Quantitative Technique 4
Further Reading 4

CHAPTER 2: BASIC PHYSICS OF MRS 5
I. Introduction 5
A. What Is an MRI? 5
B. What Is an MR Spectrum? 5
II. Measurement Protocol Choices 6
A. What Do We Mean by Localization? 6
B. STEAM and PRESS 7
C. Choice of STEAM Versus PRESS 7
D. How Does the Echo Time Affect the Spectrum? 7
E. How Does the Repetition Time Affect the Spectrum? 8

F. *Total Number of Scans and Signal to Noise: What Is Signal-to-Noise and How Does Total Number of Scans Affect It?* 8
G. *Box Size and S/N: How Does the Box Size Affect S/N and the Quality of the Spectrum?* 9
H. *Signal to Noise: A Summary of Controlling Factors* 10
III. Automated or Manual Adjustments Preceding the Scan 10
A. *Adjustments of Frequency, Transmitter, and Receiver* 10
B. *Optimizing the Magnetic Field Homogeneity: "Shimming"* 10
C. *Adjusting Water Suppression* 10
IV. Quantitation 11
A. *Numerical Analysis of the MR Spectrum* 11
B. *Ratios Relative to Creatine* 12
C. *Normal Metabolite Ratios* 12
D. *Absolute Quantitation* 12
E. *Normal Absolute Concentrations* 20
V. Summary 20
Further Reading 22

CHAPTER 3: THE CLINICAL SIGNIFICANCE OF METABOLITES 23
I. Biochemistry: Back to the Future 23
II. The New Neurochemistry 23
A. *What Is Neurospectroscopy?* 23
B. *MRS Introduces the New Neurochemistry* 24
III. Common Peak Assignments 24
A. *Pitfalls in Peak Assignment* 26
B. *Understanding Brain Chemistry* 26
C. *Brain Metabolism and the ABCs of Energy Production* 27
1. Creatine 32
2. Why Lactate? 32
3. *Myo*-inositol 34
4. N-Acetylaspartate: The *sine qua non* 34
5. Glutamate and Glutamine: Neurotransmitters 35
6. Lipids 36
7. Choline 36
IV. Classic Patterns in MRS Diagnosis 38
A. *Brain or Cerebrospinal Fluid* 38
B. *MRS of Neuronal Injury* 38
C. *MRS of Neurodegenerative Disease* 38
D. *Hypoxic-ischemic Cascade* 39
E. *Basic Neurochemistry of Tumors* 39
F. *Necrosis, Gliosis, and Other Common Patterns of Brain Injury* 41
G. *Brain Swelling, Edema, and Atrophy* 42
H. *Impact of Systemic Disease on the Brain* 42
Further Reading 43

CHAPTER 4: AN INTRODUCTION TO CLINICAL CASES 45
I. A Step-by-Step Guide to Reporting MRS 45
II. Case Example: Parietal White and Occipital Grey Matter Spectra in the Normal Adult 45
A. *Purpose of MRS* 45
B. *Summary of Patient's History* 46
C. *MRS Method, Volume of Interest, and Comments about Examination* 46
D. *Location 1 (White Matter)* 46
1. Peak Identification and Quality Assurance: Reading the Spectrum from Right to Left 46
a. N-Acetylaspartate (NAA) 47

b. Beta and Gamma Glutamine and Glutamate (β,γ-Glx) 49
c. Second NAA Peak 49
d. Creatine and Phosphocreatine (Cr) 50
e. Choline (Cho) 50
f. *Scyllo*-inositol (sI) and Glucose 50
g. *Myo*-inositol (mI) 50
h. Alpha Glx (α-Glx) 50
i. Second Cr Peak 50
j. Second mI Peak 50
2. Interpretation and Analysis 50
E. Location 2 (Grey Matter) 51
1. Peak Identification and Quality Assurance: Reading the Spectrum from Right to Left 51
2. Interpretation and Analysis 52
F. Conclusion 52
G. Summary of the Patient's History Post MRS 52
H. Other Comments 52

CHAPTER 5: CLINICAL CASES 53
5.1 HEALTHY SUBJECTS: NORMAL PROTON SPECTRA OF ALL AGES 53
5.1.1 Normal Premature, Gestational Age 35 Weeks 53
5.1.2 Normal Premature, Gestational Age 36 Weeks 57
5.1.3 Normal Baby 59
5.1.4 Normal Child 61
5.1.5 Normal Adult 63
5.1.6 Normal PROBE: Adult 65
5.1.7 Normal Young Adult 67
5.1.8 Normal Elderly 70
5.1.9 Normal MRS: Elderly Adult 72
5.2 DEGENERATIVE DISEASE: ALZHEIMER'S DISEASE, DEMENTIA, CENTRAL PONTINE MYELINOLYSIS 74
5.2.1 Alzheimer's Disease 74
5.2.2 Rule Out Alzheimer's Disease: Negative 77
5.2.3 Central Pontine Myelinolysis 80
5.2.4 Alzheimer's Disease Versus Other Dementia 83
5.2.5 Rule Out Alzheimer's Disease 86
5.3 SYSTEMIC DISEASES: LIVER DISEASE, DIABETES MELLITUS, OSMOLAR ABNORMALITIES 92
5.3.1 Liver Disease, Severe Hepatic Encephalopathy 92
5.3.2 Liver Disease 95
5.3.3 Rule Out Subclinical Hepatic Encephalopathy 97
5.3.4 Liver Disease: Transplantation 99
5.3.5 Liver Disease, Pre- and Post-TIPS, and Pre- and Postliver Transplantation 103
5.3.6 Hepatic Encephalopathy Versus Alzheimer's Disease 108
5.3.7 Liver Disease: Portal Hypertension 111
5.3.8 Korsakoff-Wernicke's Psychosis, Rule Out Hepatic Encephalopathy 113
5.3.9 Diabetic: Hyperosmolar Coma 115
5.3.10 Hyponatremia 117
5.4 TRAUMA: SIADH, HYPEROSMOLAR STATE, DIFFUSE AXONAL INJURY, HYPOXIC INJURY 120
5.4.1 Head Injury: SIADH 120
5.4.2 Head Injury: Hyperosmolar State 123
5.4.3 Head Injury: Diffuse Axonal Injury 126

5.4.4 Hepatic Encephalopathy Versus Long-Term Effect of Trauma 128
5.4.5 Head Injury: Diffuse Axonal Injury 131
5.4.6 Head Injury: Hypoxic Injury 134
5.4.7 Head Injury, Poor Outcome 140
5.4.8 Head Injury, Good Outcome 142
5.4.9 Head Injury—Hyperosmolar State Versus AIDS Dementia 145
5.5 GLOBAL HYPOXIA: NEAR DROWNING, HEART DISEASE, ASPHYXIA 147
5.5.1 Near-Drowning: Good Outcome 147
5.5.2 Near-Drowning 149
5.5.3 Near-Drowning: Fatal Outcome 154
5.5.4 Near-Drowning: Poor Outcome 158
5.5.5 Asphyxia: Moderate Hypoxic Encephalopathy 161
5.5.6 Cardiac Arrest, Rule Out Hypoxic Encephalopathy 168
5.5.7 Congenital Heart Disease 170
5.5.8 Agonal Prognosis After Hypoxic Injury 173
5.6 FOCAL HYPOXIA: VASCULAR DISEASES, STROKE, INFARCTS, AND NONVASCULAR FOCAL HYPOXIA 176
5.6.1 Hemorrhagic Infarct 176
5.6.2 Hemorrhagic Infarct, Rule Out Secondary Carcinoma: Negative Study 180
5.6.3 Asthmatic Cardiac Arrest 182
5.6.4 Apnea Versus Cerebral Vascular Accident (Part 1) and Focal Hypoxic Brain Injury (Part 2) 185
5.7 FOCAL LESIONS: ECTOPIC GREY MATTER, HIV RELATED LESIONS, TUMORS, AND TUBERCULOMA 194
5.7.1 Ectopic Grey Matter 194
5.7.2 AIDS Brain Masses, Toxoplasmosis 196
5.7.3 AIDS Brain Masses, CNS Lymphoma 198
5.7.4 AIDS Masses: Multifocal Leukoencephalopathy (PML) 200
5.7.5 AIDS Dementia 204
5.7.6 Recurrent Meningioma 206
5.7.7 Accidentally Discovered Tumor 208
5.7.8 Glioblastoma: Therapeutic Monitoring After Brachytherapy 212
5.7.9 Tumor: Postsurgical Recurrence 224
5.7.10 Untreated Glioma 233
5.7.11 Recurrent Hemangioblastoma: Negative 237
5.7.12 Tuberculoma Versus Lymphoma 239
5.8 MULTIPLE SCLEROSIS AND OTHER INFLAMATORY DISEASE 242
5.8.1 Multiple Sclerosis: Quiescent Plaque 242
5.8.2 Multiple Sclerosis 245
5.8.3 Rasmussen's Encephalitis (Adult) 252
5.8.4 Herpes Encephalitis 256
5.9 PEDIATRICS: ABNORMAL DEVELOPMENT, REYES SYNDROME, INBORN ERRORS OF METABOLISM 259
5.9.1 Hyperglycinemia 259
5.9.2 Cytomegalovirus (CMV) and Perinatal Hypoxia: Late Sequelae 264
5.9.3 Viral Encephalitis: Reyes Syndrome 267
5.9.4 Myopathy: Rule Out Mitochondrial Encephalopathy 270
5.9.5 White Matter Disease? Rule Out Canavan's Disease 273
5.9.6 Metabolic Error, Absent Creatine/Phosphocreatine 282
5.9.7 Hydrocephalus Versus Metabolic Disease 285
5.9.8 Metabolic Acidosis 289
5.10 XENOBIOTICS: ETHANOL, PROPYLENE GLYCOL, MANNITOL 293
5.10.1 Propylene Glycol in the Brain 293

5.10.2 Ethanol in the Brain 297
5.10.3 Mannitol in a Tumor 300

CHAPTER 6: MRS PATTERNS AND REFERRAL GUIDELINES 303

I. Clinical Decision Making with ^{1}H MRS 303

A. Dementia 303

B. Hepatic Encephalopathy (HE) 304

C. Hypoxic Encephalopathy and Stroke 305

D. Abnormal Development 306

E. Neonatal Hypoxia 310

F. Cerebral Tumor 311

G. Closed Head Trauma 312

H. Shaken Baby Syndrome 313

I. HIV CNS Lesions and AIDS Dementia 313

Appendix: List of Abbreviations 317

Index 319

1

Introduction to Magnetic Resonance Spectroscopy

I. A BRIEF HISTORY OF CLINICAL MAGNETIC RESONANCE SPECTROSCOPY

Magnetic resonance spectroscopy and magnetic resonance imaging (MRS and MRI), the flourishing twins (zeugma*) born out of the biological and medical application of nuclear magnetic resonance (NMR) in 1973, have since grown at unequal rates.

MRI, noninvasive imaging, is mature, and an integral part of neurology and neurosurgery, and of the radiologist's daily work. MRS, noninvasive chemical analysis, no less sturdy as an "infant," has hitherto been the daily work of only a small proportion of radiologists. Although virtually all clinical MR scanners provide the capability of performing MRS, a physicist has generally been needed to operate and interpret the MRS output.

Despite these difficulties, hundreds of peer-reviewed papers reporting thousands of clinical MRS examinations, have appeared in the radiological, neurological, and academic literature. Based upon these demonstrations of efficacy, a small number of radiologists have dedicated their efforts to identifying the clinical need for MRS. Substantial progress has been made. However, for the reasons just mentioned, access to MRS has been extremely limited. One unavoidable result has been frustration on the part of interested clinicians and radiologists, unable to perform the examinations that they require to confirm the clinical experience of their peers, or to satisfy their natural desire to help their patients. Another has been the rather small circle of physicians and radiologists who are aware of the probable diagnostic impact that MRS may have in their given specialty. As a result, clinical applications have themselves been somewhat conservative and unimaginative, confined to "sure winners," or rare and esoteric neuropathologies.

All this changed rather dramatically in 1995 with FDA approval of a fully automated, rapid, and inexpensive MRS sequence for neuro-MRS. This was the so called PROBE sequence—PROton Brain Examination—the acronym chosen for localized, single voxel ^{1}H MRS. This PROBE sequence runs without intervention by the physicist or neuroradiologist, and with minimum intervention by the MR technologist. It provides a proton spectrum from the chosen volume of interest in six minutes or less.

Suddenly, there were no longer any technical barriers, and no need for preselection of patients for a formal MRS examination. Instead, MRS could be incorporated into routine brain imaging protocols, or added at will to any still unresolved MRI examination. New clinical applications

*Greek zeugma = twins: Paul Lauterbur christened his newly discovered MRI "zeugma-tography." By the same logic, MRS might have been designated kumoscopy: kumos (Greek) = wave.

continue to emerge from this approach as neuroradiologists define the clinical impact of biochemistry on their specialty.

The need for this book? We perceive the last remaining barrier to be intellectual. After years in the shadows, we believe MRS needs an accessible, didactic source book that provides the simple ground-rules for clinical MRS, and an easy catalog of commonly encountered brain spectra. In addition, we firmly believe there is a need for a more clinically oriented manual which provides, on a case by case and patient by patient basis, the indications, preferred examination protocol, peak by peak interpretation, and clinical conclusion of the MRS examination. Also important is the identification of common clinical pitfalls and MRS artifacts, as well as discussion of the place of MRS examinations among other possible diagnostic tests used regularly.

This book has had to await the accumulation of a sufficiently comprehensive (but by no means exhaustive), patient database. As in other clinical specialties, this is different from the systematic accumulation of average data necessary to prove efficacy. This is largely available, and to be found in individual peer-reviewed articles. While we recognize the over-riding importance of that kind of information, it must await another book. The primary purpose of the present small volume is to find a place beside each practicing neurologist, neuroradiologist, oncologist, and psychiatrist who must use MRS to obtain accurate neurodiagnoses, effectively monitor therapies, and achieve an optimal outcome in their patients.

II. MOTIVATION

What the clinician needs to know in order to request, read, and interpret a brain MRS has suddenly become much more manageable. Many years experience with clinical MRI has resulted in a comfortable understanding of the MR process. The patient must be able to lie on a bed that enters a confined space where magnetic fields are switched on and off repeatedly, and tolerate the accompaniment of considerable noise. After an interval of one to two minutes, all of the radio frequency signals generated by resonating protons are mathematically mapped to produce an image. The anatomy displayed in cross-section is now the norm for all who work in medicine.

MRS is produced in the same way, with three additional steps which pass unnoticed by the patient. The image obtained at the beginning of the examination (what spectroscopists call a "localizer") is scrutinized, and a region of interest, or volume of interest (VOI) is selected for MRS. The water protons within the VOI are rendered silent by suppressing their particular frequency band (water suppression), and, using the same constellation of switched magnetic fields already familiar from MRI, a frequency map, or spectrum is produced. This is the MRS spectrum. Its profile is hardly recognizable to those who work in clinical medicine, and yet, it is as easy to "read" as an EKG. Perhaps more important for the clinician is the certain knowledge that his request (for MRS) can be fulfilled painlessly (to the patient), and effortlessly (to the MR technologist) through the miracle of automation.

MRI sequences (the histochemical stain that alters contrast in the resulting image), are fully automated, and therefore selected every day for their relevance to the clinical question at hand. As mentioned earlier, in 1995 the FDA approved clinical use of MRS sequences in their automated form (for example, PROBE for GE Medical Systems' Signa 1.5 Tesla scanner). Accordingly, the clinician's request can be fulfilled with no more trouble than an MRI series, and just as quickly. For each brain location, 3 to 6 minutes of scan time suffices.

For the clinician then, three things need to be made readily available before she/he can reasonably be expected to treat the request for MRS with the same casual ease as the request for an EKG!

A. A Working Knowledge of Brain Chemistry in Medicine

This is a simple exercise as will be seen in Chapter 3. After all, only a dozen neurochemicals need to be remembered, out of the hundreds, because the MRS filter is very restrictive—only those peaks of the spectrum which occur in health and disease can be discussed in clinical diagnosis. All the rest can safely be forgotten!

B. Clinical Indications for MRS

This is a moving target because each clinician will have an opinion. However, the body of published and confirmed clinical MRS diagnoses as applied to the brain probably doesn't exceed 50, so referral guidelines, or decision trees that need to include MRS are rather easy to visualize. This topic is covered in Chapter 6.

C. How to Read and Interpret the MR Spectrum

A cadre of "cognoscenti" at present appear to understand and interpret MRS for those of their colleagues who are interested. While the same thing happens in almost every specialty, the clinicians requesting MRS, however appropriately, will perhaps never truly feel comfortable with the result unless she/he can see it, touch it, and read it with enough confidence to debate its clinical relevance with the patient, and their medical peers. A simple flawless routine of reading, interpreting, and reporting spectra is described in Chapter 4, and used for all of the cases presented in this book.

III. A MORE DETAILED LOOK AT MRS: THE SINGLE VOLUME OF INTEREST

MRI employs water protons to create its images, and although applicable anywhere in the body, the impact of MRI has been nowhere more dramatic than in the human brain. The brain's inaccessibility, and yet almost perfect magnetic field homogeneity, combine to permit MR imaging of great sophistication. Image-contrast can be manipulated in a multitude of ways, all of which excite water protons to differentiate grey and white matter, CSF, blood, and blood-flow. But, those hydrogen nuclei (protons) that are not in H_2O are distributed throughout the hundreds of biologically significant molecules of the living brain, and are nowhere to be seen in MR images.

A relatively small proportion of these "image-invisible" molecules are present in sufficient concentrations to produce a measurable MR signal. Because each has a unique radio-frequency specific to the chemical itself, and entirely distinct from the frequency of water protons, a spectrum of brain chemistry results from which a patient's neurochemical "profile" can be constructed in which "larger" peaks result from high metabolite concentrations, and small peaks from lower concentrations, all at their unique radio frequency. This is proton magnetic resonance spectroscopy.

The clinical utility of the brain spectrum rests upon two important properties. First, the nature and concentration of brain chemicals identified is remarkably constant; hence a "normal" brain spectrum is readily recognized. Second, the particular neurochemicals concerned are of the clinical relevance in healthy and diseased brain. Included in the profile are a neuronal marker, two astrocyte markers, one energy marker, two redox markers, and one or more cell-volume markers, as well as two or more of the known neurotransmitters. Concentrations of each of these neurometabolites alter in a defined and reproducible way in diseased brain. Accordingly, abnormal brain spectra are readily recognized and classified.

Together with the abnormal anatomy and physiology identified by the MR images, the MR spectrum gives a more complete picture of disordered neurology encountered in clinical practice. On the other hand, the MR spectrum also defines neurochemical abnormality when neuro-anatomy is normal, and where no pathology was recognized by MRI.

In short, MRS is a clinical modality waiting to find its place in improving health care outcomes, adding significant diagnostic information for the benefit of patients, and providing significant cost benefits to providers.

IV. PHILOSOPHY OF SINGLE VOLUME OF INTEREST MRS

Neurology, which has been greatly enhanced by the advent of neuro-imaging, seeks to provide more and more accuracy of diagnosis by localizing deficits and pathology to specific brain structures, through the medium of finer spatial resolution and tissue contrast. Single voxel or volume of interest (VOI)* MRS takes a more unorthodox view of neurology. It has sought to define the biochemical abnormalities that underlie, cause, or are associated with neurological diseases. When accurately identifying the rather small amounts of neurochemicals (generally measured in tens of micro moles, or micrograms of the substance), the necessary precision can only be provided by the inclusion in the analysis of a relatively large volume of the brain.

Volumes of brain measuring 1 to 125 cc ($1 \times 1 \times 1$ cm^3 to $5 \times 5 \times 5$ cm^3) are selected, and isolated by localization procedures from the surrounding scalp, skull, and meninges, before acquiring specific and quantitative noninvasive measurements of a dozen or more brain metabolites. Because the biochemical analysis is the average of a multitude of cell types and structures—nuclei, grey and white matter, connections, "systems," and pathways dissected so painstakingly by generations of neurologists—there is something quite shocking to many diagnosticians about this new approach. To neuroradiologists, also, it is unsettling to reach a diagnosis after "seeing" only a small fraction of the intact brain. For them, the metabolic brain images derived by the single photon, or positron emission (SPECT or PET) technique of nuclear medicine are preferred.

For both segments of the neurodiagnostic community, therefore, single voxel MRS starts at a serious disadvantage. Indeed, for many years, single voxel MRS has been viewed as no more than an evolutionary phase of MRS until fully automated MR metabolite imaging of the brain becomes widely available.

This may indeed be a reasonable stance. But as single VOI MRS brings valuable clinical diagnostic information to the practice of neurology that cannot be provided by any of these alternative tools at the present time, and metabolite mapping by chemical shift imaging (csi) still falls short of routine implementation, single VOI MRS should be used now.

*The terms volume of interest and voxel are used interchangeably in this book.

V. METHOD OF SINGLE VOI ADOPTED IN THIS BOOK

For simplicity, we describe just one method, short echo time STEAM, together with one variant (PRESS), that impacts the final appearance spectrum only to a minor degree. In Chapter 2, we describe the basic physics of MRS and refer to some of the competing schemes for acquiring high quality MRS, but we also recommend this section be postponed for later reference, after the physician or technologist is comfortable with reading the entire range of diagnostic spectra covered in Chapters 4 and 5.

VI. THREE SIMPLE RULES FOR UNDERSTANDING AN MR SPECTRUM OF THE BRAIN

For your ease and convenience, follow these three simple rules.

A. Rule One: Spectra Are Read in Terms of Peak Ratios

Spectra may be processed by a computer and assigned the peak ratios, but peak ratios obtained by just looking at the spectrum are in most cases completely adequate. The ratio is invariably expressed relative to creatine* (Cr). In other words in understanding the peak ratios, Cr is always assumed to be equal to one (Cr = 1.00).

B. Rule Two: Uniform Clinical Methods Yield Identical Spectra

Even small changes in technique undermine diagnostic confidence. It is important to compare like with like; compare ratios with age-related normals (healthy controls), and match location to location. Tables of normal values, the reference ranges for different age groups, and brain locations, as well as graphs of how peak ratios change with age, are given throughout this book.

*NOT creat*in*ine.

C. Rule Three: MRS Is a Quantitative Technique

With only a few simple assumptions, MRS is indeed a quantitative technique because the signal obtained is directly proportional to the number of molecules. The first assumption, already mentioned, is that Cr is equal to one; peak ratios, therefore, represent tissue metabolite concentrations in a simplified form. In practice, of course, Cr as a key brain metabolite is also subject to major modification, in particular by focal brain disease. Knowing when to abandon the simplifying assumption, that Cr is always 100% of normal, is important in practice. The second assumption is that signal intensity of all metabolite peaks have approximately the same, or at least unvarying, relaxation times (T_1 and T_2, in the familiar MRI terminology).

Absolute metabolite concentrations add another useful dimension to diagnosis. Techniques for absolute quantitation (yielding concentration, e.g. [Cr] in square brackets) are discussed only briefly in Chapter 2, and results of absolute quantitation are sometimes discussed in the individual case presentations. Again, a simplifying assumption is that [Cr] is 100% of normal for age and location; with that assumption the peak ratios can be used to describe metabolite concentrations.

FURTHER READING

An Overview of MRI

Weiskoff RM and Edelman RR. Basic Principles of MRI. In: Edelman RR, Hesselink JR, and Zlatkin MB, eds. Clinical Magnetic Resonance Imaging, 2nd edition. Philadelphia: W.B. Saunders Co., 1996: 3–51.

An Overview of MR Spectroscopy

Bolinger L and Insko EK. Spectroscopy: Basic Principles and Techniques. In: Edelman RR, Hesselink JR, and Zlatkin MB, eds. Clinical Magnetic Resonance Imaging, 2nd edition. Philadelphia: W.B. Saunders Co., 1996: 353–379.

An Excellent Textbook of Clinical Neurology for the Nonspecialist

Adams RD, Victor M, and Ropper AH, eds. Principles of Neurology, 6th edition. New York: McGraw Hill, 1997.

2

Basic Physics of MRS

I. INTRODUCTION

The first part of this chapter (Sections II–III) is structured around what you see and do when you run a single volume of interest (VOI) MRS examination. First, a brief presentation of MRS, then choices of the measurement protocol are discussed (VOI, echo time, repetition time, number of averages), followed by an overview of the adjustments that precede a scan (frequency, transmitter, receiver, shim, water suppression).

The remainder of the chapter (Section IV) provides an overview of quantitation including the semi-quantitative ratio measures and the fully quantitative evaluation of an MR spectrum.

A. What Is an MRI?

Magnetic resonance imaging (MRI) is assumed to be well-known to the reader. Briefly, an MR image is acquired from a "slice" through a part of the body. A transaxial (axial) slice of the brain, for example, is illustrated in Fig. 2.1. Note the two boxes on the MR image (Fig. 2.1). One is placed in the occipital grey matter region (box below), and the other in the parietal white matter region (box above). These boxes represent two volumes of interest (VOIs) from which biochemical measurements can be obtained. In order to measure the concentration of biochemicals (or, in this case, neurochemicals) in the VOI, magnetic resonance spectroscopy (MRS) must be performed.

B. What Is an MR Spectrum?

An MR spectrum from a VOI similar to the one illustrated in Fig. 2.1 (box 2), is shown in Fig. 2.2A. A simplistic way to describe a spectrum is "a curve, or graph, with peaks." An example of a peak is shown in Fig. 2.2B.

In this book, we confine ourselves to the discussion of proton (^{1}H) MRS, so a peak can be described as the MR signal from a set of magnetically equivalent protons, for example, the three protons of a CH_3 molecular group. The position of such a peak along the horizontal axis, also called the chemical shift axis, or frequency axis, is governed by the local magnetic field at the site of the proton(s). Thus, the protons in different molecular groups of different molecules cause their corresponding peaks to appear at different positions on the chemical shift axis. In other words, each neurochemical resonates at a different frequency, or position along the chemical shift axis, in accordance with the magnetic field generated by the molecular groupings of their protons. Although peaks from nonidentical molecules may overlap, in practice, the peak position identifies the brain metabolite (Table 2.1).

In Fig. 2.2, turning our attention from the horizontal to the vertical axis, "the height" of the peak determines the

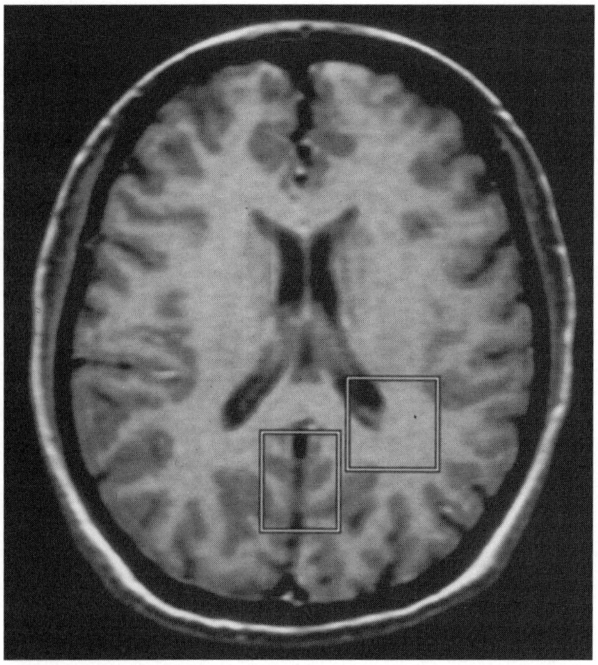

Fig. 2.1 MR images are acquired from "slices" through the body (either from the axial, sagittal, or coronal view). A typical MRI (axial) slice of the midbrain is shown, the boxes on the image indicate two typical MR spectroscopy locations.

Table 2.1 Chemical Shifts of Metabolite Resonances in MR Spectra of the Human Brain

Resonances observed in normal MRS	Chemical shift (ppm)
N-acetylaspartate (first peak, NAA_1)	2.02
β,γ-Glutamine and glutamate (β,γ-Glx)	2.05–2.5
N-acetylaspartate (second peak, NAA_2)	2.6
N-acetylaspartate (third peak, NAA_3)	2.5
Total Creatine (Cr)	3.03
Total Choline (Cho)	3.22
scyllo-inositol (sI)	3.36
Glucose	3.43
myo-inositol (mI)	3.56
α-Glutamine and glutamate (α-Glx)	3.65–3.8
Second peak of Glucose	3.8
Second peak of Cr	3.9
Second peak of mI	4.06
Other resonances (not observed under normal conditions)	**Chemical shift (ppm)**
Propylene glycol (doublet with 7 Hz splitting)	1.14
Ethanol (triplet with 7 Hz splittings)	1.16
Lactate (doublet with 7 Hz splitting)	1.33
Alanine (doublet with 7 Hz splitting)	1.48
γ-aminobutyric acid (GABA) (complex multiplet)	2.9
Glycine	3.56
Mannitol	3.8

amount of the metabolite. To be precise, it is the area under the peak that is relevant. Applying appropriate scaling and calibrations, the concentration of a given metabolite is calculated from the peak area (see Section IV of this chapter). The peak area is used because it is not influenced by a variety of physical factors (field homogeneity, relaxation, etc.) that influence the peak height. The width of the peak multiplied by the height is equal to the peak area. As long as the widths of different peaks of the spectrum do not vary, we can estimate equally well the amount of the metabolite from the peak height. Fortunately, in brain MR spectra, peak widths are usually quite similar, making it much easier to evaluate spectra directly from their appearance.

Combining the horizontal and vertical information we can see that the brain MR spectrum, by showing the presence, or absence, and the size of each of the peaks, provides a *biochemical fingerprint* from the measured brain location.

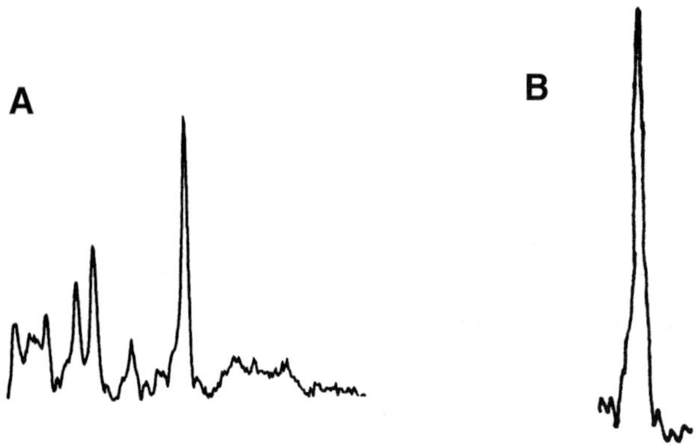

Fig. 2.2 (A) An MR spectrum of the human brain. (B) A peak of the spectrum. The *peak position identifies* the metabolite, while the *peak size is a measurement of the amount* of the metabolite. The MR spectrum shows the presence or absence of peaks, thus giving a *fingerprint of the biochemistry*.

II. MEASUREMENT PROTOCOL CHOICES

A. What Do We Mean by Localization?

Single volume MRS uses methods that allow us to measure MR signals originating only from the region of our choice. The choice of dimensions and position of a box is part of the measurement protocol (often termed "prescribing the scan").

A multitude of single volume MRS methods have been proposed; among these the "single shot" methods stimulated echo acquisition mode (STEAM) and point-resolved spectroscopy (PRESS) have proven most successful for single VOI ^{1}H MRS of the brain. The details of volume selection beyond the following summary will not be given here (see "Further Reading" for details). The basic idea is similar to slice selection in MR imaging; however, in single volume MRS, three slice-selective radio frequency (Rf) pulses are combined in such a way that the detected MR signal originates from the intersection of the three orthogonal slices (see Fig. 2.3). The success of ^{1}H MRS of brain relies very much on your choice of volume size and location (see below).

B. STEAM and PRESS

Both STEAM and PRESS are pulse sequences for single VOI MRS. The basic features are three slice selective Rf pulses that select the VOI. The result is a spectrum from the selected VOI. The acquired NMR signal from STEAM is a stimulated echo from the pulse sequence, 90°-TE/2-90°-TM-90°-TE/2-"ECHO." PRESS is a double spin echo from the pulse sequence, 90°-TE_1/2-180°-TE_1/2-TE_2/2-180°-TE_2/2-"ECHO."

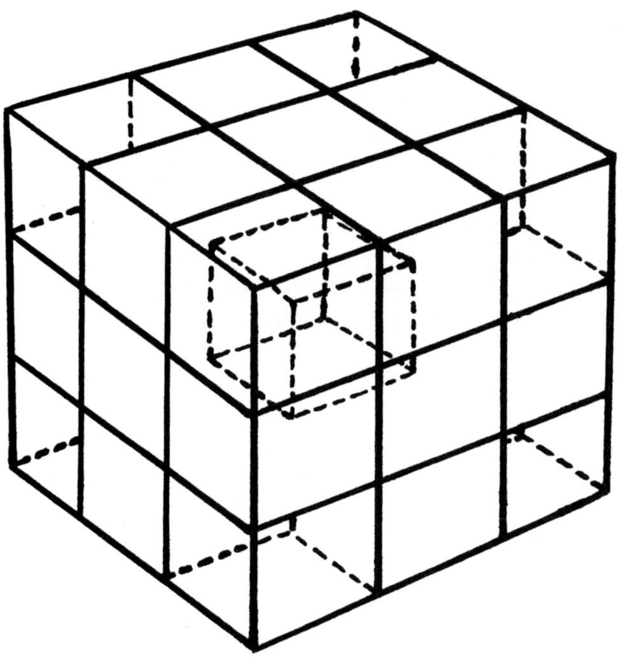

Fig. 2.3 The volume of interest (VOI) is the intersection of three orthogonal planes. Slice selective radio frequency pulses are used to measure an MR spectrum from the VOI.

The choice of STEAM or PRESS affects signal size: Given the same total echo time, the signal of PRESS is in theory a factor 2 larger than that of STEAM. The choice of STEAM or PRESS also affects the appearance of different peaks, in particular from coupled spins. Outer volume suppression is less complete with PRESS given the same total echo time and the same magnetic field gradient performance.

C. Choice of STEAM Versus PRESS

STEAM has for many been the favored choice because it is important to use a short echo time (e.g., 30 ms) in order to see some of the clinically important metabolites. Given the same hardware (mostly a matter of magnetic field gradient performance) a short echo time can better be achieved with STEAM than with PRESS, due to the problem of outer volume suppression. Therefore, the advantage of PRESS having a factor 2 larger signal can in practice not be implemented if a short echo time is also demanded. Limitations of the slice selective 180 degree pulses have also been part of the considerations. However, with the increasingly better hardware performances of MR scanners the spectra quality of STEAM and PRESS spectra measured with a sufficiently short TE become comparable, and the factor 2 signal size advantage of PRESS becomes a reason to use PRESS. When, or if, to switch from short echo time STEAM to short echo time PRESS should not depend on technical feasibility alone. A major consideration should be the need to gather thousands of spectra from normal individuals of all ages and from patients with different diseases, before we have a database that is comparable to what we have with STEAM today. This is one of the main reasons that most case studies in this book used STEAM.

The final choice between STEAM and PRESS (as well as other measurement parameters) should be a choice you stick with; the detailed appearance of the spectrum, the relative heights of the common peaks, and in particular of the coupled resonances such as glutamine, varies significantly with choice. All these details of spectral appearance assist the physician's ability to recognize pathology in a spectrum, thus should only be changed by pathology.

D. How Does the Echo Time Affect the Spectrum?

The longer one waits before sampling an MR signal, the less signal will be left. The brain spectra on Fig. 2.4A and B illustrate this. Only three peaks are left at TE = 272 ms and are all "half" the size that they were at 30 ms. In MRI, a

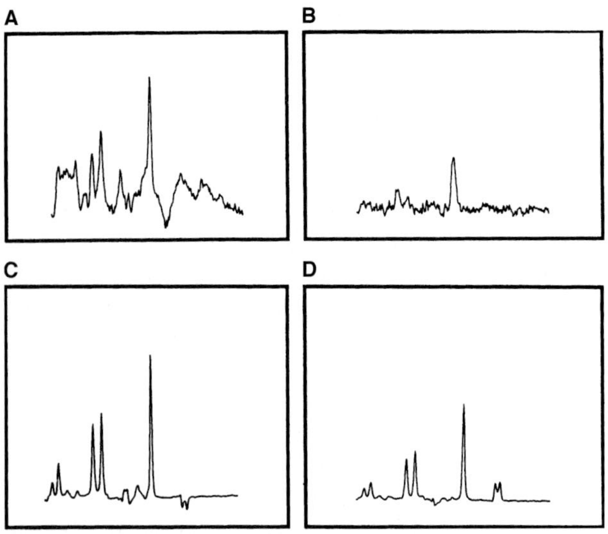

Fig. 2.4 MR spectra vary with echo time: (A) TE = 30 ms, (B) TE = 272 ms. Both are STEAM spectra of the human brain shown with the same vertical scale. A PRESS spectrum of a bottle containing normal brain metabolites plus lactate at (C) TE = 136 ms, and (D) TE = 272 ms, are both shown with the same vertical scale.

long echo time gives a T_2-weighted MRI with a different contrast. Similarly, you could say that the TE = 272 ms spectrum is more T_2-weighted than the TE = 30 ms. It results in fewer and smaller peaks as already mentioned, but also in different peak ratios, depending on the differing T_2 relaxation of the metabolites. Notice that the two smallest of the three remaining peaks in Fig. 2.4B (272 ms) have a ratio of ~1:2, where the ratio was ~2:1 in Fig. 2.4A (30 ms). So, the MRI contrast is paralleled here by MRS peak ratios. It should furthermore be noted that the spectra in Fig. 2.4A and B are shown to scale; Note the loss of signal at longer TE. This can be "disguised" by the usual operation of the MR computer, so depending on the display setting, Fig. 2.4B might have appeared to be different. The spectrum would have been scaled to fit the largest peak within the screen; the largest peak would still be the same size in the display as the largest peak in Fig. 2.4A. Then you would notice the effect of long TE as, 1) fewer peaks, and 2) an apparent larger noise level between the peaks.

An example of another echo time effect is illustrated in Fig. 2.4C and D. The spectra are sampled from a bottle containing metabolites equivalent to the normal human brain plus lactate. The lactate is seen to the right in the spectrum, at TE = 136 ms as a set of 2 peaks pointing down, but at TE = 272 ms pointing up.

The effect is a result of spin–spin coupling between protons within the lactate molecule, and is only included here to illustrate how the choice of TE can influence the appearance of coupled spin systems (see "Further Reading" for a detailed explanation of the effect).

E. How Does the Repetition Time Affect the Spectrum?

The analogy to MRI can again be used to illustrate the effect of TR on a spectrum. In MRI, a T_1-weighted image is achieved by using a short repetition time, and in MRS the result of using a shorter TR (1500 ms) is mainly just smaller peaks in the spectrum, Fig. 2.5A, because of the saturation of the signal. The apparent effect on the lipid to the right in the spectrum cannot be attributed to T_1-weighting alone, but is also caused by differences in the efficiency of the suppression of signal from outside the prescribed VOI.

F. Total Number of Scans and Signal to Noise: What Is Signal-to-Noise and How Does Total Number of Scans Affect It?

The signal-to-noise ratio (S/N) may be defined from the signal height and the peak to peak noise, as illustrated in Fig. 2.6. Unfortunately, the mathematics of S/N is not kind to the spectroscopist: The signal is proportional to the number of scans; the noise is proportional to the square root of the number of the scans, making S/N proportional to the square root of the number of scans. Therefore, to get *twice* the S/N you need *four* times as many scans. An illustration of the drastic difference in S/N when using 256 scans versus 4 scans is presented in Fig. 2.7.

In Fig. 2.7, the signal is not a factor of 256/4 = 64 larger, as it should be if "the signal is proportional to the number of scans," as mentioned. The signal is approximately of the same size. The reason is that on most MR systems the acquisitions are *averaged* instead of *summed.* A major advantage of averaging is that the spectrum can be displayed using the same scale independent of the number of

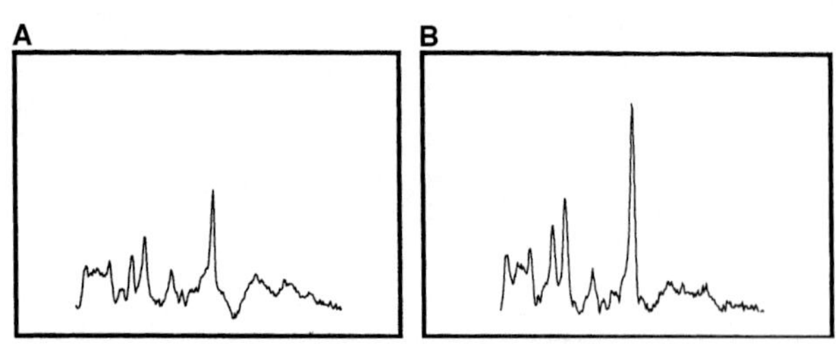

Fig. 2.5 Both spectra, (A) TR = 1500 ms, and (B) TR = 6000 ms, were acquired from the human brain with STEAM, and are shown with the same vertical scale.

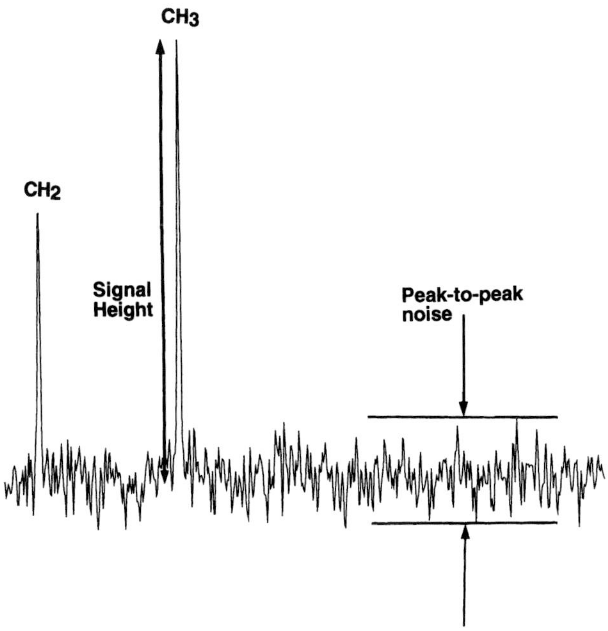

Fig. 2.6 Signal and the peak to peak noise in an MR spectrum of creatinine (sic).

acquisitions. Fig. 2.7 also clearly shows how the noise is averaged out as a result of the noise being random (incoherent), in contrast to the MR signals which are coherent and become more well-defined with increasing number of scans.

G. Box Size and S/N: How Does the Box Size Affect S/N and the Quality of the Spectrum?

When prescribing the VOI, one may be tempted to make it small to fit within a specific lesion, or anatomical region. It is also easy only to think in one or two dimensions, hereby allowing the edge lengths of a VOI to be decreased without considering the effect three-dimensionally, or in terms of S/N. The following example will illustrate how expensive, in terms of S/N, or total measurement time, it can be to select too small a VOI.

We will compare two measurements, assuming TE, TR, and all other parameters not mentioned here are unchanged. Assume that a peak's signal size from a $2 \times 2 \times 2$ cm^3 VOI is 1.00, and that it was measured averaging 64 scans, with a TR = 1500 ms (disregarding prescans), making the total acquisition time 1 minute, 36 seconds (see Table 2.2 and Fig. 2.8A). The measurement was performed with a long TE (270 ms) on a bottle containing close to normal *in vivo* concentrations of brain metabolites, plus lactate. Several peaks are clearly seen.

What would happen to the S/N if the edge length was decreased from 2 to 1 cm, and how much should the total measurement time be increased to compensate for the lost S/N? The VOI would be a factor 8 smaller; therefore, the MR signal would be 8 times smaller (Table 2.2). Fig. 2.8B shows the corresponding spectrum to scale (bottom), and enlarged by a factor 8 (top). The MR signal is in fact a factor 8 smaller, but only one of the peaks can be distinguished because of the much greater noise level.

As described above, S/N increases with the square root of the number of averages; therefore, we would need 64×64 (= 4096) averages to achieve similar S/N from the $1 \times 1 \times 1$ cm^3 VOI, and the total measurement time would be 1 hour, 42 minutes, 24 seconds.

It should now be clear that it pays to choose adequately large VOIs, and if a smaller VOI is very tempting then it is often a good idea first to use 2 minutes measuring a larger surrounding VOI to at least get one good S/N spectrum, and also to get an idea whether it is worth using a long measurement time on the smaller VOI. Dealing with the obvious problems of partial volume (e.g., focal lesion/surrounding tissue) is better than risking only getting one spectrum which might be too noisy.

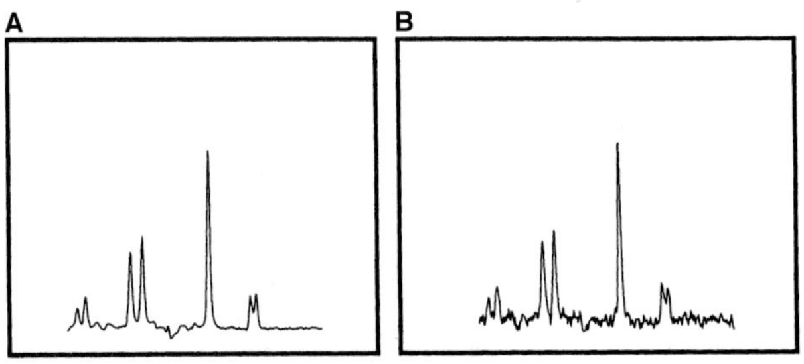

Fig. 2.7 STEAM MR spectra measured from a bottle containing normal brain metabolites plus lactate. (A) Total number of scans = 256, and (B) 4 scans are averaged.

Table 2.2 Relations Between VOI Size, Signal Size, and Number of Scans and Measurement Time Necessary to Achieve the Equivalent Quality as for the VOI Size $2 \times 2 \times 2$ cm^3 with 64 Scans

VOI size	Signal	Scans	Time
$1 \times 1 \times 1$ cm^3 (1 mL)	1/8 = 0.125	4096	1 h 42 m 24 s
$2 \times 2 \times 2$ cm^3 (8 mL)	1	64	1 m 36 s

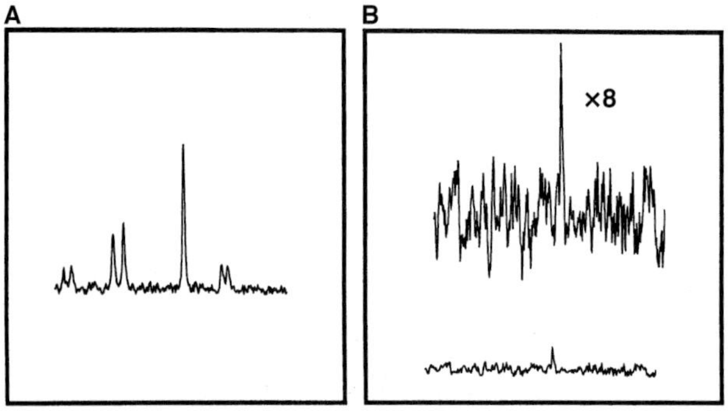

Fig. 2.8 STEAM MR spectra measured from a bottle containing normal brain metabolites plus lactate, from a (A) $2 \times 2 \times 2$ cm^3 VOI and from a (B) $1 \times 1 \times 1$ cm^3 VOI.

H. Signal to Noise: A Summary of Controlling Factors

S/N is affected by a number of factors. The factors which depend upon operator choice include volume size, total number of scans, and TR, TE, etc. Factors which cannot be changed are metabolite concentrations, relaxation times (T_1 and T_2), and other factors inherent to the measured object, or patient. Because patient movement can decrease the signal from the MRS pulse sequences, steps should be taken to avoid this. In addition, bad shim (field homogeneity) can affect S/N. In this case, check for ferromagnetic material, and air or blood near the VOI. (Field homogeneity will be explained in the following section.) Finally, bad adjustments of measurement parameters, and other technical problems, can affect S/N.

III. AUTOMATED OR MANUAL ADJUSTMENTS PRECEDING THE SCAN

A. Adjustments of Frequency, Transmitter, and Receiver

The frequency, the transmitter gain, and the receiver gain are adjusted just as it is done preceding MRI scans. These have to be adjusted because each patient affects the physical properties of the scanner slightly differently.

B. Optimizing the Magnetic Field Homogeneity: "Shimming"

The effect of shimming is to make the peak taller and narrower, as illustrated in Fig. 2.9. A "better shim" is achieved by adjusting the currents in the shim coils so that the magnetic field within the VOI becomes more homogeneous.

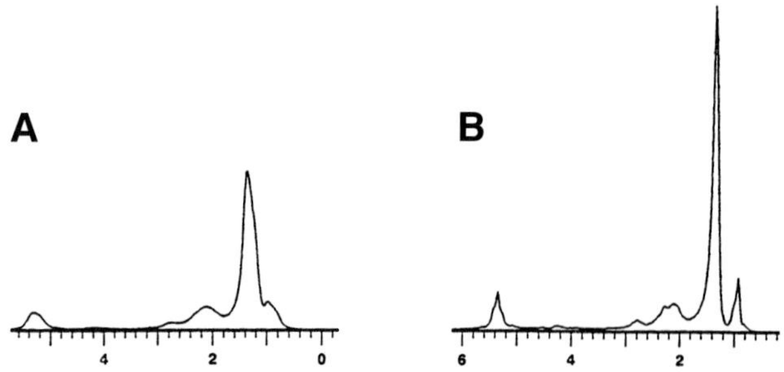

Fig. 2.9 STEAM MR spectra measured from a bottle of peanut oil before (A) and after (B) shimming.

After the shimming procedure, a fine adjustment of the frequency may have to be performed because the shimming can change the magnetic field, and therefore also the frequency.

C. Adjusting Water Suppression

When sampling a brain MR spectrum it is convenient to suppress the signal of water, which is roughly a factor 10,000 stronger than the metabolite signals. Among many methods of water suppression the chemical shift selective (CHESS) Rf pulses have proven very successful. A CHESS Rf pulse only affects a narrow band for example around the water peak. CHESS pulses followed by magnetic field gradients can be combined in a variety of ways to achieve water suppression.

The effect of water suppression is shown in Fig. 2.10. Spectra with no water suppression, medium water suppression, and with almost perfect water suppression are shown on the same scale. The latter appears as a flat line on the unamplified scale but if the vertical scale is readjusted (Fig. 2.10B) we can see the residual water peak to the left in the spectrum. The residual water peak is in this case still larger than the metabolite peaks (all seen "pointing up"), and the residual water peak is out of phase (pointing mostly down compared with the metabolite peaks). The water suppression is quite good, but it could have been better: a) It is possible to make the water peak smaller than the metabolite peaks, and b) it is possible to achieve a residual water peak that is in phase with the metabolite peaks (here termed "avoiding over suppression," it will become clear in the following chapters that being able to recognize "over suppression," is important when reading the clinical MR spectra).

Fig. 2.10C shows the same spectrum, but zoomed in further, so that now only the brain metabolites (from a solution in a bottle) are shown. The effect of the "over suppressed" water is seen in the form of a downward sloping baseline to the left in the spectrum.

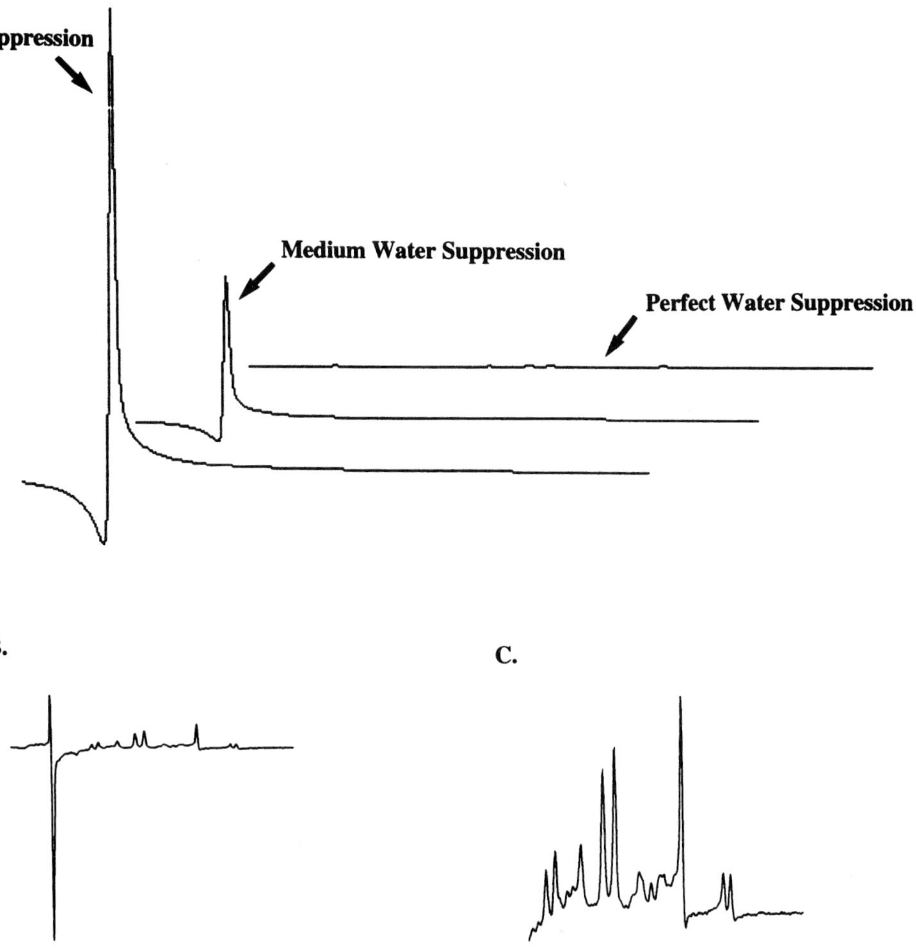

Fig. 2.10 (A) Various degrees of water suppression; (B) water suppressed spectrum of brain metabolites "zoomed" to include the residual water peak, and (C) "zoomed" to show the metabolites only.

The spectrum in Fig. 2.10C is displayed as spectra are often displayed in publications or on the screen right after acquisition. The residual water peak and peaks on the other side of water (few and small) are usually not displayed. But it can often be helpful to consider how the water suppression and the residual water peak affect the spectrum. In particular, recognizing its effect on the baseline may be essential to a correct interpretation of the spectrum.

It is possible to further eliminate the residual water signal by manipulating the spectrum in the postprocessing procedures. This book contains spectra shown both with and without this final elimination of water.

IV. QUANTITATION

In Clinical practice there is often a sharp dichotomy between laboratory results which are recorded and understood numerically, and "images" (x-ray, CT, nuclear medicine, MRI, PET and even EKG) which are intended to be read visually. MR spectra are intrinsically visual, and can be read as any MR image. Metabolite ratios, although "numerical" are not really quantitative: they simply reinforce the impression and add precision and reproducibility to the MRS exam.

Quantitation on the other hand, records concentrations of metabolites, and permits a more profound clinical interpretation at the level of cellular pathology.

A. Numerical Analysis of the MR Spectrum

We describe three types of numerical analysis: Visual inspection of the spectrum can provide information such as "peak A is about twice the size of peak B," in itself a numerical analysis, which can be refined using a ruler to measure the peak heights. The truly relevant measure is the peak area, but in vivo the widths of the individual peaks are similar, simplifying the issue. The third type of numerical analysis is computer analysis. In the simplest form a crude baseline is chosen and the peak heights or peak areas (integrals) are calculated. A multitude of more advanced methods of fitting the spectrum to an ensemble of idealized peaks are also used by MR spectroscopists. The computerized numerical analysis developed at HMRI has been used consistently since 1990 for the purpose of creating a large database of identically processed spectra. (Differences in data processing methods, rather than real differences, are probably the most important contributor to deviations between different clinical results.) HMRI's data analysis has been applied to a majority of the spectra presented in this book. It includes removing the residual water baseline using a filter, and obtaining numbers both for peak heights and areas. A number of different fitting routines are applied, and the unsuppressed water signals and the signal from an external reference are analyzed to obtain the numbers needed for the absolute quantitation, (see "Further Reading" for details).

B. Ratios Relative to Creatine

The basic numerical analysis involves getting a number for the "size" of each peak. The scaling of the spectrum depends on the sensitivity of the coil for the conditions of each examination, as well as on the receiver gain and other analog or digital amplifications of the signal before the spectrum appears on our screen. Therefore, the number for the "peak size" is in a sense arbitrary. However, all peaks have been scaled by the same factor, thus ratios will be unaffected, making the peak height ratios, interchangeable with the visual impression. Ratios relative to Creatine (Cr) are most common, because Cr is the metabolite among those visible that has the most stable concentration. However, Cr does change in certain pathologies, and one must always keep in mind that the metabolite ratio only "quantifies" the metabolite assuming that Cr is normal. In other words the ratios measure the metabolites on a scale that assumes that the measure of Cr = 1.

Many other peak ratios, e.g. Cho/NAA, as well as ratios of metabolites relative to internal water have been employed by various research groups, but throughout this book only ratios to Cr are used for interpreting the MR spectra to "enhance" and standardize diagnostic sensitivity.

C. Normal Metabolite Ratios

The metabolite ratios are useful because a set of VOI location dependent and age dependent normal values are known, which differ significantly from the values obtained in various diseases. Normal adult metabolite ratios are found in Tables 2.3 and 2.4. These normal values are different in HMRI's standard grey and white matter locations, Cho/Cr in particular is about 30% larger in the white matter location. Other locations or VOI sizes in principle require their own set of normal values, but as a first very rough estimate, one can use the values of the most similar standard location. Note however that the ratios in e.g. putamen are significantly different from those of HMRI's standard locations, Table 2.5.

Metabolite ratios will be the same measured in any group of normals with any scanner and anywhere in the world providing all measurement parameters are the same. Significant variations may however be seen if the average age differs. This important influence is a continuous process, but for simplicity we can consider just 4 age groups; infants up to about 18 months, children, adults, and elderly: The greatest changes are seen during the first year of life. These changes mainly reflect myelination and of course also take place in utero. MRS ratios from premature infants therefore need to be compared to normal ratios of the same gestational age (see Fig. 2.11).

Our second age group, children, can usually be treated as adults, with the exception that NAA/Cr is 10–20% higher in children, peaking at the age of 10–14 years. The third age group is young adults, our "normal," and the fourth group is the group of the elderly. In the elderly population (60–90 years) the main differences are that NAA/Cr is 10–20% lower than in the adult population and the average Cho/Cr is elevated about 10%. Keeping in mind that the changes are continuous it is not surprising that very precise measurements with 10 or more volunteers in each group can show small but significant differences of NAA/Cr in between a group of young adults and one of middle-aged adults (see Chapter 4). But for simplicity, and because in practice only NAA/Cr is affected we use the adult set of normal values in most cases, even for children and for elderly. For children we then remember that NAA/Cr is 10–20% higher, and for elderly 10–20% lower, and apply that to our analysis. In newborn and very young infants, individual results are interpreted only with the aid of age-related curves (see Fig. 2.11).

D. Absolute Quantitation

The metabolite ratios are most useful when the assumption of normal Cr is valid. Methods of absolute quantitation have been developed to provide independent measures of all metabolites including Cr. The method developed at HMRI employs an external reference that is measured once for each patient, the signal from the reference combined with calibration values provides a scaling factor for the peak area. Other scaling methods such as the methods of using the optimized transmitter amplitude, which relies on the

Table 2.3 Metabolite Ratios (Averages ± Standard Deviations)

	Grey matter		White matter	
	A	B	A	B
NAA/Cr	1.32 ± 0.04	1.38 ± 0.07	1.40 ± 0.07	1.58 ± 0.16
Cho/Cr	0.59 ± 0.09	0.62 ± 0.07	0.83 ± 0.05	0.83 ± 0.07
mI/Cr	0.57 ± 0.06	0.57 ± 0.05	0.61 ± 0.06	0.60 ± 0.07
α-Glx/Cr	0.48 ± 0.06	0.40 ± 0.06	0.46 ± 0.07	0.42 ± 0.06
β,γ-Glx/Cr	0.35 ± 0.03	0.36 ± 0.04	0.26 ± 0.05	0.28 ± 0.06
$Glucose_1/Cr$	0.55 ± 0.09	0.40 ± 0.07	0.48 ± 0.07	0.43 ± 0.05
$Glucose_2/Cr$	0.12 ± 0.04	0.10 ± 0.02	0.13 ± 0.03	0.15 ± 0.08

A: Obtained in a group of 12 healthy adult volunteers, 43 ± 4 years old (NAA/Cr, Cho/Cr, mI/Cr, α-Glx/Cr, and β,γ-Glx/Cr ratios are from: Ross, Jacobsen, Villamil, Korula, Kreis, Ernst, Shonk, and Moats, Subclinical hepatic encephalopathy: Proton MR spectroscopic abnormalities, Radiology 1994; 457–463; and the $Glucose_1/Cr$ and $Glucose_2/Cr$ ratios are from: Kreis and Ross, Cerebral metabolic disturbances in patients with subacute and chronic diabetes mellitus by proton magnetic resonance spectroscopy, Radiology 1992; 184:123–130). The $Glucose_1/Cr$ and $Glucose_2/Cr$ ratios are from a similarly aged but not identical volunteer group. B: Obtained with a later software version, but with an identical pulse sequence (except for the prescan Gaussian water suppression Rf pulses which were changed from (90° + α,90° + α,90° + α) to (67°,103°,119° + α)), in a group of 11 younger healthy volunteers (27 ± 4 years old). The values clearly overlap, with the exception of NAA/Cr (significantly different $p < 0.05$), but this can easily be understood as a result of the age difference in between the two groups. The grey matter values of α-Glx and $Glucose_1/Cr$ also differ in between the two groups; this is understood as a small effect of the differing water suppression schemes.

principle of reciprocity for a transmit-receive coil, have been used by other research groups, (see "Further Reading" for details).

The scaled metabolite peak areas are then corrected for a number of NMR parameters including effects of T_1 and T_2 relaxation and of J-coupling. For each new group of patients one needs to establish that these NMR corrections are not significantly different from healthy volunteers. That has proven to be the case for most patient groups until now. Thanks to the fact that we use short echo time, T_2 changes

Table 2.4 Metabolite Ratios and Normal Aging

Age range	N	NAA/Cr	Cho/Cr	mI/Cr
Grey matter				
16–25	10	1.41 ± .08	0.56 ± .05	0.60 ± .04
26–37	10	1.36 ± .08	0.61 ± .08	0.60 ± .04
40–78	12	1.26 ± .09	0.60 ± .05	0.59 ± .06
White matter				
16–25	10	1.54 ± .09	0.77 ± .05	0.59 ± .05
26–37	10	1.49 ± .07	0.78 ± .06	0.60 ± .04
40–78	12	1.41 ± .12	0.82 ± .08	0.63 ± .05

Source: Shonk and Ross, Role of Increased Cerebral myo-Inositol in the Dementia of Down Syndrome. Magn. Reson. in Medicine 1995; 33: 858–861.

N is the number of healthy volunteers in each group. The data were obtained with the exact same software as the data listed in Table 2.3 column A. The agreement with Table 2.3 is clear, especially when the slight continuous changes with age are considered.

Table 2.5 Metabolite Ratios for Basal Ganglia (Mainly Putamen)

NAA/Cr	Cho/Cr	mI/Cr
1.13 ± 0.06	0.73 ± 0.06	0.48 ± 0.04

A group of 8 healthy volunteers were measured. The data were obtained with the exact same software as the data listed in Table 2.3 column B. It is obvious that the normal ratios from these locations are different from the ratios from the two standard locations.

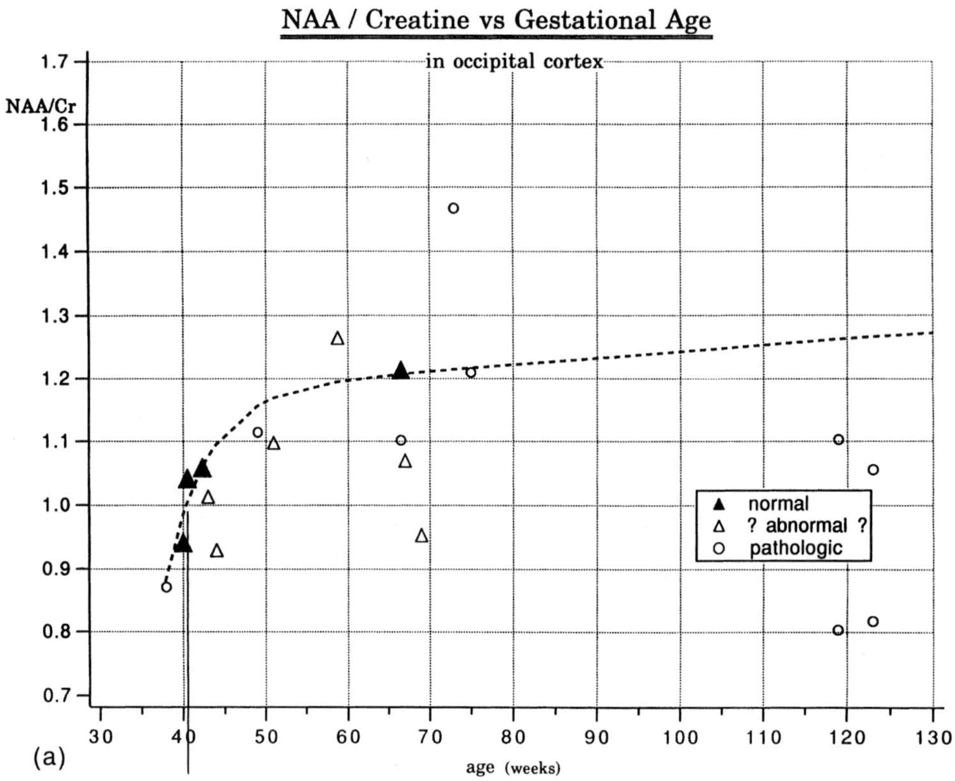

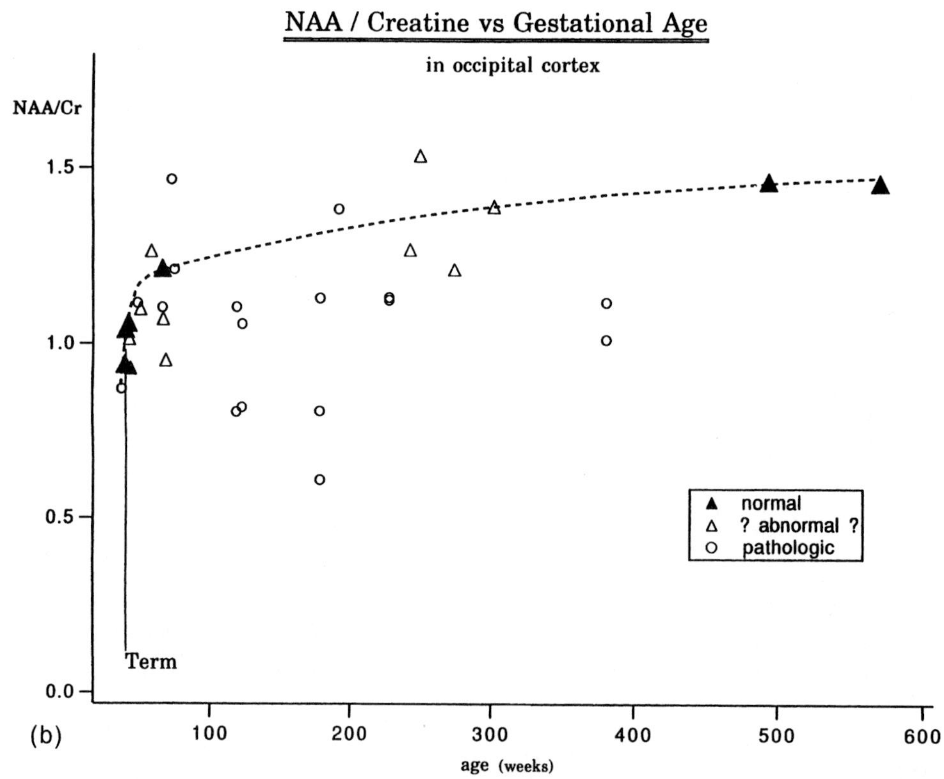

Fig. 2.11 Age dependence of metabolite ratios in HMRI's standard locations (in infants grey and white matter is not yet differentiated so the standard locations are: occipital cortex, in adult named grey matter, and parietal cortex in adults named white matter): (a) NAA/Cr in occipital cortex 0–130 weeks gestational age. (b) NAA/Cr in occipital cortex 0–600 weeks gestational age. (c) Cho/Cr in occipital cortex 0–130 weeks gestational age. (d) Cho/Cr in occipital cortex 0–600 weeks gestational age. (e) mI/Cr in occipital cortex 0–130 weeks gestational

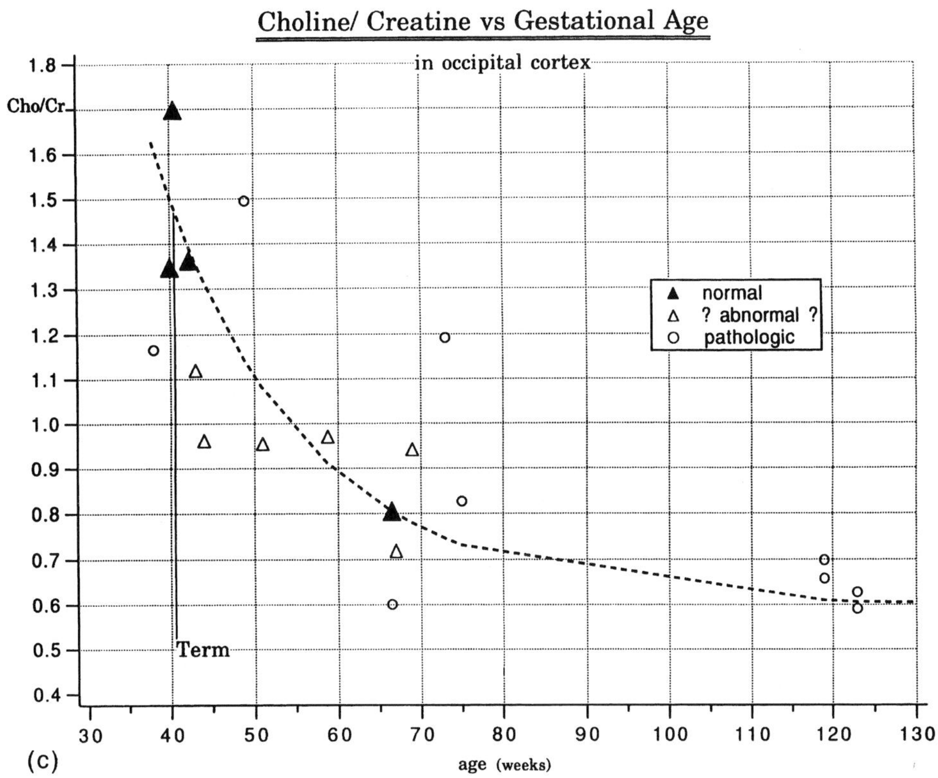

(c)

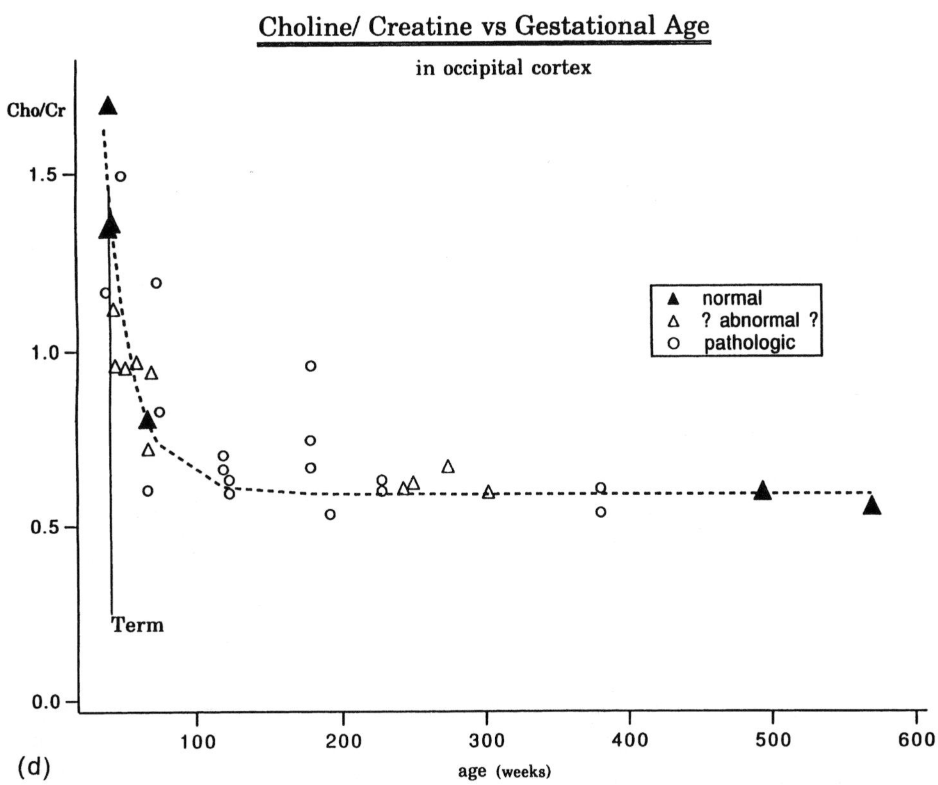

(d)

age. (f) mI/Cr in occipital cortex 0–600 weeks gestational age. (g) NAA/Cr in parietal cortex 0–130 weeks gestational age. (h) NAA/Cr in parietal cortex 0–600 weeks gestational age. (i) Cho/Cr in parietal cortex 0–130 weeks gestational age. (j) Cho/Cr in parietal cortex 0–600 weeks gestational age. (k) mI/Cr in parietal cortex 0–130 weeks gestational age. (l) mI/Cr in parietal cortex 0–600 weeks gestational age.

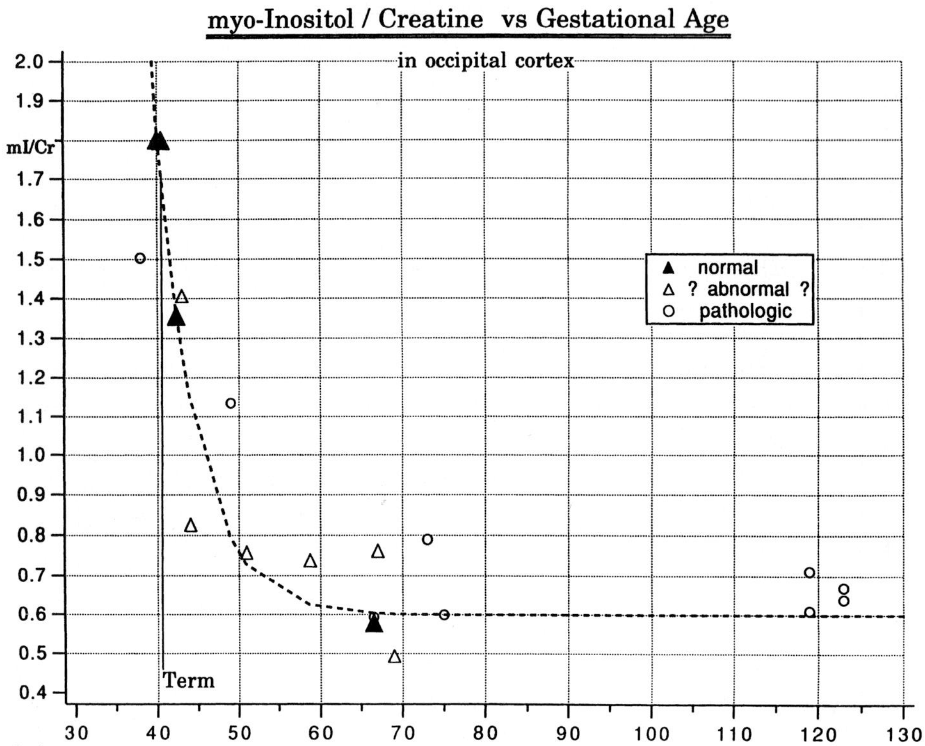

(e)

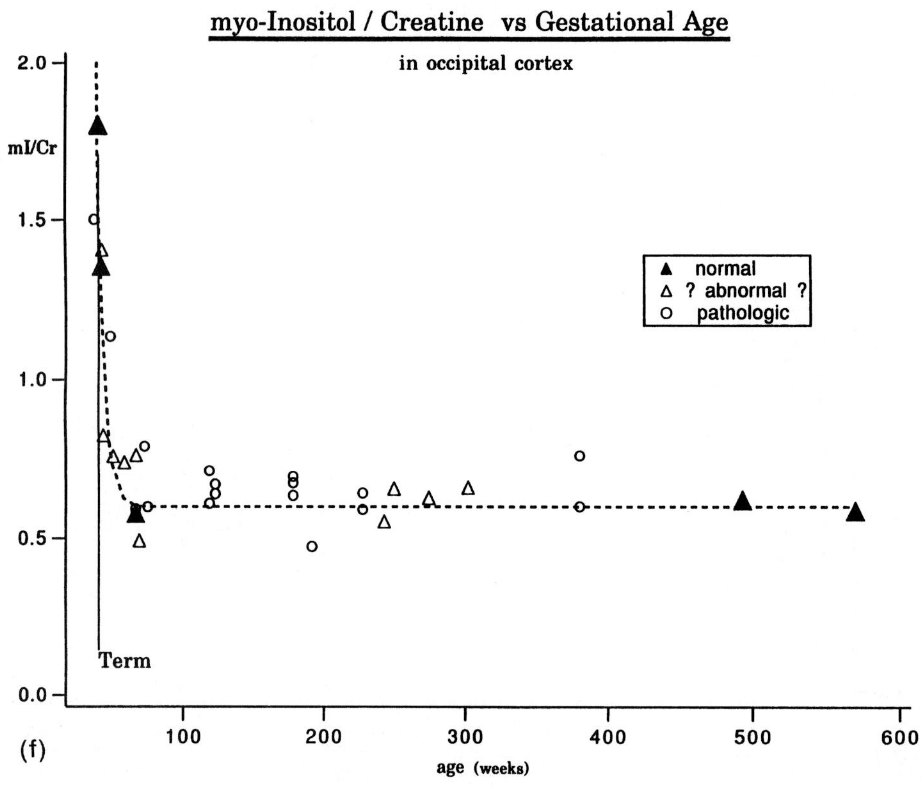

(f)

Fig. 2.11 Continued

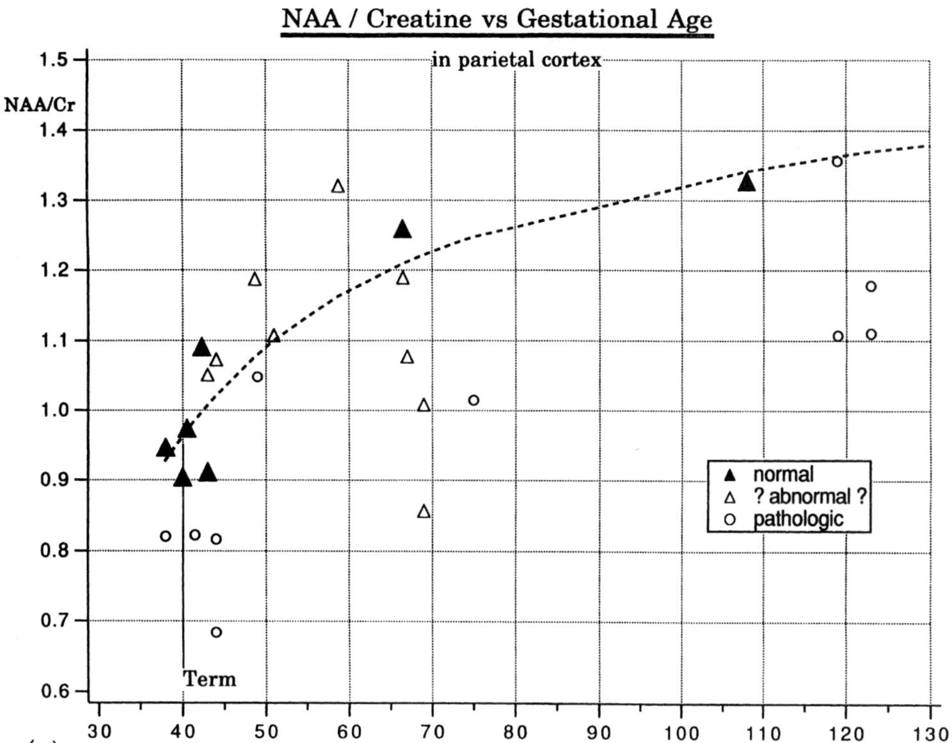
NAA / Creatine vs Gestational Age
in parietal cortex
NAA/Cr
1.5
1.4
1.3
1.2
1.1
1.0
0.9
0.8
0.7
0.6
30 40 50 60 70 80 90 100 110 120 130
age (weeks)
Term
normal
? abnormal ?
pathologic

(g)

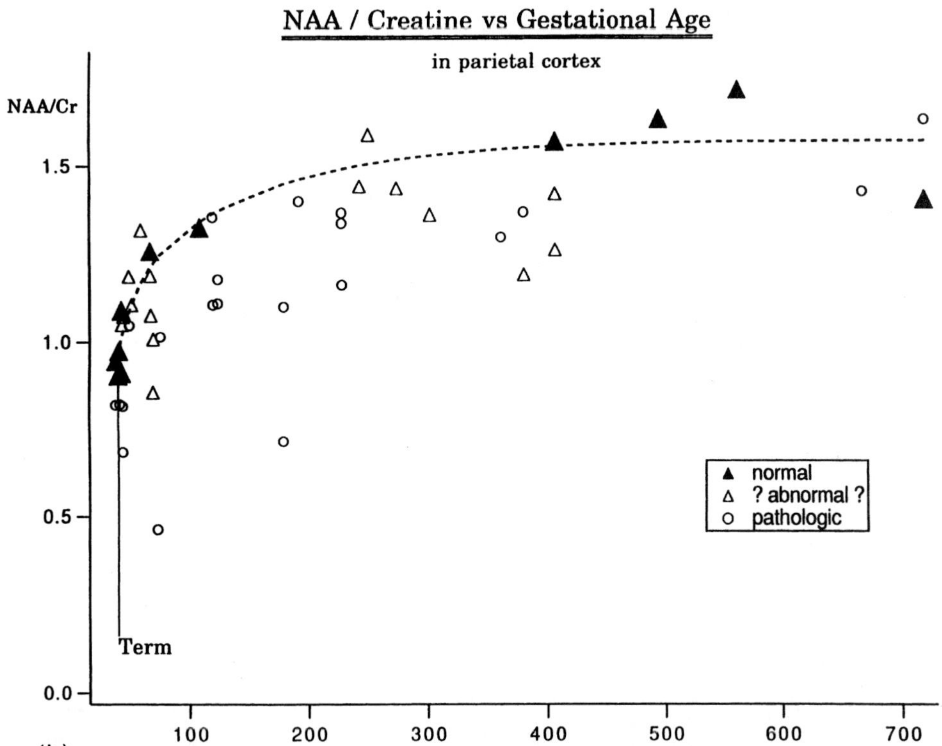
NAA / Creatine vs Gestational Age
in parietal cortex
NAA/Cr
1.5
1.0
0.5
0.0
100 200 300 400 500 600 700
age (weeks)
Term
normal
? abnormal ?
pathologic

(h)

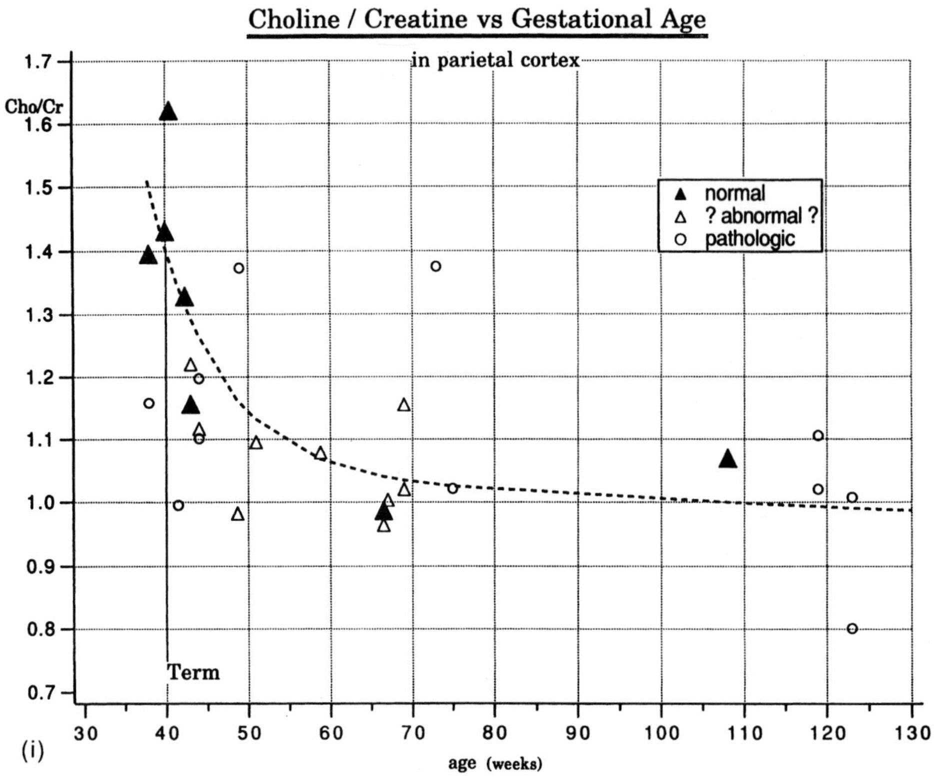

(i)

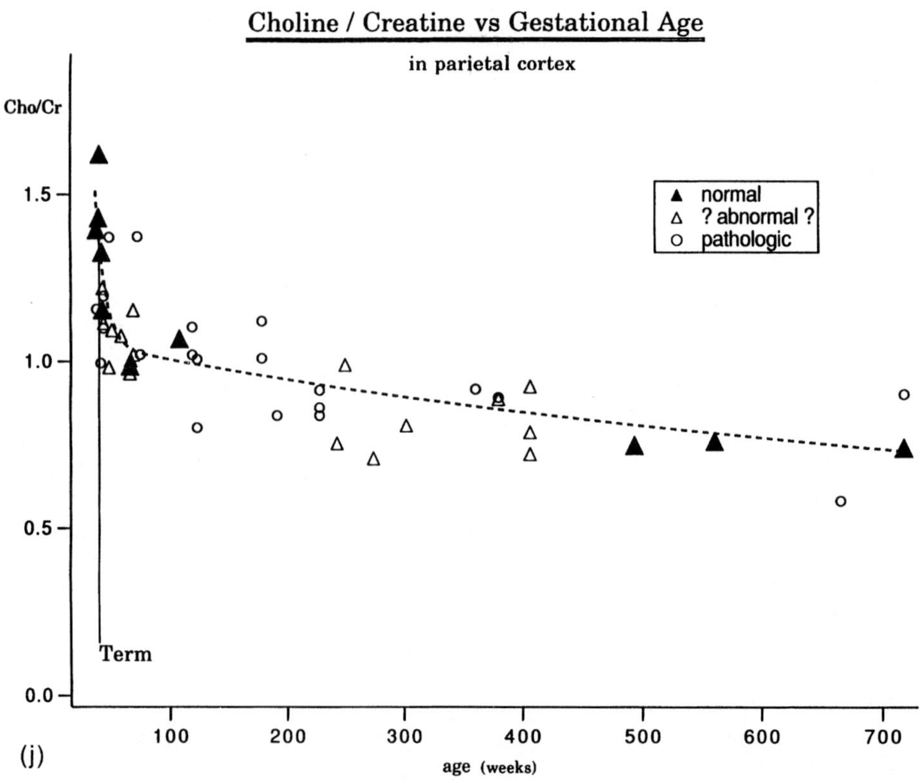

(j)

Fig. 2.11 Continued

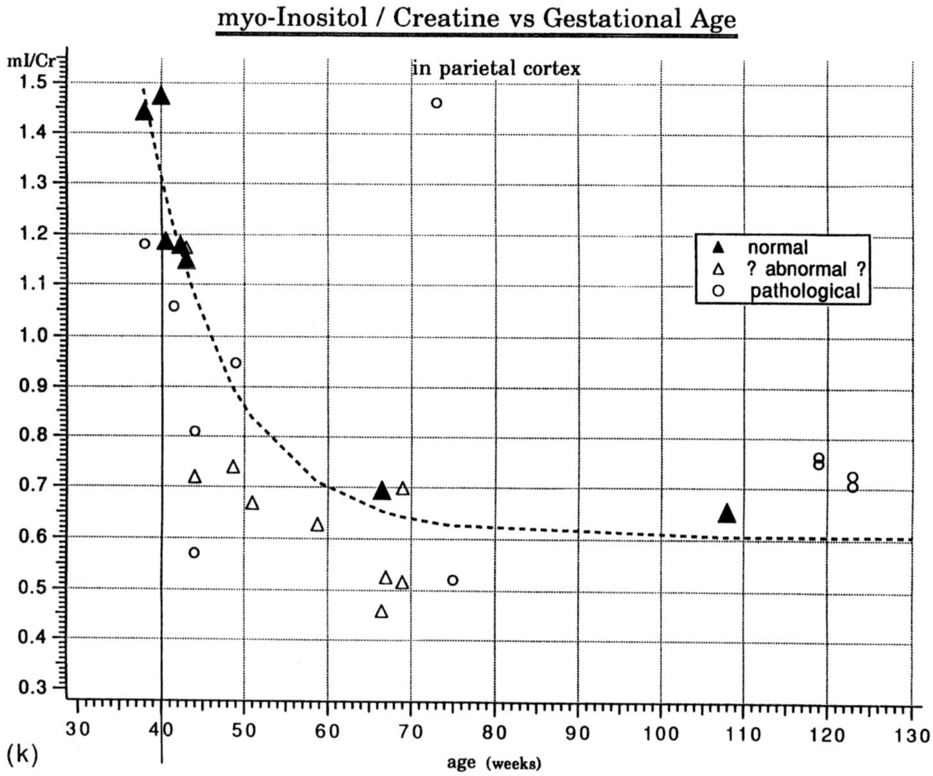
myo-Inositol / Creatine vs Gestational Age
in parietal cortex
mI/Cr
1.5
1.4
1.3
1.2
1.1
1.0
0.9
0.8
0.7
0.6
0.5
0.4
0.3
normal
? abnormal ?
pathological
30
40
50
60
70
80
90
100
110
120
130
age (weeks)
(k)

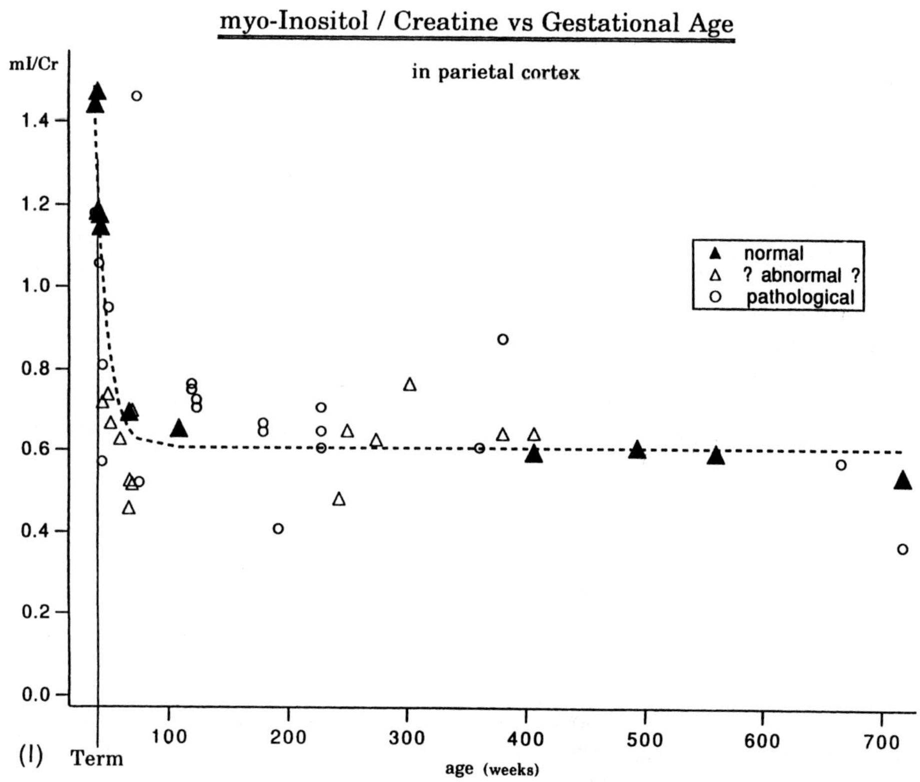
myo-Inositol / Creatine vs Gestational Age
in parietal cortex
mI/Cr
1.4
1.2
1.0
0.8
0.6
0.4
0.2
0.0
normal
? abnormal ?
pathological
Term
100
200
300
400
500
600
700
age (weeks)
(l)

need to be dramatic before any effect on the relaxation signal reduction factor occurs. Similarly, repetition times have been sufficiently long that T_1 saturation factors have been unaffected. Coupling signal reduction factors for short echo times are believed not to be significantly affected by disease. Altogether, in practice this means that we have established the NMR corrections for healthy volunteers, and the values are not measured in each individual patient. In each "new" disease the validity of the values is confirmed before use.

Without further corrections the concentrations would be calculated as mmol/liter in the total VOI. But it makes sense to calculate the concentrations only per volume of brain tissue, because most of the metabolites are found only in brain tissue, not in cerebrospinal fluid (CSF). Therefore the unsuppressed water signals are analyzed and quantified to obtain relative size of the three compartments: the volume of brain tissue water, of CSF water, and of dry matter (also includes invisible water), and the CSF correction is applied. The differentiation between CSF and tissue water relies on the approximately 10 fold different T_2 relaxation times. Further corrections include conversion of the volume concentrations (mmol/liter) to mass concentrations (mmol/g), (see "Further Reading" for details). Although this sounds rather complicated, the overall effect is to allow a VOI to be selected which also includes part of the ventricle or even a fluid-filled cyst. Subsequent data processing can relate metabolite content to only the proportion of the VOI which was truly "brain."

Table 2.6 Normal Concentrations in HMRI's Standard Locations for Grey Matter (GM) and White Matter (WM)

	GM	WM
NAA	9.24 ± 0.26	8.81 ± 0.18
Cr	8.04 ± 0.32	6.33 ± 0.14
Cho	1.43 ± 0.08	1.58 ± 0.02
mI	7.26 ± 0.52	6.68 ± 0.44

Source: Kreis et al. Journal of Magn. Reson. B 102, 9–19 (1993).
The result was obtained from measuring N = 10 healthy volunteers (22–34 years old, 5 males/5 females). The values of the tissue concentrations (mmol/kg tissue) are presented as mean ± "standard error of the mean." (Standard deviations are obtained by multiplying with the square root of N).

E. Normal Absolute Concentrations

Normal metabolite concentrations are given in Table 2.6, but for convenience we present the concentrations relative to the normal value for the location of interest in healthy adult volunteers. All numbers in Table 2.6, would hereby become 100% or = 1.00, in what is also called Institutional units. This way normal values become very easy to remember, and μmol/g concentration can readily be obtained by multiplying with the relevant value from Table 2.6.

Age dependence is treated in a similar fashion as described above for the metabolite ratios, see also the graphs in Fig. 2.12 for age dependence in infants.

V. SUMMARY

An MR spectrum as the one shown in Fig. 2.13 from a VOI in the brain provides information about the concentration and the presence or absence of metabolites in the VOI.

Before acquiring the spectrum the operator prescribes the measurement choosing VOI location, and position, echo time, repetition time, number of scans to be averaged, and possibly other parameters. These choices affect the appearance and diagnostic quality of the spectrum, and consistency in choosing affects reproducibility in a clinical setting.

Prior to sampling the spectrum, a number of adjustments are carried out either automatically or manually. These include frequency, transmitter and receiver adjustments, and shimming and optimization of the water suppression.

The peaks of the spectrum can be measured "by eye," more systematically with a ruler, or by a computerized analysis; all three methods lead to peak heights or areas. The spectrum is analyzed by this peak measure, either in a semiquantitative form as peak height or area ratios, or in the form of scaled and corrected absolute quantitative numbers; metabolite concentrations. The scaling takes care of differences in how much the signal is amplified in each examination; other corrections include those of relaxation, and tissue compartmentation. Both the peak ratios and the absolute numbers need to be related to normal values from the same brain location in healthy age matched volunteers. Variation with age is particularly important during the first year of life.

It must be by now obvious, but bears repeating: The secret of successful clinical MRS is standardization, without which, the user must cope with every one of the multiple variables described in this chapter.

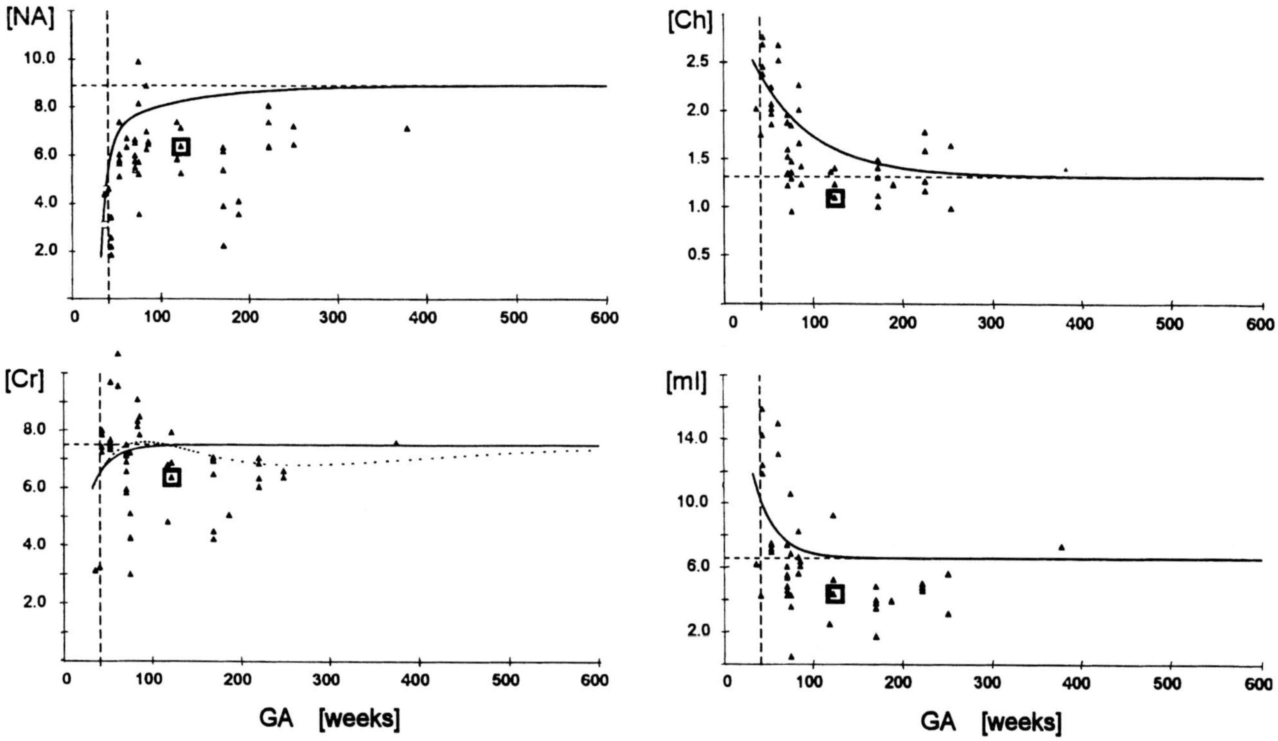

Fig. 2.12 Absolute metabolite concentrations as function of gestational age, determined by ^{1}H MRS. The concentrations are expressed as millimoles/kg brain tissue and are corrected for T_1 and T_2 alterations, underlying spectral features and voxel composition (see "Further Reading" for details). The filled symbols represent data from occipital, grey matter, the light markers stand for data from parietal, (developing) white matter. Exponential curves were fitted from all occipital data and the parietal data of subjects younger than 80 weeks gestational age. The parietal data of children older than 80 weeks are included in the figure, but not used in the fitting. The plotted lines are biexponential for N-Acetyl residues (NA, mainly N-acetyl aspartate, therefore usually referred to as NAA), and monoexponential for Cr, Ch and mI. The dotted curve in the Cr plot, which describes the data somewhat better has triexponential character. The full lines indicate best fitting double exponential curves for both the occipital and the parietal locations. The horizontal lines identifies the value determined for occipital grey matter (dashed lines) and parietal white matter (broken lines) in a group of adults with a mean age of 28 years. The vertical dashed line marks term at 40.5 weeks gestational age. The filled square indicates data from a group of infants; three who were truly normal, and the remainder had no primary cerebral pathology, but who had to undergo an MRI examination for some other than neurologic reason. The filled diamonds indicate data from a group of infants with pathology considered immaterial to the cerebral metabolite profile, including recovered infants after successful Extra Corporeal Membrane Oxygenation (ECMO) treatment, the apparently unaffected hemisphere in well localized unilateral disease, such as middle cerebral infarct, seizures etc. This second group was attributed less weight in constructing the normal curves.

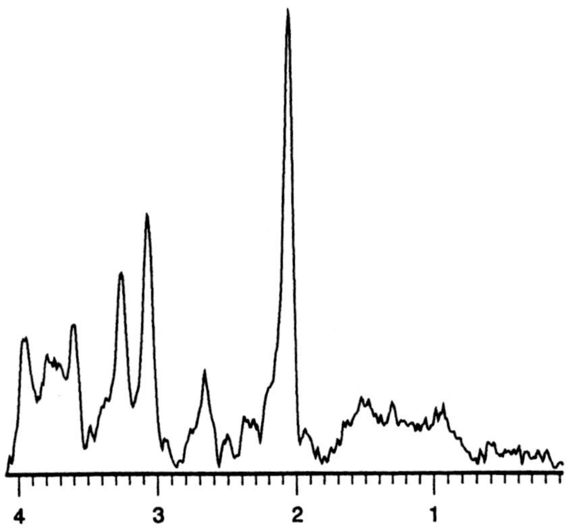

Fig. 2.13 A typical MR spectrum from mostly white matter of the human brain.

FURTHER READING

This three-volume book describes all features of In Vivo Magnetic Resonance Spectroscopy with a focus on the more technical aspects.

Diehl P, Fluck E, Günther H, Kosfeld R, and Seelig J, eds. In-Vivo Magnetic Resonance Spectroscopy. I: Probeheads and Radiofrequency Pulses, Spectrum Analysis. II: Localization and Spectral Editing. III: In-Vivo MR Spectroscopy: Potential and Limitations. Berlin: Springer-Verlag, 1992.

An excellent introduction to in vivo NMR. The book explains the principles and methods with great clarity and precision. It is often the first recommendation to the "In vivo NMR novice," and it is the basic reference book to many "In vivo NMR spectroscopists."

Gadian DG. NMR and its applications to living systems, 2nd edition. Oxford: Oxford University Press, 1995.

Original References Describing Details of Quantitative Localized Human MRS

Ernst T, Kreis R, and Ross BD. Absolute Quantitation of Water and Metabolites in the Human Brain. I. Compartments and Water. J. Magn. Reson. 1993; B 102: 1–8.

Kreis R, Ernst T, and Ross BD. Absolute Quantitation of Water and Metabolites in the Human Brain. II. Metabolite Concentrations. J. Magn. Reson. 1993; B 102: 9–19.

Kreis R, Ernst T, and Ross BD. Development of the Human Brain: *In Vivo* Quantification of Metabolite and Water Content with Proton Magnetic Resonance Spectroscopy. Magn. Reson. Med. 1993; 30: 424–437.

Danielsen ER, Michaelis T, and Ross BD. Three Methods of Calibraton of Quantitative Proton MR Spectroscopy. J. Magn. Reson. 1995; B 106: 287–291.

Michaelis T, Merboldt K-D, Bruhn H, Hänicke W, and Frahm J. Absolute concentrations of metabolites in the adult human brain in vivo: Quantification of localized proton MR spectra. Radiology 1993, 187; 219–227.

3

The Clinical Significance of Metabolites

I. BIOCHEMISTRY: BACK TO THE FUTURE

Biochemistry, which is so vital if we are to understand MRS, is a lost art. Its principles are enshrined in several key textbooks, of which useful examples are given below, but few practitioners knowledgeable in the area of intermediary metabolism are found any more in biochemistry departments. MRS is the most powerful and versatile technique that we have for defining chemical species, molecular structure and the *in vivo* chemical environment. In a period of ever expanding interest in molecular diagnosis, MRS provides quantitative, noninvasive assays of metabolites from which many such diagnoses can be made.

Biochemistry, as defined by MRS, is limited to those metabolites which can be assigned and quantified by the methods at hand, or revealed by ingenious editing sequences. Therefore, we first present the common assignments available *in vivo* from the brain using proton MRS:

Amino-acids: N-acetylaspartate, alanine, glutamate
Amines: glutamine, choline, creatine
Sugars: *myo*-inositol, glucose, *scyllo*-inositol
Others: lactate, long-chain fatty acids, triglyceride, ketones
Xenobiotics: ethanol, mannitol, propanediol

II. THE NEW NEUROCHEMISTRY

At its inception, MRI carried with it the potential not only for exquisitely detailed neuroanatomical imaging, but also for noninvasive analysis of neurophysiology and neurochemistry. One technique, proton MRS (^{1}H MRS) also known as neurospectroscopy, has matured to the point of providing clinical neurophysiologic and neurochemical information. MRS, we would suggest, is indispensable for state-of-the-art neurodiagnosis. This chapter will aid the practicing neurologist to confidently request, perform, and interpret the results of proton MRS either alone, or as an adjunct to neuro MRI/CT.

A. What Is Neurospectroscopy?

We define neurospectroscopy as the field of study resulting from MRS examination of the human brain. Diseases and pathologies of the brain are commonly classified as: (1) structural (including degenerative, tumor and embryogenic defects); (2) physiological (essentially interruption of blood supply); and (3) biochemical or genetic. Of the latter, some are receptor and neurotransmitter-related (e.g. dopamine in Parkinson's disease), but many are directly or indirectly related to disturbances of the pathways of oxidative, anabolic and catabolic intermediary metabolism, the TCA cycle, glutamine/glutamate turnover, glycolysis, ketogenesis, or fatty acid metabolism. Genetic diseases involve mutations of nuclear or mitochondrial DNA. PET, and to a lesser extent SPECT, MRA, fMRI and "diffusion-imaging" address blood flow, glucose turnover and oxygen consumption, and PET and SPECT are uniquely able to "image" targeted receptor ligands. However, until the advent of MRS, no direct noninvasive assay of the cerebral metabolites, was available. There was no neuronal marker, no astrocyte marker, and no technique to directly determine energy metabolism. These gaps are now filled by neurospectroscopy, and with increased clinical experience a diagnostic "need" for magnetic resonance spectroscopy (MRS) of the brain emerges.

B. MRS Introduces the New Neurochemistry

Fig. 3.1 depicts some of the neurochemical pathways which have become more relevant since the advent of neuro-MRS. The major metabolites of the ^{1}H MR spectrum, N-acetylaspartate (NAA), total creatine (creatine plus phosphocreatine; Cr), total choline (as reflection of phosphoryl choline and glycerophosphoryl choline; Cho), *myo*-inositol (mI), and glutamate plus glutamine (Glx), were only infrequently encountered in neurochemical discussion of physiology or

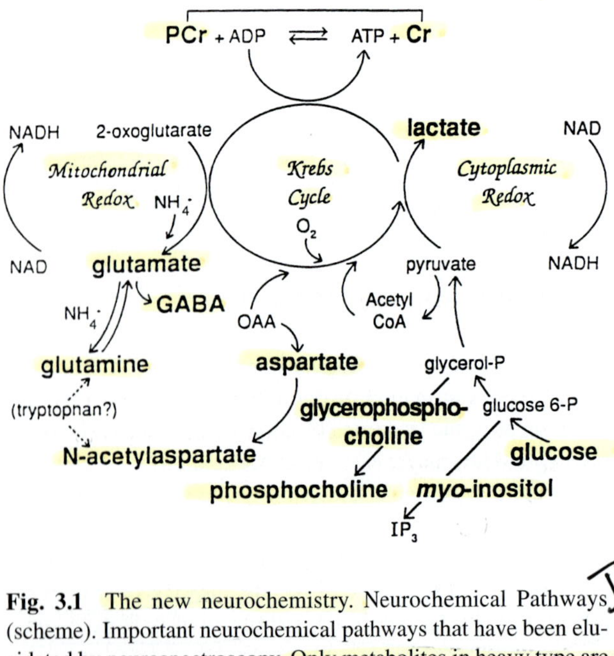

Fig. 3.1 The new neurochemistry. Neurochemical Pathways (scheme). Important neurochemical pathways that have been elucidated by neurospectroscopy. Only metabolites in heavy type are readily visualized in ^{1}H MRS, but information gathered from ^{1}H MRS extends to the other metabolites shown.

disease, before the advent of MRS. They now join glucose uptake and oxygen consumption as the most easily measured neurochemical events, and will become increasingly important in neurological discussion.

Because the concentration limit of protons is about 0.5–1.0 mM for MR-detection, virtually all true neurotransmitters, including acetyl-choline, norepinephrine, dopamine, and serotonin (the exceptions are glycine, glutamate, glutamine and GABA), are currently beyond detection by conventional neuro-MRS. Similarly, the hormone messenger inositol polyphosphates and cyclic AMP are not detected. This leaves important gaps in the New Neurochemistry.

Another evident shortcoming of MRS is the inaccessibility of most macromolecules, due to their limited mobility. Accordingly, phospholipids, myelin, proteins, nucleosides and nucleotides, as well as RNA and DNA, are effectively "invisible" to this family of methods.

III. COMMON PEAK ASSIGNMENTS

Although the brain spectrum will be described in detail in the following chapter, before we address the relevant clinical pathobiochemistry of the metabolites observed in proton spectra, it is instructive to remind ourselves of the basis upon which each peak or complex was assigned its chemical identity.

A composite of a human ^{1}H brain spectrum identifies 17 peaks, including the residual water and lipid (Fig. 3.2). Simplification of the spectrum, by procedures generally described as "editing" can be achieved. Metabolites not at present visualized in the spectrum shown might be seen if yet other "editing" procedures were to be applied during initial acquisition of the signal. Another form of "simplification" reduces a proton spectrum of brain to only the most prominent three resonances; this spectrum is a commonly encountered profile. All this leads us to artificially subdivide the spectrum for the ensuing discussions of biochemical and clinical significance of ^{1}H MRS.

Spectrum I (Fig. 3.2) is the simplest grouping of the major resonances observed in the brain: NAA, Cr, and Cho. NAA actually has three resonances, and creatine has two, so that the simplest analysis nevertheless ascribes six peaks to three metabolites. The metabolites are chemically unrelated and the clinical significance of each is discussed elsewhere. Spectrum II is in fact not a spectrum, but an area in which lactate, alanine, and a number of other "methyl" protons, including ethanol, would resonate. Lactate (and alanine) are normal and important metabolites that existing techniques can demonstrate only at pathologically elevated concentrations. The metabolic significance of lactate is discussed

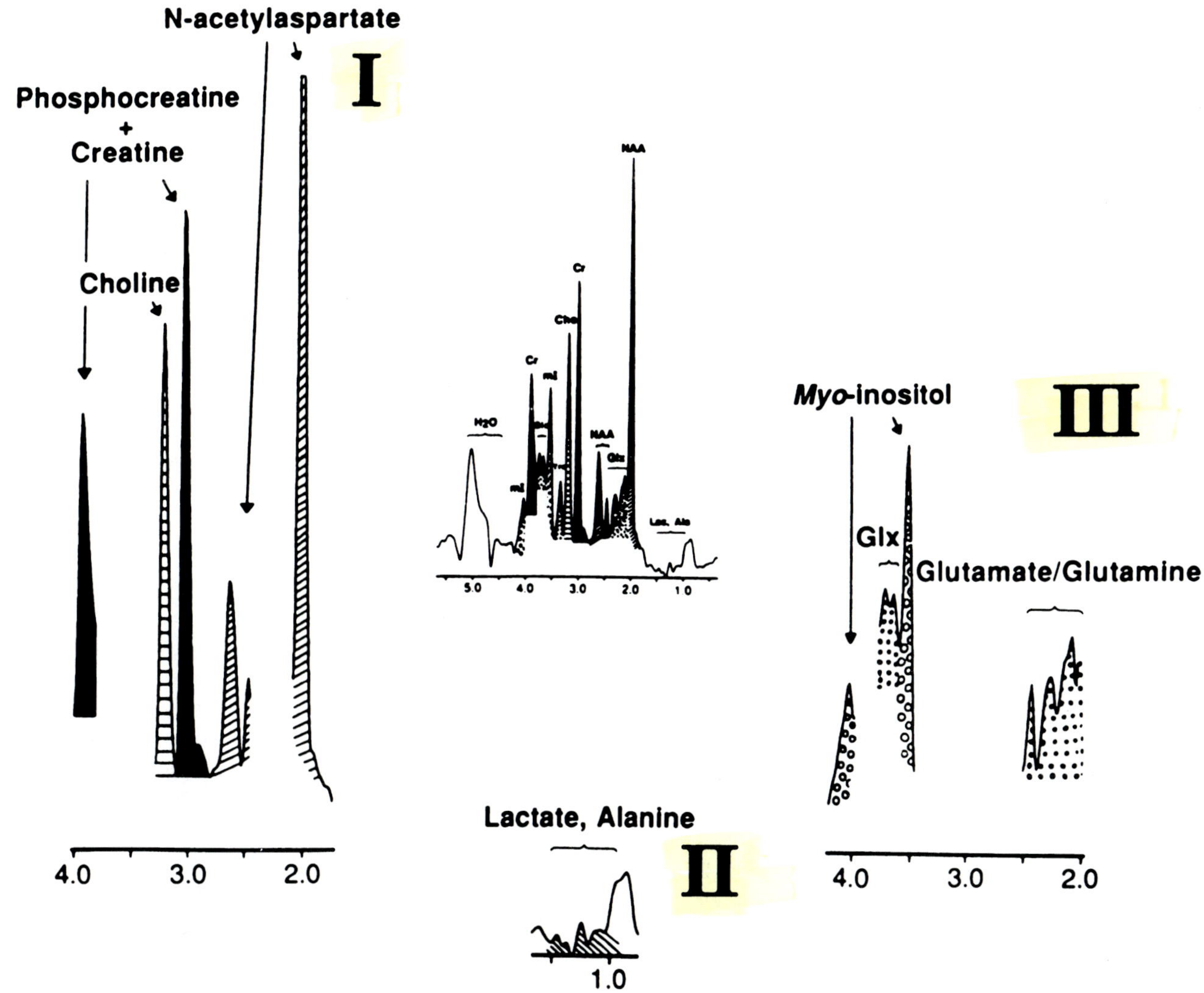

Fig. 3.2 Proton spectrum of human cerebral cortex. The central spectrum was obtained at 1.5 T after water suppression, and with a relatively short echo time (30 ms). The insets illustrate the main metabolites of a proton spectrum: N-acetylaspartate, creatine, and choline (I); lactate and alanine (barely perceptible at physiological concentration) (II); and glutamate plus glutamine, and *myo*-inositol (III). N-acetylaspartate (NAA); phosphocreatine plus creatine (Cr); choline complex (Cho); *myo*-inositol (mI); glutamate plus glutamine (Glx); *scyllo*-inositol (sI)/glucose/taurine; lactate (Lac); alanine (Ala). (From Ross NMR in Biomedicine 1991; 4(2):i–ii.)

later. Spectrum III selects from the remaining peaks those associated with three further metabolites: glutamate (Glu) plus glutamine (Gln) (together labeled Glx), and mI. The six or more protons of Glx yield several peaks in two distinct regions of the spectrum, while mI, an unrelated cerebral metabolite, yields two. However, the pathophysiological importance of these compounds more than repays the efforts made to analyze them successfully. Biochemistry of Glu, Gln, and mI are described below.

Structural chemistry determines the spectrum, as we see with the simple molecule ethanol (alcohol) (Fig. 3.3). Two principal peaks at a fixed frequency represent the CH_3 and the CH_2 protons. Each peak is rendered more complex (a triplet and a quartet of peaks, respectively) in a pattern which is different in fine detail, depending upon the play of magnetic field gradients and the field strength used for MRS. This is called the coupling pattern. Provided the same method of MRS is adhered to, this pattern can be relied upon to appear in the proton brain spectrum, just as it does in the "bottle" (Fig. 3.4). The "artificial" brain spectrum on the right (Fig. 3.4E) comes from the mixture of metabolites added in appropriate proportions, several of which are

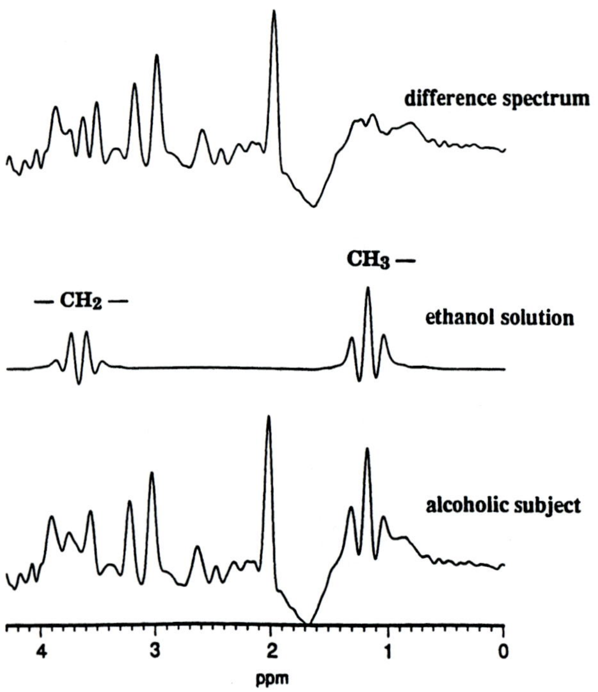

Fig. 3.3 The ^{1}H MR Spectrum of alcohol (Ethanol). The middle spectrum identifies ethanol—C_2H_5OH on the basis of the its coupled resonances CH_3 and CH_2. In the lower spectrum, the familiar human brain spectrum is dramatically altered after the subject imbibes an alcoholic beverage, by the addition of the CH_3 triplet from alcohol now within the brain. The upper spectrum in a computer-generated difference spectrum proving the triplet to be that of ethanol. Subtraction of the two spectra restores the "normal" brain spectrum.

shown as spectra of single components (Fig. 3.4B–D, F–H). Subtle differences between these model spectra and the "real" brain (Fig. 3.4A) are the peaks of chemicals in *in vivo* human brain that were not added to the artificial brain mixture; for example, glutamine.

A. Pitfalls in Peak Assignment

Assignment is readily accomplished by generating spectra of model solutions. A word of caution, however. When new peaks must be assigned, likely candidates can be selected from published chemical-shift tables, and then be *rigorously* confirmed in a three-step process: 1) Acquire a spectrum under identical conditions, including localizing gradients, echo times, etc., to ensure that couplings (if any), correspond to the *in vivo* peak(s). 2) Quantitate the *in vivo* peak(s), at least to within 10–20% to avoid assigning a likely candidate if its concentration is orders-of-magnitude improbable. 3) In an extract, "spike" with the candidate compound and run MRS, or better still, run the MRS, then incubate the extract with specific degradative enzyme under conditions designed to remove only the candidate compound (say AMP), and rerun MRS.

A very good exercise in peak assignment is illustrated by a brain tumor spectrum containing mannitol (see Fig. 3.5 and inset of Fig. 3.6); the crucial proof of identification is provided by the echo time. The spectra obtained from a phantom solution of mannitol at three different TEs (Fig. 3.6), demonstrates that its detection is difficult at the longer TE of 270 ms. This emphasizes the importance of the first step in our three-step confirmation process: *acquire under identical conditions to ensure that couplings correspond to the* in vivo *peak.* A similar exercise is demonstrated for confirmation of glycine in the brain spectrum of an infant (see Case 5.9.1).

B. Understanding Brain Chemistry

The more than 100 brain spectra assembled in this Handbook can be understood in many ways. One is to learn complex metabolic maps of classical intermediary metabolism (for example, the inositol pathway; Fig. 3.7), in which case many pages of maps would be required. Another is to recognize patterns of common neuropathology (Fig. 3.8), and pigeon-hole each new patient. This is a powerful method in practice, and we suspect will be that preferred by radiologists who are adept at pattern recognition. Most of the case studies in Chapter 5 follow this approach.

A third approach, represented by the cartoons in Fig. 3.9, takes the spectrum one stage further and reminds us of the basic function of the four principal metabolites in the proton spectrum, so that loss of NAA is immediately equated with neuronal (or axonal) damage. A decrease in Cr indicates an energy deficit; change in Cho marks membrane changes; and an increase, or decrease in mI could point to astrocyte (glial) pathology. This approach leads to a useful classification of spectra and ultimately to a handy table (Table 3.1) of differential diagnoses which can be expanded to include all pathologies published in the MRS literature and all cases experienced by the neuroradiologist.

A fourth approach is the one taken in this chapter. To be truly at ease "reading" clinical MRS, we suggest that the clinician will want a greater depth of knowledge of what each of the metabolites contributes to normal brain function, and to know something of the many pathological processes, which in practice, alter their turnover, metabolism, and may even result in their disappearance. In this we can stop short of the complex metabolic maps because for the most part MRS in the human brain is two orders-of-

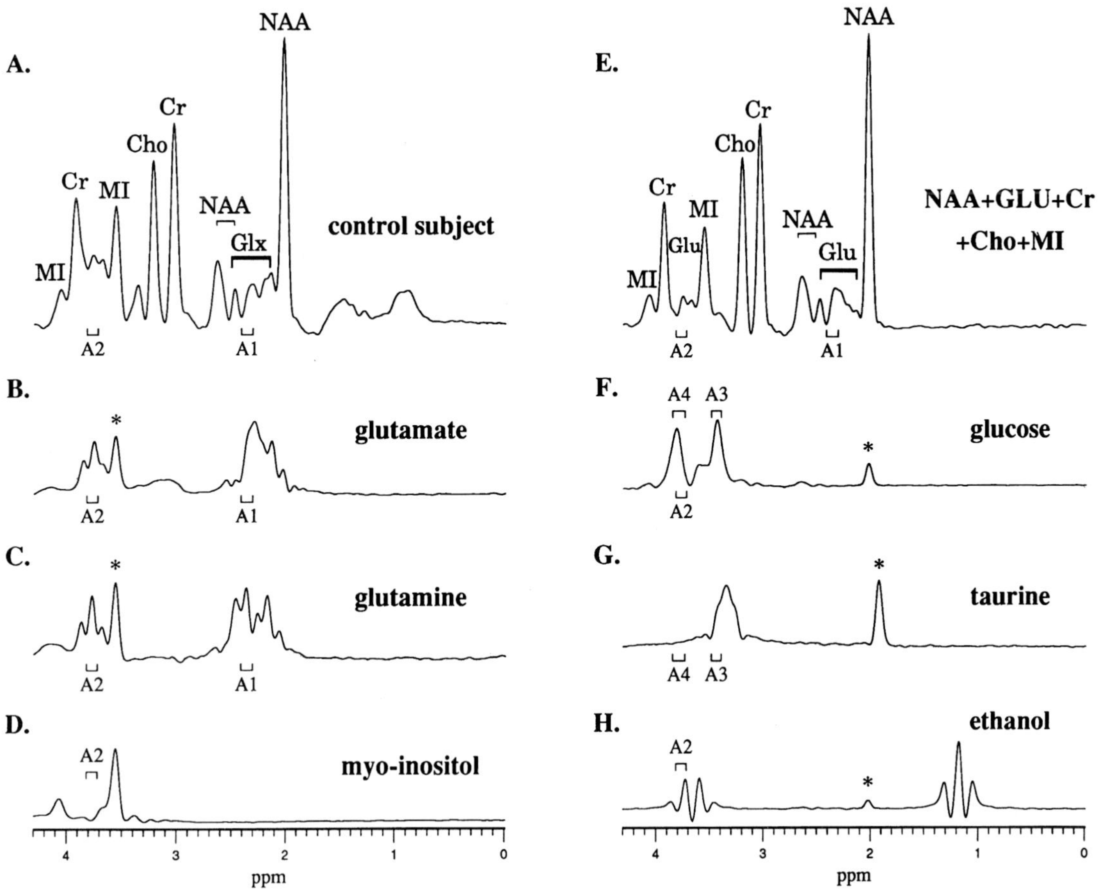

Fig. 3.4 Assigning the peaks of the *in vivo* brain spectrum to its constituents. A is the *in vivo* brain spectrum, while B–H are the spectra of model solutions containing one or more metabolites as indicated. Spectrum E, which looks remarkably like the *in vivo* brain spectrum, is a solution containing only the 5 chemicals indicated. They are added in appropriate proportions so that the relative peak intensities are essentially those of human white matter *in vivo*. From the individual spectra shown it can be readily seen that a single chemical not only has more than one peak, but that each has a characteristic and recognizable shape, which is repeated in the mixture and in the *in vivo* spectrum. Peaks marked * are added as chemical shift reference points; glycine for B and C, and acetate for F, G, and H.

magnitude less sensitive than the biochemical tools used to create those maps. Instead, we will start with simplified maps, and finish again with the checklist of factors (Table 3.1) which should go through the clinician's mind when abnormalities appear in either individual peaks, or in patterns of peaks in the MRS.

C. Brain Metabolism and the ABCs of Energy Production

Perhaps the most remarkable thing about brain spectra is their constancy! The simple explanation for this is the balance of energy production and consumption, and the stable ionic and osmotic equilibrium of living cells (Fig. 3.10). A pivotal fact, which is useful in interpreting some pathological spectra, is that this equilibrium (described by R. L. Veech with the term Gibbs-Donnan equilibrium, an extension of the more familiar Donnan equilibrium), also maintains brain cell volume. Osmotically active metabolites (the "osmolytes"), include some very familiar brain metabolites of the MR spectrum: Cr, PCr, mI, and Cho. Cerebral energy state, brain cell volume, and the MR spectrum are intimately related.

There are three steps to converting brain fuels into adenosine triphosphate (ATP) and phosphocreatine (PCr), the actual sources of chemical energy for thought, electrical transmission, and all of the biosyntheses required to maintain brain function.

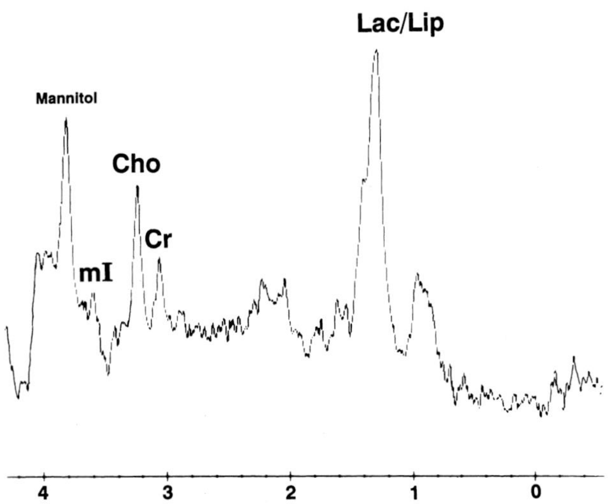

Fig. 3.5 A potential pitfall in peak assignment. A rare *in vivo* brain tumor spectrum exhibiting an unexpected peak (mannitol). The prominent peak of mannitol was initially overlooked because of the similarity to the spectral appearance of *myo*-inositol. Accurate reading of the chemical shift avoids the erroneous assignment. This patient received mannitol to control intracranial hypertension.

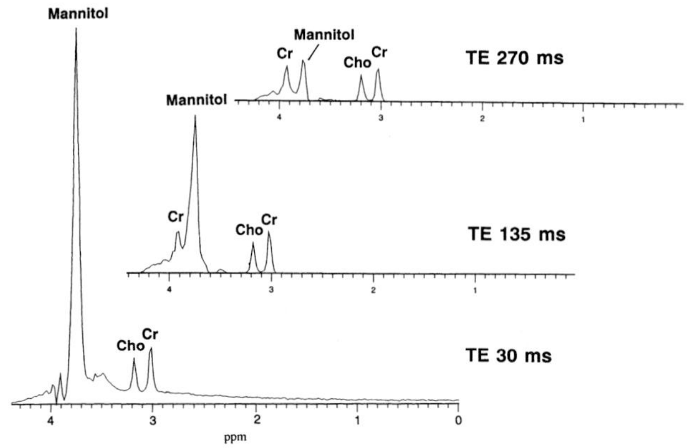

Fig. 3.6 Positive assignment of an MRS peak to mannitol. Three spectra of the same phantom solution are shown, acquired with long echo times (TE = 270 and 135 ms) and short (TE = 30 ms). The latter TE is that used throughout this Handbook. At TE 30 ms mannitol dominates the spectrum. Loss of signal at this chemical shift with long TE is useful confirmation that the peak arises not from a normal brain metabolite, but from a strongly coupled metabolite, probably a xenobiotic. Mannitol fits this description (Courtesy of Drs. Hungerford, Blüml and Norfray).

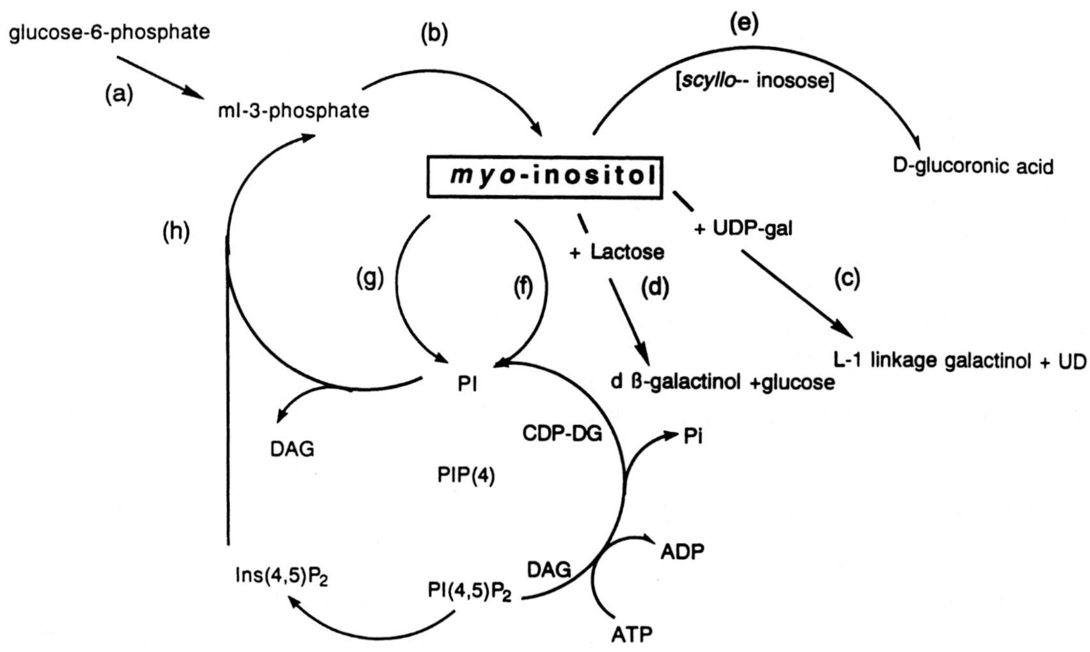

Fig. 3.7 Reactions involving *myo*-inositol.

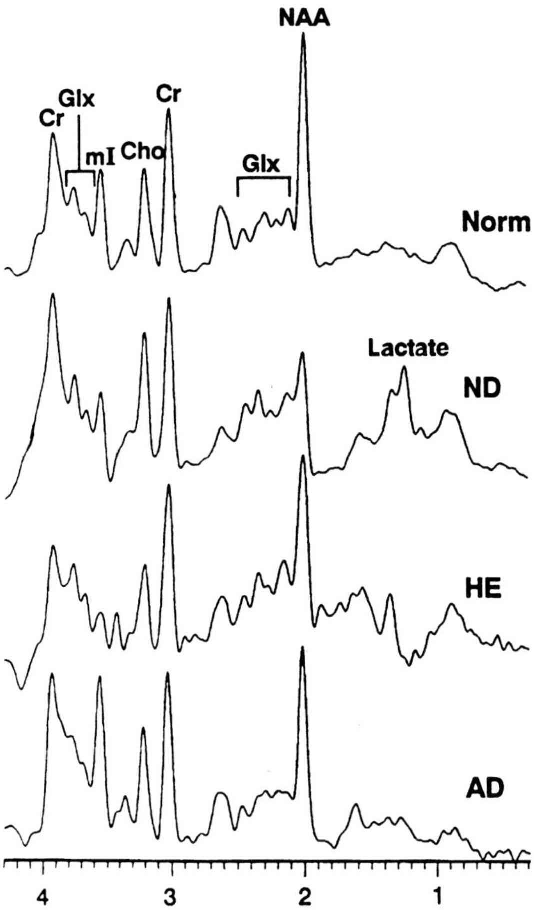

Fig. 3.8 Normal brain MRS and three commonly encountered patterns of abnormality. Normal peaks are identified in the upper MRS (Norm). Additional peaks and alterations in other peak-heights relative to Cr are recognizable in the brain MRS of ND—a child after near drowning, HE—an adult with clinical hepatic encephalopathy, AD—a female aged 63 years with dementia of the Alzheimer type.

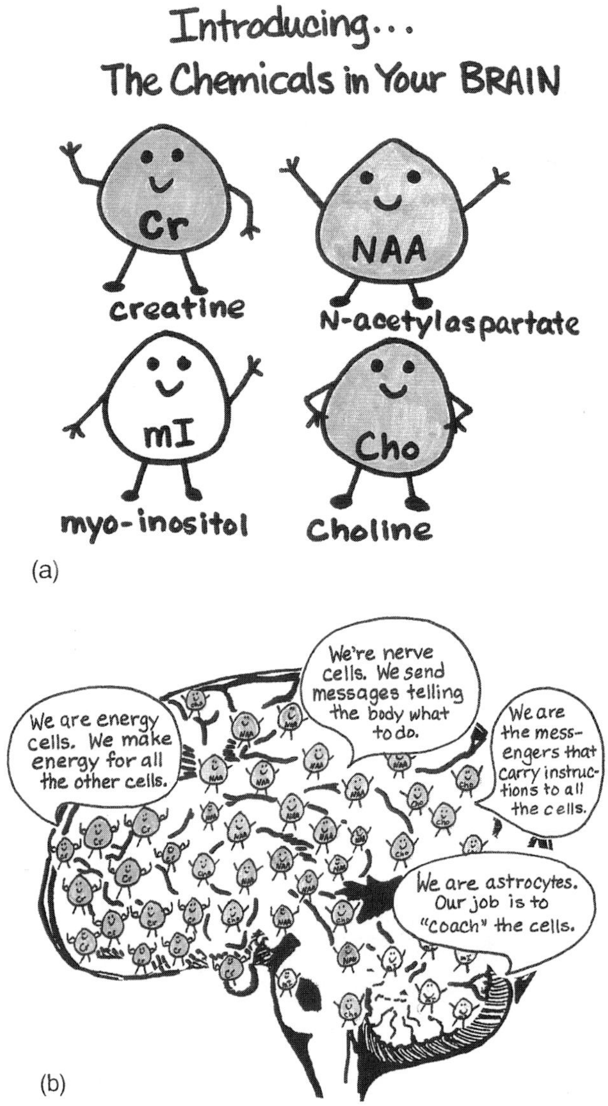

Fig. 3.9 A memory aid to reading and understanding clinical brain spectra. (a) Peaks of MRS reflect individual metabolites. (b) Metabolites reflect brain cell-type or function.

Step 1

Transport and metabolism of the primary fuels—glucose, ketones (in infants), amino acids, and lipids—to two-carbon fragments (Ac CoA; Fig. 3.11).

Step 2

Oxidation of 2-carbon fragments in the Krebs cycle (TCA; Fig. 3.11), with the formation of ATP by oxidative phosphorylation within brain mitochondria ("36"; Fig. 3.11). Failure at this step deprives the brain of ATP, some of which is made up by diversion of pyruvate to lactate (not shown in Fig. 3.11).

Step 3

Creatine kinase, a reversible enzyme (Fig. 3.12) permits synthesis of high-energy PCr from ATP; conversely PCr provides an additional source of brain ATP, releasing free creatine. The substrate, PCr and its product Cr, are maintained in nearly equal proportions. Metabolic adaptations in the brain (Fig. 3.13) may also need to be considered.

Table 3.1 Differential Diagnostic Uses of MRS

Metabolite (normal cerebral concentration)	Increased	Decreased
Lactate (Lac) (~1 mM; not "visible")	(Often) Hypoxia Anoxia Near-drowning ICH Stroke Hypo- and hyperventilation Inborn errors of TCA, etc. Canavan's Alexander's Hydrocephalus Lymphoma (lac + lipid) PML (small increase of lac + lipid) Toxo (15–20 fold increased lac + lip)	(unknown)
N-Acetylaspartate (NAA) (8–9 mM)	(Rarely) Canavan's Axonal recovery e.g. MELAS, MS Infant development Hyperosmolar states Post head injury (hyperosmolar response)	(Often} Developmental delay Infancy Hypoxia Anoxia Ischemia Dementia including Alzheimer's disease ICH Near drowning Alexander's Epilepsy Neoplasm MS Stroke Diabetes mellitus Closed head trauma Toxo Cryptococcoma Lymphoma PML Tumor Herpes II encephalitis Hyponatremia SIADH
Glutamate (Glu) and/or glutamine (Gln) (Glu = 10 mM; Gln = 5 mM)	Chronic hepatic encephalopathy (HE) Acute HE Hypoxia Near-drowning OTC deficiency	Hyponatremia SIADH Post trauma Possible Alzheimer's disease
Myo-inositol (mI) (5 mM)	Neonate Alzheimer's disease Renal failure Diabetes mellitus Recovered hypoxia Hyperosmolar states PML	Chronic HE Hypoxic encephalopathy Stroke Tumor Toxo Cryptococcoma Lymphoma Low grade malignancy

Table 3.1 Continued

Metabolite (normal cerebral concentration)	Increased	Decreased
Myo-inositol (continued)		Hyponatremia SIADH
Total creatine (Cr) (8 mM)	Trauma (hyperosmolar response) Hyperosmolar Increasing with age	Hypoxia Stroke Tumor Infant Trauma Toxo PML Lymphoma Cryptococcoma Hyponatremia SIADH
Glucose (Glc) (~ 1 mM)	Diabetes mellitus ? Parenteral feeding (G) ? Hypoxic encephalopathy	Not dectectable
Choline (Cho) (1.5 mM)	Trauma Diabetes "White" vs. "grey" Neonates Post liver transplant Tumor Chronic hypoxia Hyperosmolar Stroke (common) Elderly normal ? Alzheimer's disease PML	Asymptomatic liver disease HE Stroke (rarely) Nonspecific dementias Toxo Cryptococcoma Hyponatremia SIADH
Lipids and low molecular weight proteins (broad resonances at 0.8–1.5 ppm)	Contamination from outside the VOI e.g. the scalp if the VOI is placed too close to the skull Cryptococcoma Active growing tumor Lymphoma PML Toxo Hypoxic injury and trauma with hypoxic injury (especially in babies) SBS In rare cases true lipid droplets contribute e.g. CPM Inflammatory diseases	Not detectable
Aceto-acetate Acetone Phenylalanine Glycine Aromatic amino acids Xenobiotics (Ethanol; Propanediol; Mannitol)	Detectable in specific settings	

ICH: intracranial hemorrhage; OTC: ornithine transcarbamylase; MS: multiple sclerosis; HE: hepatic encephalopathy; PML: progressive multifocal leukoencephalopathy; Toxo: toxoplasmosis; SBS: shaken baby syndrome; MELAS: myopathy, encephalopathy, lactic acidosis with strokelike lesions; SIADH: syndrome of inappropriate antidiuretic hormone secretion; CPM: central pontine myelinolysis.

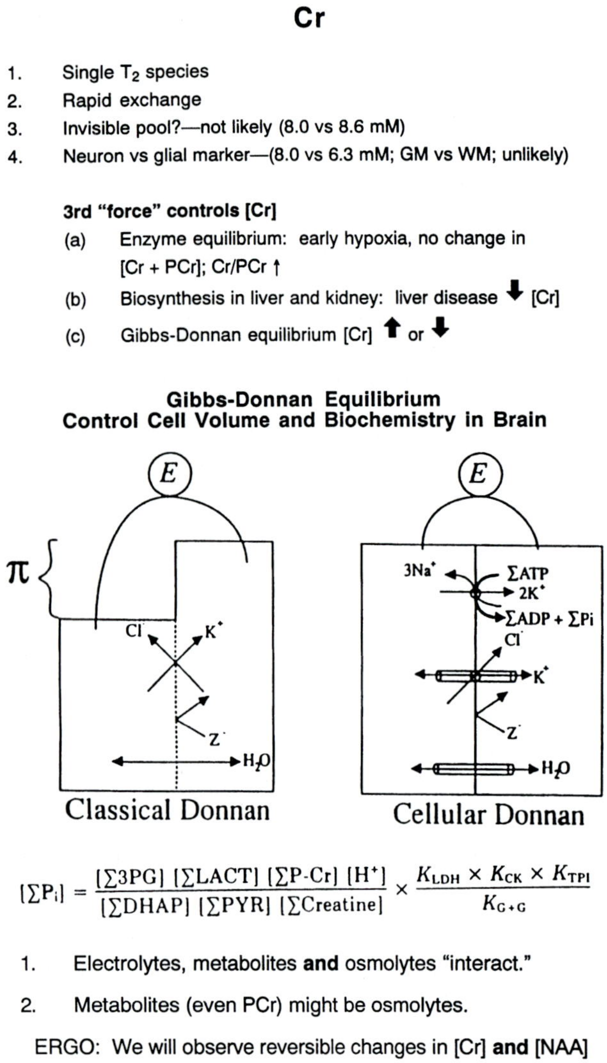

Fig. 3.10 Significance of brain creatine. Diagramatic representation of homeostatic forces maintaining a constant brain spectrum. Cr as the pivotal marker of cerebral energy metabolism alters only when the homeostatic mechanisms are severely compromised. (Thanks to R. L. Veech).

1. Creatine

The sum of Cr and PCr appears as the central peak in the proton brain spectrum, and is an extremely reliable marker of intact brain "energy metabolism" (refer to Steps 1–3, above). Accordingly, it is reasonable to assume that Cr = 1 in the expression of all ratios (see Chapter 2, Section. IV. B), and this assumption will help us considerably in interpreting many pathological MR spectra. Another useful reminder is that creatine itself is imported to the brain from extracerebral sources of synthesis (Fig. 3.14). Hence, systemic disease may impact Cr in the brain, and the effects seen in the MR spectrum. A striking and rare example which proves this point is the complete absence of a Cr peak in a child with a metabolic defect in the kidney (Fig. 3.15); kidney is one of the links in the chain of Cr biosynthesis (see Fig. 3.14).

The most commonly encountered instance of the absence of Cr in the brain spectrum is in secondary brain tumor, from a primary tissue that lacks creatine kinase (indeed most tissues that metastasize to the brain lack a Cr peak). Other thoughts that should be triggered when lower than normal Cr is observed in a brain spectrum are summarized in Fig. 3.10.

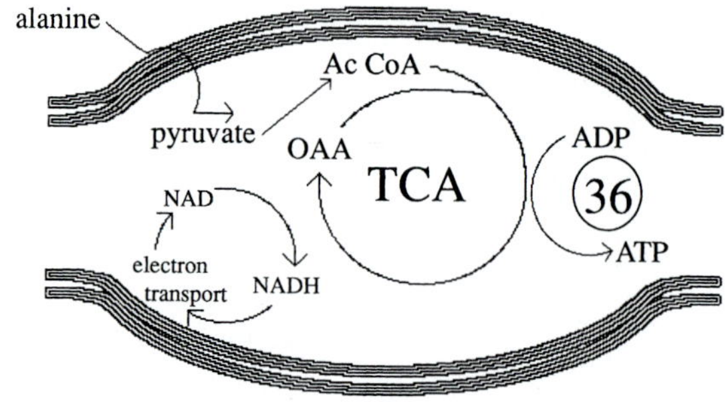

Fig. 3.11 Sources of metabolic energy in the brain. Oxidative pathways generating ATP (adenosine triphosphate). (See also the two following figures.)

2. Why Lactate?

The end-product of failed oxidative metabolism in the brain is lactate, so that the appearance of lactate in an MR spectrum is correctly perceived as evidence of ischemia, i.e., stroke. However, it is more useful to have the fuller picture of the origins and fates of cerebral lactate (Fig. 3.16) if the

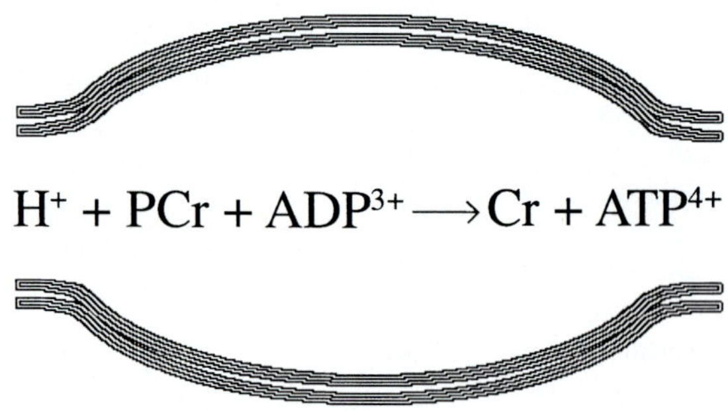

Fig. 3.12 Sources of metabolic energy in the brain. Link between Creatine, PCr and ATP. (See also the previous and following figure.)

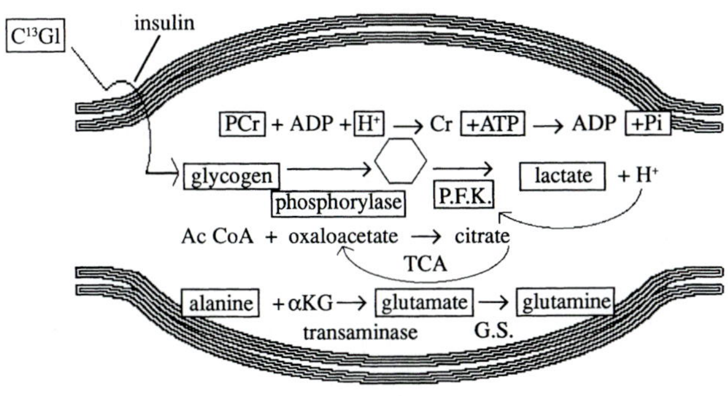

Fig. 3.13 Sources of metabolic energy in the brain. Major pathways of substrate metabolism. (See also the two previous figures.)

clinical utility of MRS is to be maximized. Interruption of the Krebs cycle (by mitochondrial damage), absence or inhibition of pyruvate dehydrogenase, the enzyme which finally generates the 2-carbon fragments that are to enter the TCA, and most frequently of all, acceleration of glycolysis itself (in tumors and "recovering" hypoxic tissues), are more common causes of lactate in the brain spectrum than is simple hypoxia.

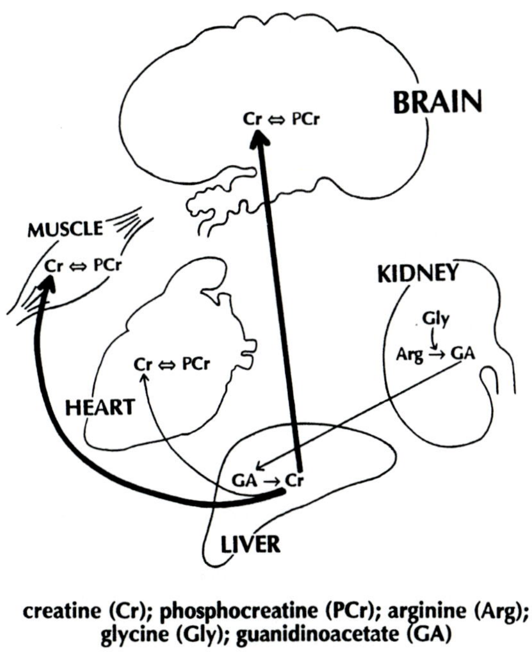

Fig. 3.14 The creatine pool. Several organ-specific metabolic pathways are illustrated. Each is required to maintain brain Cr, which is probably not synthesized in that organ, de novo.

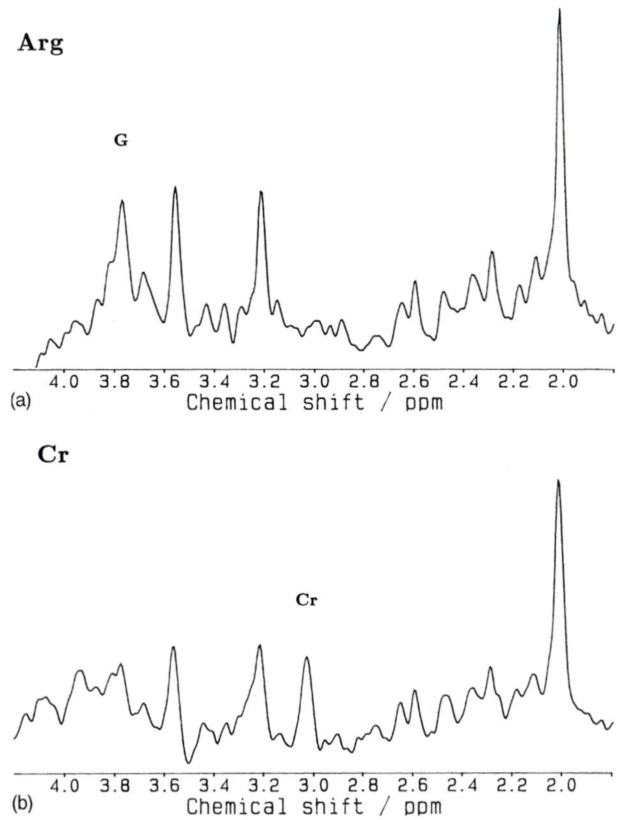

Fig. 3.15 MRS of brain lacking Cr. (A) This MRS is remarkable for the absence of the normal peak at 3.0 ppm. Dietary supplements with arginine (Arg) did not restore creatine, but a peak of guanidinoacetate (G), the precursor of creatine, was detected. (B) Dietary supplements with creatine (Cr) partially restore the Cr resonance at 3.0 ppm.

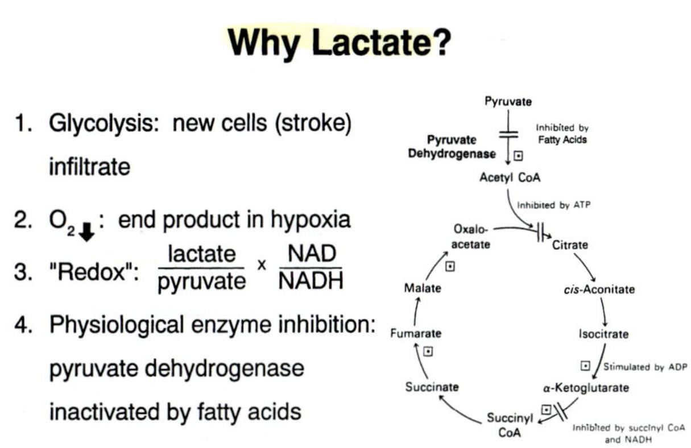

Fig. 3.16 Diagramatic representation of sources of brain lactate.

3. *Myo*-inositol

Myo-inositol is a simple, sugar-like molecule, with a complicated spectral pattern that is explained by the asymmetry of its six hydrogen atoms, easily remembered by referring to the famous turtle cartoon (Fig. 3.17). *Myo*-inositol appears to be almost exclusively located in astrocytes where it is now recognized as the most important osmolyte, or cell-volume regulator. Listed in Fig. 3.17 are some of the discoveries about human brain mI which have been brought to light by clinical proton MRS. It is now apparent that these discoveries make the most sense when mI is considered in its role as an *osmolyte* and *astrocyte marker*, rather than as the neural messenger, inositol polyphosphate ($PI(4,5)P_2$) (Fig. 3.7), admittedly, a much more exciting role.

What Have We Learned

***Myo*-inositol (*scyllo*-inositol)**

10-fold Fluctuations

1. Astrocyte marker; little evidence
2. Osmolyte; hyper (x3)+ hyponatremia (x 1/2); newborn x2 post-natal
3. Myelin "product" in MS and HIV
4. Metabolic "marker" (IPP_3)
   AD—Down syndrome
   Diabetes
   Hepatic Encephalopathy:
   (a) SCHE?
   (b) Neuropsych. marker?

Fig. 3.17 Stereochemistry of *myo*-inositol and *scyllo*-inositol. A complex coupled peak arises for *myo*-inositol (mI) because of the asymmetry of C2 (head of turtle up). *Scyllo*-inositol (sI) is symmetrical (differs from mI in that C2, head of turtle down). Cartoon turtle with the carapace as the inositol ring, and each of its 6 hydroxyls represented by an appendage (after B. Agranow).

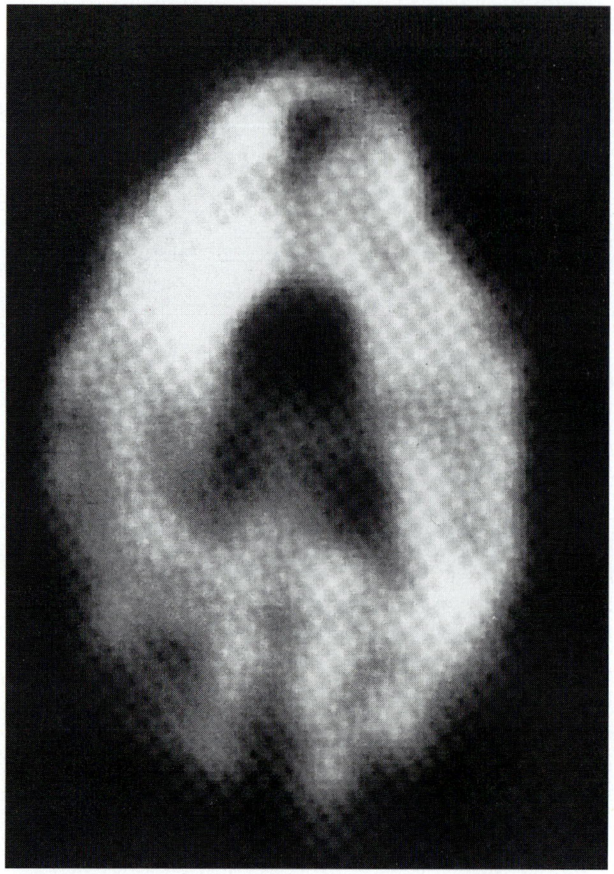

Fig. 3.18 Density map of NAA.

4. N-Acetylaspartate: The *sine qua non*

"Function unknown, but place assured." This phrase sums up NAA, the tallest peak in the brain spectrum. As a neuronal marker, the accurate determination of this peak justifies all but a small handful of clinical MRS examinations. No amount of juggling with MRI sequences, with PET, SPECT or with CT, can yield the central piece of information: How many viable brain cells are there? It is hard to see how neurology, the medicine of neuronal function, existed before MRS, and how it can evolve in future without MRS.

NAA, in its role as neuronal marker, is evenly distributed throughout the cerebral cortex (Fig. 3.18). NAA, as well as the less common analog N-acetyl aspartylglutamate (NAAG), is proven by immunohistochemistry to be within the bodies of neurons (Fig. 3.19). Its pathways of biosynthesis and breakdown, as far as they are accurately known, are shown in Fig. 3.20.

As an axonal marker, NAA arrives by slow transport along the axon from the site of synthesis in the neuronal

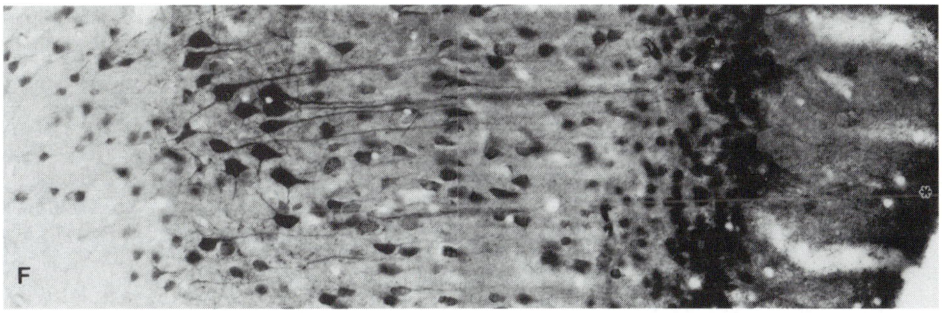

Fig. 3.19 Immunocytochemistry proof of NAA (and NAAG) localization within neurons. Dark staining is confined to neurones and their dendrites. In addition to the photomicrograph, postulated pathways of synthesis and breakdown of NAA and NAAG are illustrated. NPM = neuron plasma membrane; GPM = glial plasma membrane. (From Moffett JR, and Naamboodiri MAA, J. Neurocytol 1995; 24:409–433 with permission).

mitochondria. Thus, NAA is equally relevant in the white matter spectrum. Disappearance of NAA is a sign of neuronal death (presumably, irreversible), and of axonal injury (often recoverable), so that a clear understanding of the pathobiology of NAA is required if every spectrum is not to be dismissed as nonspecific brain damage! Mostly, this has come from purely clinical observation and extrapolation. Thus, the "half-life" of NAA in the grey matter spectrum after severe hypoxia (Fig. 3D in Chapter 6) may reflect the time required for "death" of neuronal nuclei, some 48 hours. Other causes of altered NAA in MRS are listed in Fig. 3.21; see 4a and b in the figure under Pathobiology.

Finally, the reappearance of NAA in MR spectra from which NAA was lost or reduced may have several different pathological explanations, all of which should be considered in differential diagnosis (Table 3.2).

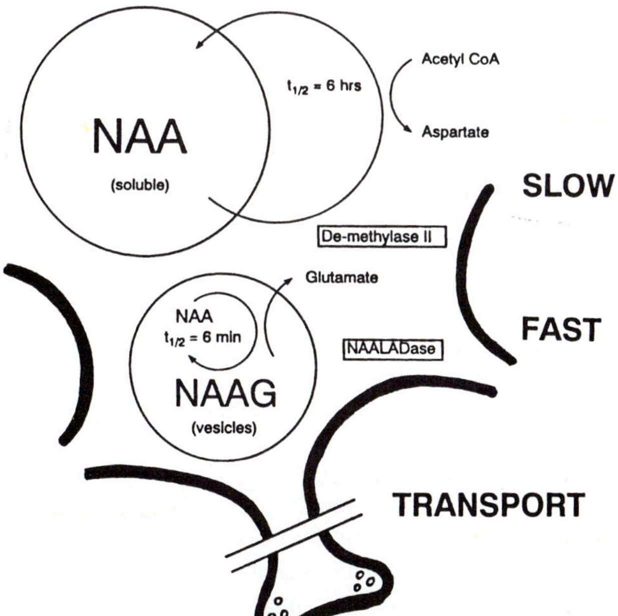

Fig. 3.20 NAA biosynthetic pathways.

5. Glutamate and Glutamine: Neurotransmitters

The small, complex peaks in the MR spectrum, designated Glx (Fig. 3.22), belie the relatively large concentrations of glutamate (>10 mM) and glutamine (>5 mM) that are present in all brain regions. This is the *unhelpful* effect of J-coupling. Furthermore, their very similar chemical structure paradoxically means that the Glx profile we see in spectra acquired with STEAM (TE = 30 ms) reflects principally glutamine in the β,γ- region, and both glutamate and glutamine in the alpha (α-) region. Glutamine, like mI, is an astrocyte marker, and the principal factors that govern its appearance in pathological brain spectra are outlined in Fig. 3.22. Glutamate, which is viewed as an important neurotoxin when its concentration exceeds that needed for neurotransmission, is also a participant in the redox cycle (Fig. 3.1) that governs lactate accumulation. Accordingly, there is a common paradox in clinical MRS (Fig. 3.23). In hypoxic, ischemic, or recovering brain, Glx is elevated. This is not, unlike lactate, a grave prognostic finding. Instead, it is probably a reflection of the protective function of the astrocyte, where glutamine synthetase removes the potentially toxic excess of glutamate to accumulate as the relatively harmless product glutamine. At this point, we should remind the reader that glutamine and glutamate are routinely

Biosynthesis (slow) precedes myelination

Oligodendroglia/neurones

Location — neurones and axons

GM ≈ WM
(spinal cord and cerebellum NAAG)

Pathobiology — 1. Loss of NAA ≡ loss of neurones

- glioma ↓
- MS ↓
- stroke ↓
- dementias ↓
- hypoxia ↓

2. Axonal marker—WM NAA after GM

- white matter disease ↓
- MS ↓
- near-drowning ↓

3. Recovery of NAA?

post hypoxia MS and MELAS

- ≠ neuronal recovery
- = axonal recovery
- = artefact of atrophy
- = another metabolic role?

"NAA responds to:

4. Other metabolic events
   (a) Metabolically heterogeneous "peak"
   diabetes ↓
   (b) Reversible osmolyte? (Lien, Chan et al)
   hypernatremia ↑ ↓
   hyponatremia ↓

Fig. 3.21 NAA location/pathobiology.

identified as Glx in the MR spectrum, to emphasis the point that without spectral "editing," we are mostly looking at changes in glutamine, on top of an unchanging brain glutamate signal.

6. Lipids

Often reported, the presence of large, broad, in-phase resonances at 0.9 ppm and 1.3 or 1.4 ppm (Fig. 3.24) is definitely pathological, but unless the resonances are pronounced, they are rather difficult to distinguish from the normally elevated baseline in this region of the spectrum. Little has been done to identify precisely which long-chain fatty acids appear under what circumstances, but it is probably sufficient here to note that over 20% of the dry weight of the brain (and much more in newborns) is lipid, in the form of membrane associated myelin, phospholipids, sphyngomyelins, and lecithins. These macromolecules do not appear unless some severe pathological process liberates the MR-visible triglyceride and long-chain fatty acids from them; lipid may therefore dominate the MR spectrum in infections, inflammations, necrosis of tumor, and stroke. It is our experience that lipid signals are an almost universal accompaniment of brain injury (trauma, hypoxia, or viral infection) in very young infants as in Fig. 3.24.

Table 3.2 An Incomplete List of Routes to Recovery of Neuronal N-acetylaspartate (NAA)

Resynthesis	Canavan's disease
Regeneration of neurons	Neural transplant
	Infants?
Regeneration of axons	Multiple sclerosis
	MELAS
	Other injuries
Re-entry	Dialysis
	Osmosis
	Hyponatremia
	Hydrocephalus
Reorganization	Masses
	Atrophy

Source: Ross and Blüml, New Aspects of Brain Physiology, NMR Biomed 1996; 9:279–296.
MELAS: myopathy, encephalopathy, lactic acidosis with stroke-like lesions.

Are lipids ever nonsignificant? Most tissues other than brain contain triglycerides in huge amounts. Contamination of the VOI will often detect these extracerebral signals and confuse the diagnosis. The most common pitfall is a VOI which has drifted too close to the skull; lipid from outside the volume is readily recognized by its inverted MR signals.

7. Choline

The "gold mine" of clinical information that is to be gleaned from observation of alterations in the choline peak in MRS is left to last because it is still the least well-understood in terms of its neurochemistry. Several attempts have been made to change the name—to trimethylamines, for example—but the misnomer choline (abbreviated Cho) is now firmly established in the literature. What we see is approximately the sum of phosphorylcholine and glycerophosphorylcholine, with less than 5% of the peak as free choline, and none as acetylcholine. More important in understanding the significance of alterations in Cho is the fact that phosphatidylcholine, the major choline metabolite of brain under normal circumstances (Fig. 3.25), is MR-invisible in myelin, cell membranes, and other complex brain

1. Glutamine (5 mM) glutamate (10 mM)
   ARE distinguishable at 1.5 T, 2 T and 4 T
2. ΔGlutamine: common ΔGlutamate: rare

↑ (ΔGlutamine)	↓ (ΔGlutamine)	↓ (ΔGlutamate)
HE	AD	AD
Reyes	hyponatremia	
OTC def	(post-transplant)	
hypoxic-ischemic	liver	
hyper-osmolar		
valproate (NH_3 ↑)		

3. GABA
   ↑ "vigibatrine"

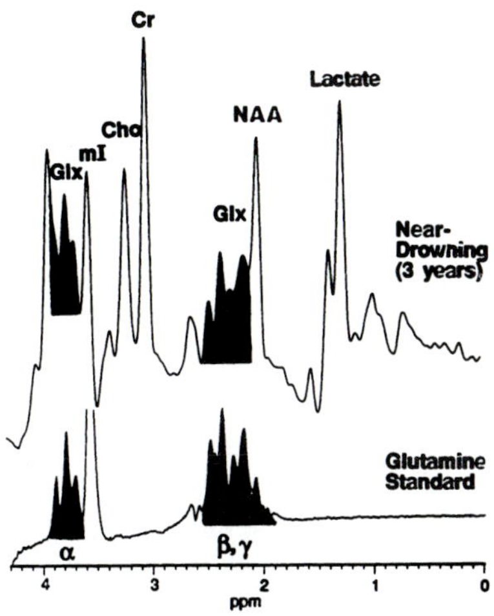

Fig. 3.22 What have we learned: Glutamine/Glutamate. The spectra illustrate how closely the brain spectral region identified as Glx, reflects the spectrum of glutamine, suggesting that under pathological circumstances, it is glutamine accumulation that occurs.

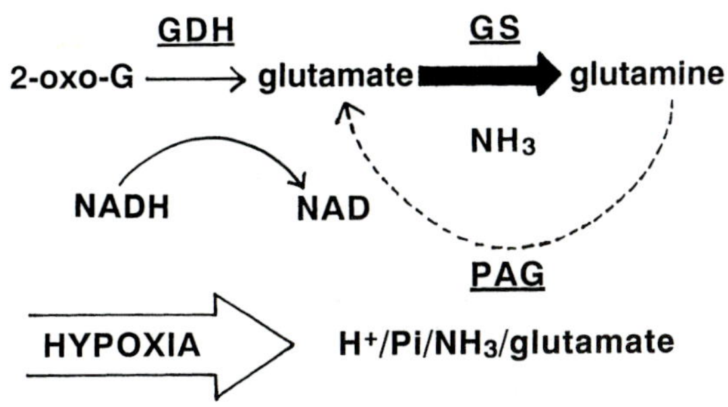

Fig. 3.23 Anomaly: Near-drowning and stroke. The expected glutamate excess is avoided by its rapid conversions to glutamine.

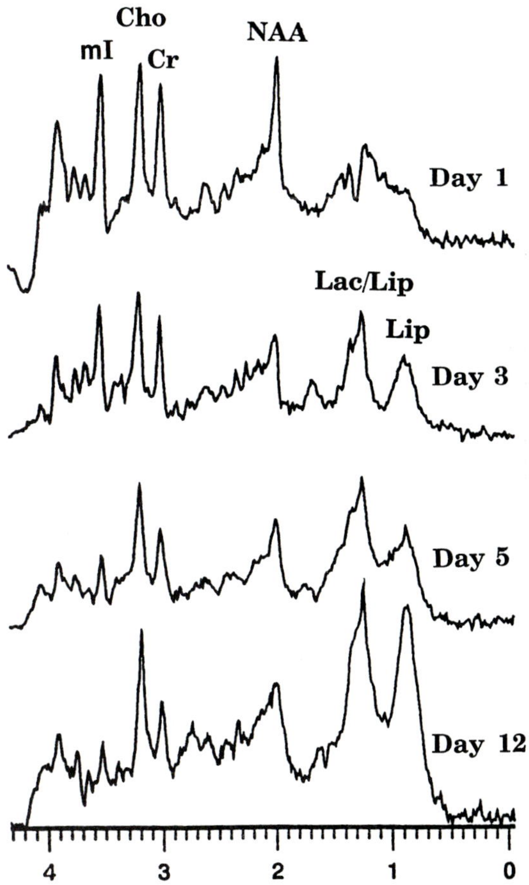

Fig. 3.24 Lipids in MRS as an abnormal metabolite. A series of spectra from the brain of a child with shaken baby syndrome (SBS) shows little or no visible lipid on day 1 (as in normal infant). However, injury is marked by progressive increase in signal at 0.9 and 1.3–1.4 ppm, which have been assigned to mobile lipids (day 3–day 12).

lipids (see Lipids, above). Under many pathological conditions, however, visible Cho may be released from this pool.

The ultimate source of choline (and glycerol) for brain metabolism is probably systemic and dietary, so the neurologist should think holistically when reading this peak.

In contrast to glutamate, where the unkind trick of J-coupling deprives us of a potentially valuable neurochemical marker, cholines (as trimethylamines) contain nine protons per molecule with virtually all of the signal contributing to the single narrow peak at 3.2 ppm. Pathological changes in choline are amplified almost tenfold; increases of only ±0.1 mM appear as major differences from normal in MRS. Some observers have abandoned the more general convention that expresses metabolite ratios to Cr (NAA/Cr and Cho/Cr), and use the Cho/NAA ratio to take

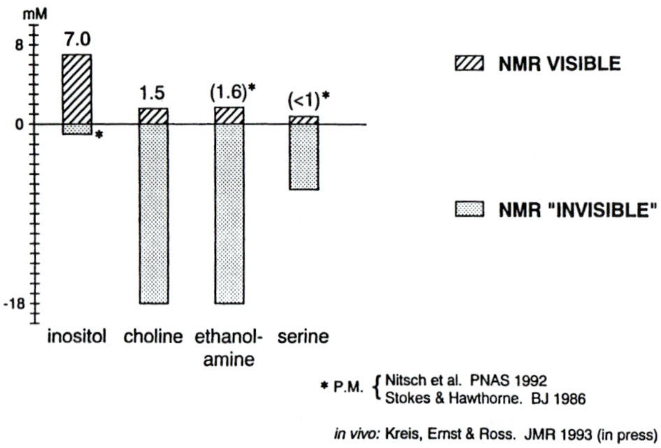

Fig. 3.25 Bound and free (lipids of human brain). Brain phospholipids are sources of excess choline(s) in common diseases. Because macromolecules are the norm, only the small fractions of choline, ethanolamine, serine and inositols shown above the line contribute to the brain spectrum.

advantage of this amplification. MRS of choline is therefore altered in response to a host of pathological processes, some of which are listed in Fig. 3.26.

IV. CLASSIC PATTERNS IN MRS DIAGNOSIS

There is a neurochemical logic to the several patterns of pathological MRS which comprise this book. Much of this is summarized in Chapter 6.

1. A myth dispelled by QMRS
   Cho WM **NOT** > Cho GM
   1.6 vs 1.4
   (impression due to GM Cr > WM Cr)

2. No visible head-groups of Pt Cho

3. Local and systemic; regulated by development; aging

Local	**Systemic**
Tumor	Diabetes
MS	Liver
(AD)	Renal failure
(depression)	Dialysis
Head Injury	Osmotic events
Stroke	Post-transplant (liver)
Hypoxia; brain death	
WM disease	
(ALD etc)	
Radiation	
HIV infection/inflammation	

4. PC vs GPC decoupled ^{31}P

Fig. 3.26 What have we learned: Cholines.

A. Brain or Cerebrospinal Fluid

It is useful to remember that cerebral metabolites are virtually absent from CSF. Thus, a VOI in which CSF is unavoidably included, e.g., cortical atrophy, cyst, etc., will show two noteworthy features: 1) Reduced S/N of the major cerebral metabolites. 2) Excess of lactate (peaks at 1.33 ppm), and glucose (peaks at 3.43 and 3.8 ppm; upper spectrum) (Fig. 3.27).

Lactate is an important negative marker in neurologic disease, and will be encountered throughout this book. And yet, because lactate is a normal constiuent of CSF, errors of voxel placement, registration, or even in the method of localization used, remain the most frequently encountered sources of lactate in clinical MRS.

B. MRS of Neuronal Injury

Generalized loss of neurons, followed by axonolysis is more common than at first believed likely. Thus, although MRS is unable to distinguish one neuron type from another, and one layer of cortex from the next, reduced NAA/Cr ratio can be used to "map" neuronal loss over time (Fig. 3.28).

C. MRS of Neurodegenerative Disease

NAA is a neuronal and an axonal marker. By definition, therefore, it is inescapable that reduction in NAA/Cr should be sought in diagnostic workup of patients with neurodegenerative disease (i.e., Alzheimer disease). MRS can go

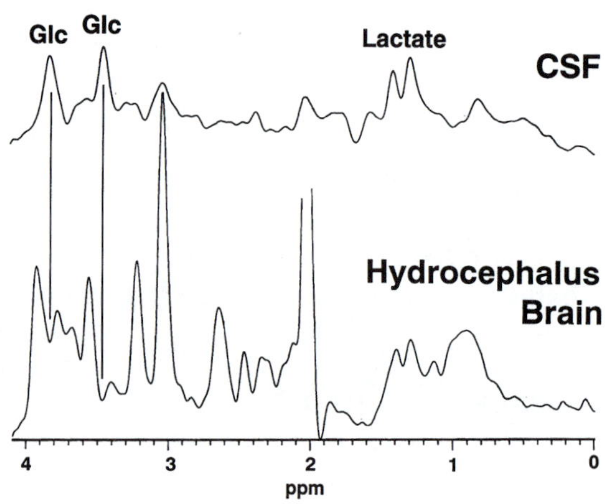

Fig. 3.27 CSF and hydrocephalus brain. Note that the recognized brain metabolites are not measurable by *in vivo* MRS in cerebrospinal fluid (CSF). The CSF spectrum reflects the somewhat higher concentration of lactate and glucose in CSF compared with intact brain.

NEUROPATHOLOGY

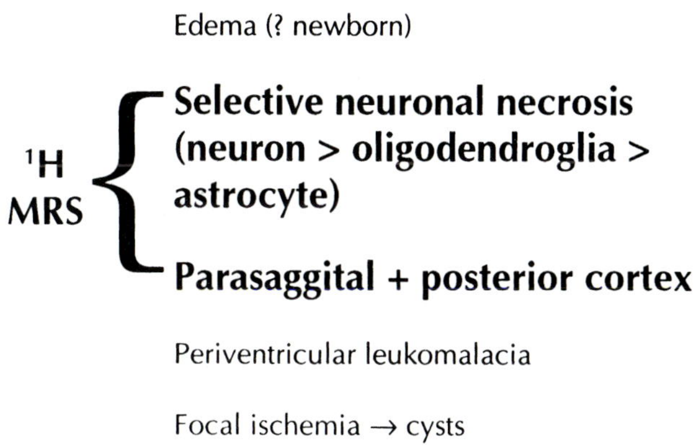

Fig. 3.28 Hypoxic encephalopathy: Role of ¹H MRS in early diagnosis.

far beyond this, however, and also diagnose underlying neuropathologies (Fig. 3.29); amyloid, gliosis, or plaque formation in Alzheimer disease, may account for an increase in the astrocyte marker, mI.

Vascular and inflammatory disorder (examples are multi-infarct, and AIDS dementia complex), may more readily increase Cho/Cr (see below).

D. Hypoxic-ischemic Cascade

Loss of NAA, appearance of lactate, increase in Glx, and ultimately loss of Cr, are logically anticipated in brain injury precipitated by the well-established hypoxia-ischemia

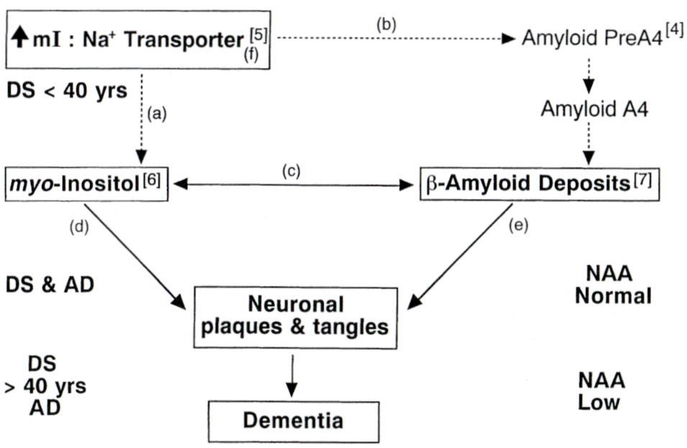

Fig. 3.29 Theory of preventable AD. ¹H MRS recognizes two neurochemical changes of AD. Because that of mI precedes the change in NAA in the related disorder of Down's syndrome (DS) we speculate that a similar sequence occurs in AD. If this were so, interruption of the sequence may "prevent" loss of NAA.

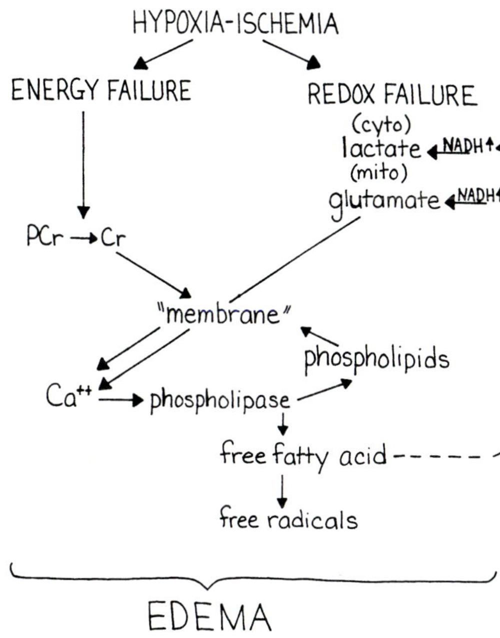

Fig. 3.30 Hypoxic-ischemic cascade. A simplified scheme of the metabolic events which follow ischemia and/or hypoxia in the infant brain. Some events, including free-radical formation, phospholipase activation, with release of MRS-visible lipids, are more popularly associated with restoration of oxygen delivery after a period of ischemia.

cascade (Fig. 3.30). Empirically, excess lipid is a frequent accompaniment in MRS. The order of lipolysis in the cascade, whether early or late, is unknown, but has been given a central place in Fig. 3.30, for convenience.

E. Basic Neurochemistry of Tumors

There are *five* biochemical defects common to the majority of brain tumors, as measured by MRS (Fig. 3.31). They are as follows: NAA is decreased; lactate is increased; lipid is increased (see inset); creatine plus phosphocreatine is decreased (as is PCr determined alone); and choline is increased. This increase in choline is readily recognized in ¹H spectra, as an increase in the ratio Cho/Cr. However, it should be remembered that in ¹H spectra all metabolite concentrations are often reduced so that this peak ratio correctly reflects relative changes rather than absolute increments or decrements. This aspect of tumor diagnosis and analysis becomes clearer when quantitative MRS is applied.

Quantitative MRS, either performed systematically in absolute terms, or by inference comparing voxels within tumors with those from surrounding normal brain, takes account of heterogeneous anatomy of tumors, including

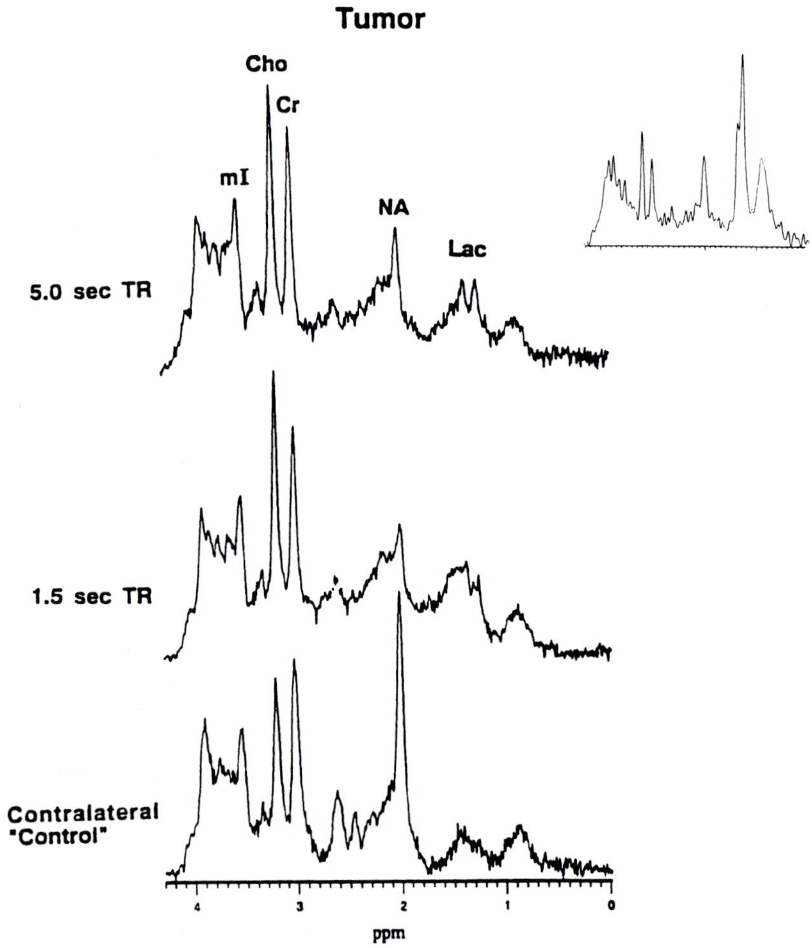

Fig. 3.31 An obvious primary tumor showing absent NAA; increased (Cho/Cr), and particularly at longer TR, increased lactate (all echo times 30 ms). Inset is a second patient showing the now classical profile of lipid plus lactate; probably recurrence post-therapy.

regions of necrosis or cysts. The most thorough studies have used metabolite imaging (CSI or other techniques) to throw entirely new light on the metabolic heterogeneity of brain tumors. These, when combined with the insights from MRI significantly alter management.

Significance of changes in tumor spectra: Most of what has been defined is self-explanatory. Reduction in or absence of NAA expressed the absence of neurones and axons from most tumors, which are histologically derived from astrocytes, which entirely lack NAA (gliomas), connective tissue (meningiomas), ectoderm (craniopharyngioma) or from remote tissues of nonneuronal origin (virtually all secondary tumors). Lactate, normally undetectable in brain, accumulates in cysts, necrotic tissues or within active tumors because of the high rate of glycolysis (glucose to lactate) within even aerobic tumors of all types. Heterogeneous distribution of lactate reflects its accumulation in cysts, regions of high tumor "activity" or conversely, areas of tumor necrosis. Alanine, an alternative "reduced" partner of pyruvate derived from glycolysis has been systematically reported in meningiomas, an observation which at present is interesting rather than helpful.

Lipid signals commonly associated with tumor spectra have usually been ascribed to necrotic regions, in untreated tumors, or to treatment responsive necrosis in treated tumors, and this is most probably correct for *in vivo* tumor spectra. On the other hand, in high resolution spectra of brain tumor biopsies, rather more specific lipids are identifiable (Fig. 3.32). These are likely contributors to the unresolved lipid peaks of the *in vivo* tumor itself.

Decreased creatine (plus PCr) or of PCr itself, is an inconstant finding in tumors. When present it has usually been ascribed to the low-energy status of glycolysing tumors in general. In the case of secondary tumors the expla-

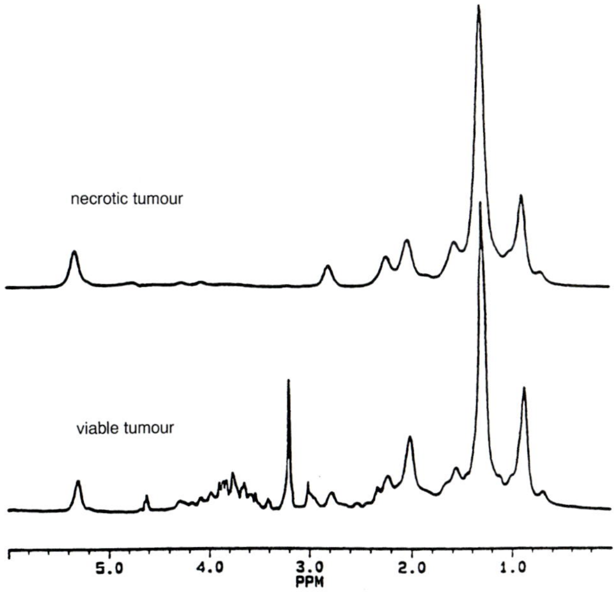

Fig. 3.32 Absolute intensity ex vivo 1-D ^{1}H MR spectra at 37 degrees C of samples from two biopsies of a subependymal giant cell tumor in a patient with tuberous sclerosis: (top) histologically 99% viable tumor; (bottom) histologically 99% necrotic tumor. From Kuesel et al. with permission.

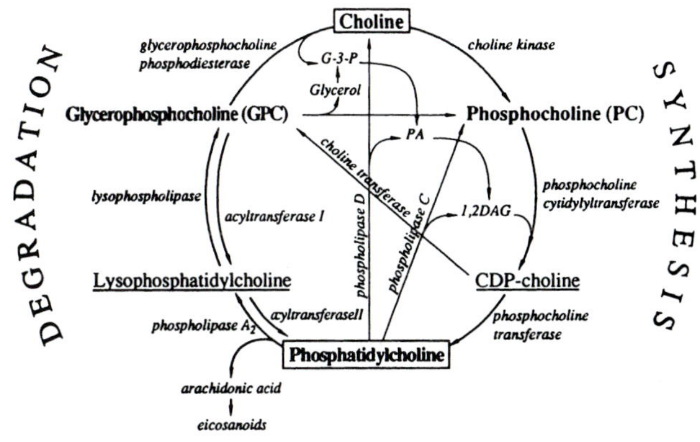

Fig. 3.33 Anabolism and catabolism of choline or ethanolamine, phosphomono- and di-ester.

nation is more straightforward, since most tissues that commonly metastasize to brain (kidney, lung, breast, lymphoma, prostate) lack both PCr and creatine kinase. On the other hand, elevated choline is nearly always observed and has been substantiated by in vitro chemical assays on tumor biopsies. Proliferation of cell membranes, the explanation usually offered, is too vague; breakdown of NMR-invisible phosphatidylcholine, to release NMR-visible cholines is another possibility.

Thus, in summary, there is a good theoretical basis for all of the neurochemical information concerning brain tumors now available with ^{1}H MRS. Attempts have been made to refine this information into a better tool for differential diagnosis, grading and staging.

Choline metabolism is too complex to be fully described here, but Fig. 3.33 is a memory aid that explains the origin of elevated Cho/Cr in tumor, and in many inflammatory diseases.

F. Necrosis, Gliosis, and Other Common Patterns of Brain Injury

Broadly speaking, the huge lipid content of normal human brain is not visible to ^{1}H MRS. Lipid peaks, when they do appear are therefore of the greatest clinical importance. Necrosis of any cause results in appearance of broad so-called lipid peaks at 0.9 ppm and 1.3–1.4 ppm. Triglycerides, a form of lipid, is frequently encountered in the MR spectra of active growing tumors, and could be derived from membrane or myelin-lipid breakdown without cell necrosis. Necrosis of growing tumors can result in the appearance of lipid peaks. However, an important result of necrosis, as its name implies, is loss of viable cells. The result is reduction in MRS signals from all of the normal brain metabolites. This varies from a notable reduction of the S/N, with preservation of a "normal" appearing MRS, to a flat spectrum with no metabolite peaks apart from the lipids at 0.9 and 1.3–1.4 ppm (Fig. 3.34, Spectrum A).

Immature brain (Fig. 3.34, Spectrum D) and the brain recovered from severe hypoxic insult is notable for the excess of Cho/Cr (Fig. 3.34, Spectra B). Absence of NAA in Spectra B (upper and lower, Fig. 3.34) is proof that neurones are largely replaced by astrocytes and the spectrum is therefore that of gliosis. The excess Cho in immature brain may reflect growth: Excess Cho in a tumor spectrum is also taken to represent growth that is persistent or recurrent tumor. The differential diagnosis of tumor recurrence from tumor necrosis, and of both from radiation necrosis is relatively straightforward with ^{1}H MRS. Based upon the principles outlined above, and with high probability, excess Cho/Cr, reduced NAA/Cr, with or without excess lipid/lactate is consistent with tumor. Lipid/lactate excess without excess Cho/Cr is consistent with radiation necrosis.

New growth, other than neoplasm is also now occasionally encountered in Clinical Practice (neurotransplant is the solitary example offered). The presence of near normal NAA and Cr peaks on MRS differentiates such entities from malignant tumors with certainty.

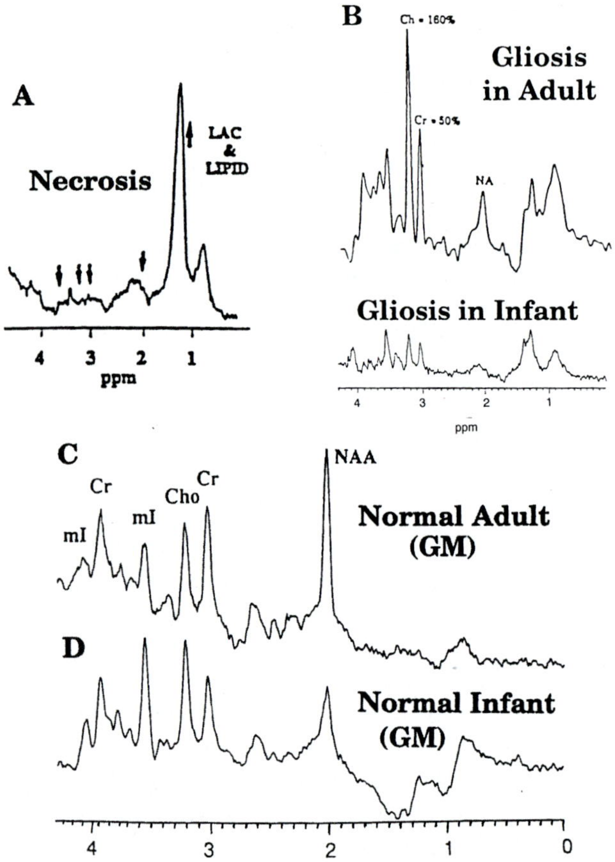

Fig. 3.34 Simplified interpretation of MRS responses to common neuropathologies. MRS is rarely "disease-specific." Instead, common patterns reflect the limited possibilities the brain possesses in responding to many different noxious or destructive events: (A) Necrosis—loss of major metabolites; liberation of membrane bound lipids: (B) Gliosis—the relative increase in astrocytes after neuronal or axonal loss, as well as the MRS pattern of "scarring." (C) Normal adult, in which NAA is dominant and (D). Immature brain in which "myelin" precursors mI + Cho predominate over unformed neural elements (low NAA).

G. Brain Swelling, Edema, and Atrophy

Without having heard of the "idiogenic osmoles" of brain, some common MRS patterns make little sense. Remember mI is the "lost osmolyte" and that brain defense in dehydration, rehydration, and incipient edema involves swelling and shrinkage (Fig. 3.35).

Ultimately, a remarkable paradox becomes clearer if this concept is pursued to its logical end. The brain severely affected by cortical atrophy, may have a normal appearing

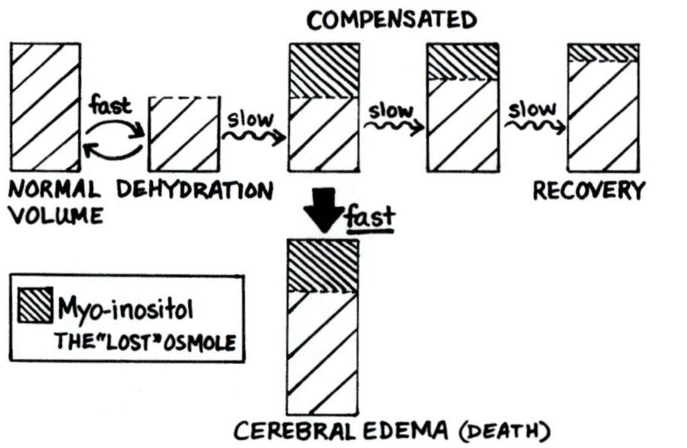

Fig. 3.35 Cartoon representing brain cell volume changes in dehydration and recovery. The upper panel indicates reduction in brain volume, followed by synthesis of brain osmolytes, principally mI (shaded). Safe recovery is accompanied by removal of mI and maintenance of normal volume. Over-enthusiastic salt replacement (lower panel) is believed to precipitate cerebral edema as intracellular sodium is replaced before removal of the balancing intra cellular osmolyte, mI.

spectrum. Early MRS is recommended because late in the course of many "destructive" brain disorders, the MRS may return to normal.

This is certainly true for hydrocephalus. Fig. 3.36 illustrates a theoretical basis for such findings which are common in clinical practice. Thus, cell swelling or shrinkage can occur, leaving metabolite concentrations normal (osmotic equilibrium restored) or abnormal (underlying brain injury). MRS can be helpful in segregating these two general categories.

H. Impact of Systemic Disease on the Brain

Why perform brain MRS in the patient with liver disease or diabetes mellitus? Ammonia arising from the bowel is normally converted to urea in the liver. A remarkable constellation of neurospectroscopy changes accompanies any failure of this mechanism. Excess ammonia diverted to the brain is well known to produce hepatic encephalopathy. Excess Glx (primarily, glutamine), and depletion of Cho and of mI appear to be closely linked, even though the pathogenesis is unclear. A "diagnostic" brain MRS occurs, and explains the neurological abnormality after transjugular intrahepatic portasystemic shunt procedure (TIPS), hepatic encephalopathy and its imitators, urea cycle disorders, valproate toxicity, Reye's syndrome. This is a striking example of how systemic disease influences the brain spectrum. Several such diagnostic patterns are described in this

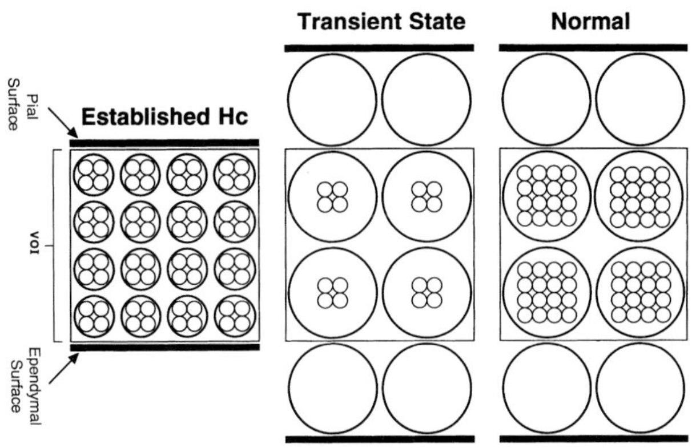

Fig. 3.36 Cortical thinning in slowly-progressive hydrocephalus. In contrast to the scheme in the previous figure, physiological adaptation allows brain water to be lost, at the same rate as the intracellular osmolytes (shown as small circles within the larger circles, each of which represents a single brain cell). Cell volumes contract as water is lost. The brain "shrinks" until normal osmolality is restored, and appears as cortical thinning.

book. Hyponatremia similarly affects the brain, resulting in subtle changes in MRS. Diabetes mellitus, which affects several peaks of the brain spectrum and alters brain glucose, as well as diabetes insipidus, appear in this Handbook because of their effects on the brain MRS.

FURTHER READING

An Overview for Physicians of Human Biochemistry and Regulation of Metabolism

Newsholm EA and Leech AR. Biochemistry for the Medical Sciences. Chichester, Sussex UK: John Wiley & Sons, 1983.

A Valuable Source Book for Human Biochemistry and Metabolism

Scriver CR, Beaudet AL, Sly WS and Valle D, eds. Metabolic Basis of Inherited Disease, 6th edition. New York: McGraw-Hill, 1989. (previous eds.: Stanbury JB, Wyngaarden JB, and Fredrickson DS).

Tables of Normal Concentrations of Common Metabolites in Different Tissues Are Given in:

Lentner C, Lentner C, and Wink A, eds. Geigy Scientific Tables, Vol 4. Biochemistry, Inborn Errors of Metabolism, Pharmacogenetics and Ecogenetics. Basle, Switzerland: Ciba-Geigy Ltd., 1986.

Beyond Proton MRS: A Review of Newer Magnetic Resonance Spectroscopy Methods Available.

Bachelard H, ed. Magnetic Resonance Spectroscopy and Imaging in Neurochemistry, vol. 8 of Advances in Neurochemistry. New York: Plenum Press, 1997.

4

An Introduction to Clinical Cases

I. A STEP-BY-STEP GUIDE TO REPORTING MRS

There are three simple rules, or principles, to remember when interpreting MR spectra, and generating clinical reports: 1) Spectra are read in terms of peak ratios; 2) MRS is a quantitative technique; and 3) Absolute metabolite concentrations add another useful dimension to diagnosis. (For a more detailed discussion of the principles of quantitation, refer to Chapter 2.)

The format of the MRS case reports on the series of individual patients presented in Chapter 5 is as follows: Purpose of MRS; Summary of Patient's History; MRS Method, VOI and Comments about Examination; Location of VOI(s) (which includes peak identification, quality assurance, interpretation, and analysis of the spectrum obtained in each location); Conclusion; Summary of Patient's History Post MRS; Other Comments. Each case is preceded by a title (i.e., diagnosis), and grouped according to pathology. Each case is completely self-contained, and can be read as an exercise. Alternatively, by referring to the index, cases relevant to a particular question can be rapidly identified and reviewed. Unless otherwise indicated, spectra for the clinical cases presented in Chapter 5 were acquired using short echo time STEAM.*

To illustrate the type of information contained under each heading, the first case report presents the average of normal adult spectra in two standard locations—parietal white and occipital grey matter—obtained from a group of healthy volunteers.

II. CASE EXAMPLE: PARIETAL WHITE AND OCCIPITAL GREY MATTER SPECTRA IN THE NORMAL ADULT

A. Purpose of MRS

MRS is typically performed to 1) evaluate if a condition is consistent with a certain disease; 2) evaluate outcome in a critically ill patient; or 3) obtain further information about a patient's condition. A large range of clinical questions may lead to the request for an MRS examination, as is demonstrated by the cases in Chapter 5.

*Some spectra were acquired with PRESS, a sequence that changes the final appearance of the spectrum only to a minor degree.

B. Summary of Patient's History

This section is self-explanatory. It is important to know the patient's age because the normal MR spectrum varies with age, particularly in the first year of life (see Chapter 2). Various clinical facts can be helpful in giving a diagnosis and listing possible differential diagnoses. The case reports present a clear picture of what to look for in different types of pathologies. Hypertension, renal failure, diabetes, or abnormal serum, drug history, and liver function tests, are all clinical factors that need to be known.

C. MRS Method, Volume of Interest, and Comments about Examination

In order to interpret the MR spectrum correctly, one needs to know how it was acquired. It is therefore a good habit always to note the MRS method (or pulse sequence) and measurement parameters. For example: Localized STEAM, TE = 30 ms, TM = 13.7 ms, and TR = 1500 ms. This tells you that the pulse sequence used was stimulated echo acquisition mode (STEAM), single voxel localization, and states the measurement parameters: echo time (TE), delay, or mixing time (TM), and repetition time (TR). Different measurement parameters, as for example TE = 20 ms and TM = 30 ms as found in some of the case reports, change the metabolite ratios enough that a different set of normal values must be used. The appearance of glutamine and other coupled resonances are particularly affected by changes in pulse sequence and measurement parameters. Most spectra presented here are measured using the standard localization STEAM method, TE = 30 ms, TM = 13.7 ms (with various TRs), reflecting HMRI's consistent use of the same method over many years.

This book shows MR spectra acquired with single VOI MRS methods. Many studies were done using the HMRI's quantitative version of STEAM, with the measurement parameters as mentioned above. These spectra were all acquired with a GE Signa 1.5 Tesla MR scanner. In some cases, measurement parameters were different mainly because they were acquired with a Siemens SP 1.5 Tesla MR scanner, using its default short echo time parameters.

As explained in Chapter 2 the metabolite ratios acquired with PRESS are sufficiently different from those acquired with STEAM, to unnecessarily complicate clinical reading. In addition an extensive clinical database for PRESS spectra is currently unavailable; therefore PRESS is not represented among the clinical cases in Chapter 5.

Both single VOI methods, STEAM and PRESS, are available in an automated version called PROBE-STEAM or PROBE-PRESS (GE Medical Systems). A large number of case studies were performed using PROBE-STEAM with the standard HMRI measurement parameters, TE = 30 ms, TM = 13.7 ms. The appearance of these PROBE spectra is not the same as that of the spectra measured at HMRI, because the manipulation of the spectra and the display is different.

We also present examples of PROBE-Q spectra, which are basically automated PROBE spectra measured and displayed as described above. Although the display is unaltered, PROBE-Q also provides (peak height) metabolite ratios obtained using postprocessing identical to that used at HMRI.

The position and the size of the VOI should be characterized because the appearance of normal spectra varies accordingly, primarily due to different relative amounts of grey and white matter in the VOI. The best strategy for nonfocal diseases is to measure in standardized locations where results from healthy subjects are well-defined, facilitating better distinction from disease.

In order to get the best results, it is important to be careful in prescribing the VOI in a consistent and reproducible fashion. In the case reports, the wording "HMRI's occipito-parietal white matter location" describes a 25 × 25 × 20 mm^3 VOI, and "HMRI's occipital grey matter location" describes a 21 × 27 × 20 mm^3 VOI, as illustrated in Fig. 4.1. The occipito-parietal white (Fig. 4.2) and occipital grey matter (Fig. 4.3) spectra are averaged spectra from a group of healthy adult volunteers. The locations of the volume of interests (VOI) are standard for this Laboratory (HMRI). Notice that the top, middle, and bottom slices are shown (Fig. 4.1); it is important to check the location in all slices covering the VOI.

Comments about the examination may include the administration of a sedative, or the use of other medical interventions such as ventilation, MRI contrast agents, or other drugs used before or during the examination. It is also important to notice if the patient moved their head during the examination. Small movements are usually benign, but in more severe cases artifacts or loss of spectral resolution may result. In the worst case, the VOI may have been totally shifted. In cases of restless patients it is often valuable to reimage after the MRS examination to confirm that the VOI location is still valid.

D. Location 1 (White Matter)

1. Peak Identification and Quality Assurance: Reading the Spectrum from Right to Left

Quality assurance and identification of the peaks is the first step of the MRS reading, followed by interpretation and numerical analysis.

Quality assurance consists of evaluating 1) the size of the signal compared to the size of the noise, or signal-to-noise ratio (S/N); 2) the resolution of the spectrum, or in other words, how narrow the peaks are; 3) whether or not

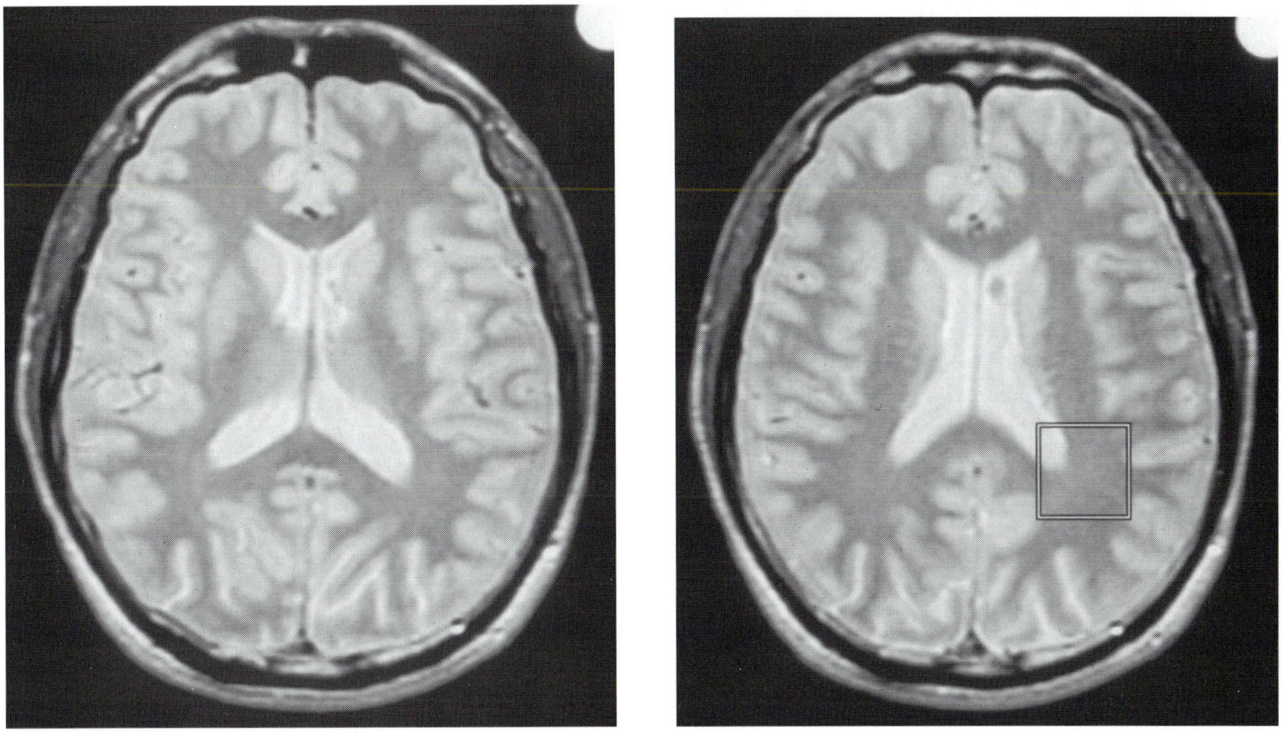

A B

Fig. 4.1 (A–F) The top, middle, and bottom slices (slice distance is 10 mm) of the typical grey matter and white matter VOIs. It is advisable to check all slices of the VOI before prescribing the location. In this way, best reproducibility is achieved, and getting too close to the scalp can be avoided. (G) The greatest source of variation between repeat MRS in individual patients is angulation of the head, since graphic prescription selects only truly orthogonal planes. Therefore, patient positioning, attention to MRI landmarks, and dimensions are all critical to achieving the highest quality in the ensuing MRS.

the water suppression worked; 4) whether or not the baseline is as flat as it should be; 5) whether or not any peaks, humps, or spikes originating from technical or patient compliance problems (artifacts) are present. In the rare cases when the spectrum scores negatively on the quality account, it should be decided if the spectrum is still of diagnostic quality or if a repeat examination should be recommended. It is better to reject the spectrum than misinterpret a potentially crucial clinical finding.

First we notice that an imaginary line through the tops of the peaks slopes down from right to left. This is normal for an adult white matter spectrum, where the left peak at 3.56 ppm is usually the smallest and the peak to the right at 2.02 ppm is the tallest peak. The angle which this imaginary line makes with the horizontal axis is called "Hunter's angle," (Named after Dr. Hunter Shelden, who initially suggested this straight forward method of pattern recognition in the evaluation of MRS). "Hunter's angle" applies to white matter where all four major peaks follow the line; for grey matter only NAA, Cr, and mI establish the imaginary line with Cho much below.

An alternative method of interpreting the spectrum is the analysis of each individual peak, as described below. The identification of peaks is best done systematically, so that no details are overlooked. Our habit is to read the spectrum from right to left, from 0 ppm and towards water at 4.7 ppm along the chemical shift scale. The white matter spectrum in Fig. 4.2 would be read as follows:

a. N-Acetylaspartate (NAA)

The first noticeable peak is the N-acetylaspartate (NAA) peak at 2.02 ppm. To be more precise, one should remember that the peak originates *mainly* from NAA, but a smaller contribution may be from other molecules with an N-acetyl molecular group. (Note: You may have noticed the small humps in the region of the spectrum from 0.8 to 1.5 ppm. The baseline is rarely completely flat in this region of the spectrum; the humps are mainly from lipid resonances, and include low molecular weight proteins. When the humps are of this insignificant size, they are not usually worth mentioning in the reading of the spectrum.)

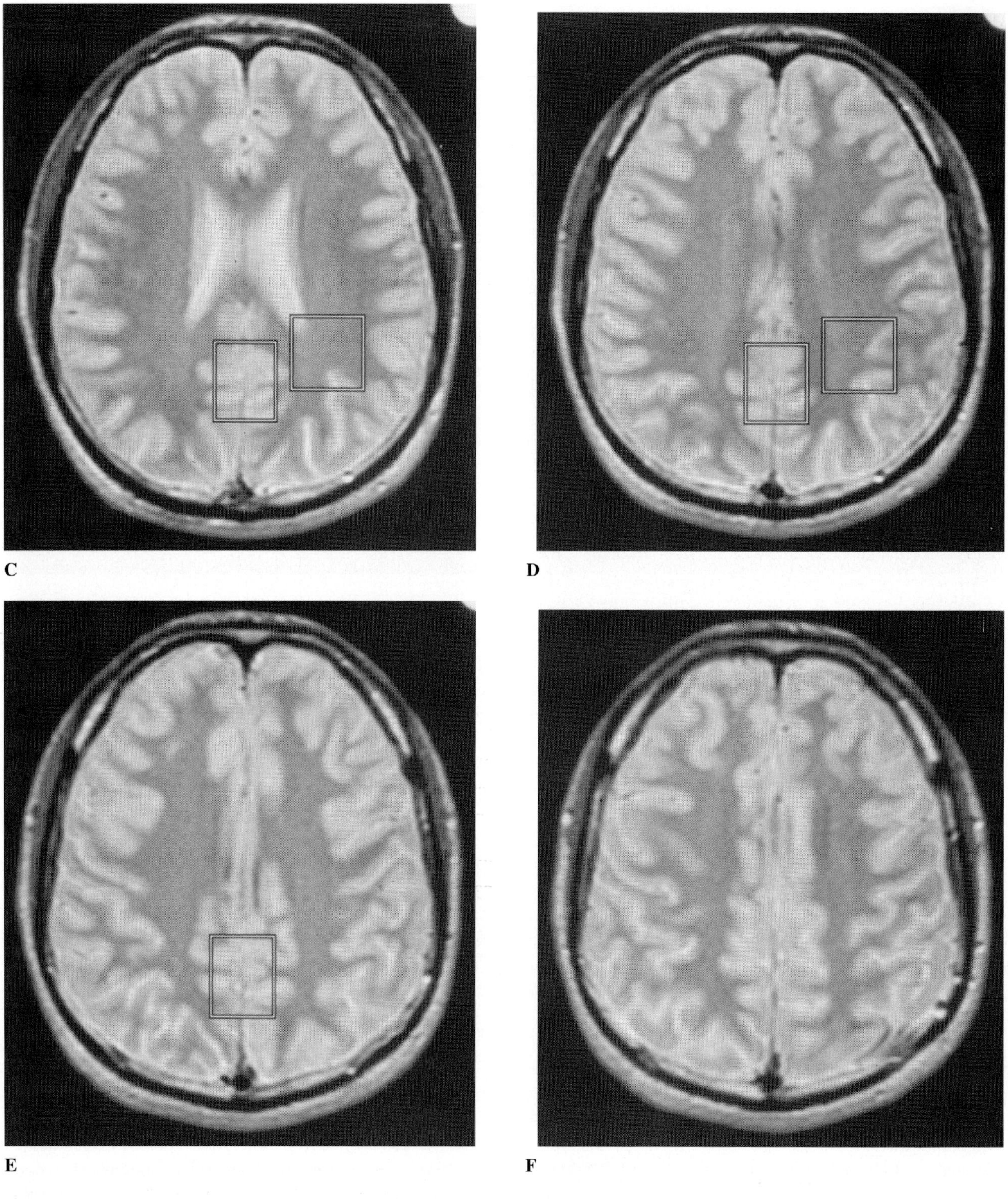

Fig. 4.1 Continued.

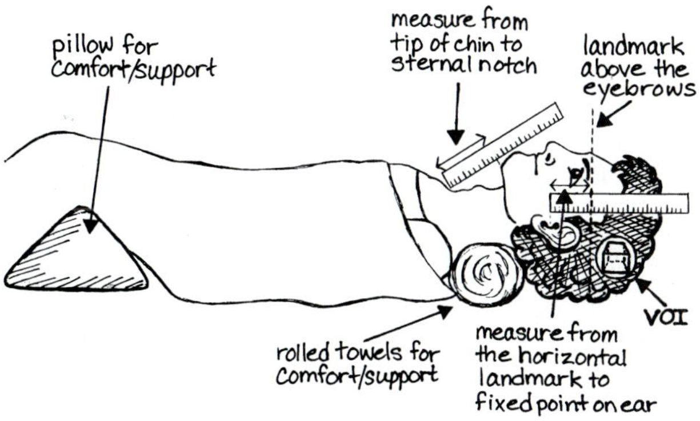

G

Fig. 4.1 Continued.

b. Beta and Gamma Glutamine and Glutamate (β,γ-Glx)

After NAA at 2.02 ppm, the next resonances seen are from the β- and γ-resonances of glutamine and glutamate (Glx). These are affected by strong spin–spin coupling, and the resonances are therefore collapsed into the pattern of resonances seen between 2.05 and 2.5 ppm. The β- and γ-resonances of glutamine and glutamate partially overlap, so the two are often referred to as β,γ-Glx.

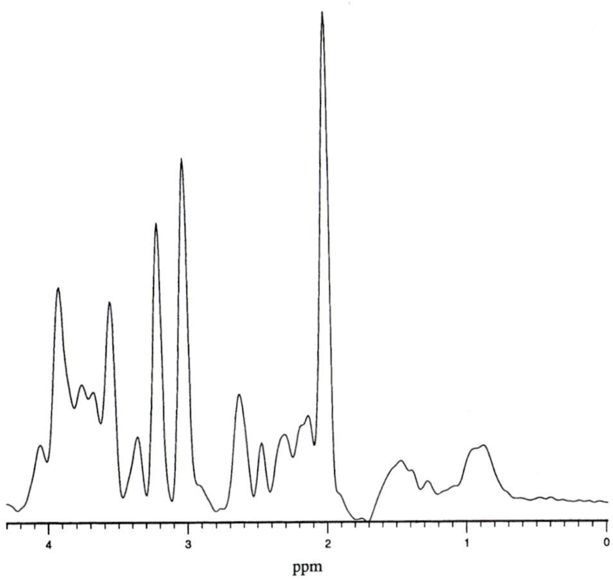

Fig. 4.2 The averaged white matter spectrum from a group of adult volunteers.

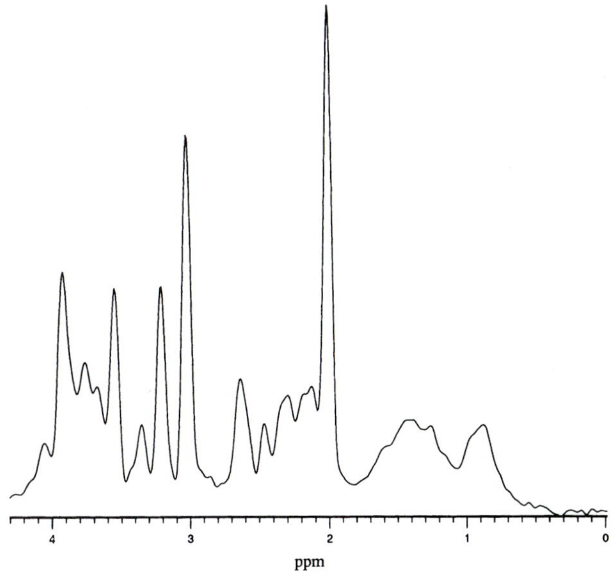

Fig. 4.3 The averaged grey matter spectrum from a group of adult volunteers.

c. Second NAA Peak

The second NAA resonance seen at 2.6 ppm is a little larger than the Glx resonances. The first NAA peak at 2.02 ppm results from the CH_3 group of the molecule, and is a singlet; in other words, it is a narrow noncoupled peak. This second NAA peak results from a CH_2 group, and is strongly coupled to its neighboring CH group; therefore this peak (which at higher resolution is shown to be a collapsed multiplet consisting of multiple peaks), is not as tall as it would have been if it were a singlet. A singlet would have been 2/3 the height of the first NAA peak's height (2 protons versus 3 protons in the two molecular groups), but this collapsed peak is smaller. The relative size of the two NAA peaks will, however, always be the same (if measurement parameters are not changed drastically), simply because the relative amount of CH_2 and CH_3 groups in the NAA molecule obviously is not going to change. This can be used to double-check the interpretation: if the relative sizes of the two NAA peaks are not as expected, something is "wrong," perhaps indicating that some other molecule has contributed to one of the peaks, or there may be a technical problem. (Several brain metabolites generate secondary resonance peaks: Glx, creatine and *myo*-inositol; see below.) A third peak of NAA, referred to in some case studies, is found at 2.5 ppm, overlapping Glx. A trick to estimating the Glx is to determine if the ratio of the third "NAA" to the second NAA peak is above 0.5, then Glx is increased compared to normal. This "pattern method" only works with TE 30 ms and TM 13.7 ms; other measurement parameters give different Glx patterns.

d. *Creatine and Phosphocreatine (Cr)*

The next tall peak at 3.03 ppm originates from creatine and phosphocreatine. The two resonances are so close that they totally overlap when acquired at 1.5 T, so the peak is usually referred to as "total creatine" and abbreviated "Cr." This Cr peak is a narrow singlet peak originating from the CH_3 group of the molecule. (Please note that the well known metabolite "creatinine" is not the metabolite observed here, but is its breakdown product.)

Adjacent to Cr, at 2.9 ppm, a very reproducible small resonance assigned to GABA (γ-amino-butyric acid) is not discussed in routine MRS. A separate "editing" technique is essential for reliable demonstration of this important neurotransmitter amine.

e. *Choline (Cho)*

At 3.22 ppm a resonance from choline-containing compounds (mainly phosphorylcholine and glycerophosphorylcholine), is found. The usual abbreviation is "Cho," but like Cr the different cholines can not be separated, and in referring to Cho it should be remembered that the resonance is the combined total choline peak. The Cho peak is a narrow singlet peak originating from, in NMR terms, nine identical protons: $(CH_3)_3$. Everything else being equal, a Cho and a Cr peak of the same height would not correspond to the same metabolite concentrations due to threefold more protons contributing to the Cho peak. The fairly large Cho peak, therefore, is produced by a smaller concentration than one might anticipate, because "choline" is only 15–20% the concentration of Cr. Accordingly, clinically significant variation in Cho is a frequent occurrence.

f. *Scyllo-inositol (sI) and Glucose*

The next small peak is not necessarily seen in all normal spectra, but the S/N in this summed spectrum is very good so we can clearly see the *scyllo*-inositol (sI) peak a 3.36 ppm. The sI peak is a singlet arising from the six magnetically identical CH protons. One should be careful with the identification, because if glucose is elevated, a glucose peak is found very nearby at 3.43 ppm. To identify the peak's chemical shift precisely, place a piece of paper with one edge along the chemical shift scale (the horizontal ruler under the spectrum), and align the perpendicular edge with the peak, then read off the chemical shift (in ppm) at the corner of the paper. This method will allow precision within 0.01 to 0.02 ppm.

g. *Myo-inositol (mI)*

The next larger peak is the *myo*-inositol (mI) peak at 3.56 ppm, and appears as a narrow singlet peak, but is in reality a complex collapsed peak from CH groups (H1, H3, H4, and H6) of the molecule.

h. *Alpha Glx (α-Glx)*

Before we get to the next large narrow peak, we find the α-resonances of glutamine and glutamate at 3.65 to 3.8 ppm. Here they are seen as two peaks, but if the resolution is even better, as for example if measured from a solution in a bottle, then an apparent triplet (three peaks) can be seen. As with the β,γ- resonances, the α- resonances are also coupled. The α- resonances of glutamate and glutamine overlap, and are often referred to as α-Glx. Similarly to the first and second resonances of NAA, the two sets of resonances of Glx should appear in appropriate relative proportions. Rare examples of pitfalls are: 1) if α-Glx appears too large, there could be underlying signal from a second glucose peak at 3.8 ppm; or 2) apparently increased and distorted β,γ-Glx could be caused by other underlying signals from, for example, ketones.

i. *Second Cr Peak*

The large narrow peak at 3.9 ppm arises from a second set of protons in Cr, a CH_2 group. The second Cr peak is a singlet like the first, so the sizes of the first and the second Cr peaks can easily be compared. The relative peak heights should be approximately 2/3, the ratio of the number of protons. This is a good quality check on the spectrum in this region so close to water (at 4.7 ppm), where artifacts sometimes occur.

j. *Second mI Peak*

The last small peak at 4.06 ppm is a second mI peak originating from the H2-CH group of mI. The peak is not always seen because it may be small compared to the noise, and because it may be affected by the residual water signal.

2. Interpretation and Analysis

In this section, metabolite ratios and other relevant numbers are listed, compared to normal values, and discussed. If possible the numbers are compared to normal adult numbers obtained with the same method (pulse sequence, measurement parameters, MR scanner manufacturer, software release, VOI location, etc.). For simplicity, the age-related corrections, are applied afterward. Age dependent corrections may not be explicitly calculated in each case, but appear as "normal for age" or "NAA is 20% low compared to normal adult, and even lower considering the patient's age of 10 years."

The metabolite ratios which may be listed are NAA/Cr, Cho/Cr, mI/Cr, β,γ-Glx/Cr, α-Glx/Cr, $Glucose_1$/Cr and $Glucose_2$/Cr. The normal ratios relevant for most *of the case studies are given in Tables 2.3 and 2.4. These normal values were obtained by computer analysis of the spectra, but the ratios NAA/Cr, Cho/Cr and mI/Cr can in most cases be obtained almost as precisely by visual inspection of peak height ratios. The ratios β,γ-Glx/Cr, α-Glx/Cr, $Glucose_1$/Cr and $Glucose_2$/Cr were also obtained by computer analysis of the spectra, as ratios of the integrals of appropriately chosen chemical shift ranges, not necessarily reflecting apparent peak height ratios. They may be more difficult to reproduce by visual inspection, but with experience you will get used to recognizing if the Glx or glucose resonances are elevated or not. Ratios depend on TE, TM, TR, pulse sequence, VOI location, and other measurement parameters. The numbers listed here are obtained using our standard parameters. Provided that the measurement is precisely reproduced, and a similar group of healthy volunteers is included, one should find the same values (within the margin of variation) anywhere in the world.

Ratios cannot always give all of the available information; knowledge of an absolute reference is often also valuable. We usually list the absolute Cr as [Cr]. Ratios to Cr together with [Cr] in principle should give full information about the metabolite levels in the VOI. However, to be more complete, sometimes absolute values for other metabolites are also given. The absolute values, as we list them, are obtained by first calculating the metabolite concentration in the brain tissue of the VOI. Most metabolites are absent from cerebrospinal fluid, CSF, so especially in cases of major atrophy, it becomes important that the concentration "in the brain tissue" of the VOI is used, that is correcting for the partial volume of CSF.

With normal values for many metabolites, which even differ depending on the VOI location, it quickly becomes impractical to memorize them all. We therefore normalize the absolute metabolite concentrations for each metabolite with respect to the normal value found in the given brain location in healthy adult volunteers. This is the so-called Institutional Unit. Thus [Cr] equal to 1 means that Cr is normal; [Cr] equal to 0.75 means that Cr is 25% reduced, and so forth. If one wants to know the absolute concentration in mmoles per kg brain tissue, it is readily obtained by multiplying by the relevant normal metabolite concentration. See Chapter 2 for further details on absolute quantitation.

The water concentration is calculated in a similar fashion based on the absolute water concentration, but normalized slightly differently, namely with respect to the concentration of pure water. The value [H_2O] equal to 0.75 hereby indicates a 75% water content in the brain tissue compared to pure water. Normal values for [H_2O] are 78% ± 1.5% for HMRI's occipital grey matter VOI, and 69% ± 1.5% for HMRI's parietal white matter VOI. The water content can therefore often be used to verify the validity of the absolute metabolite concentrations. For example, if [H_2O] is equal to 0.40 (obviously too low), the lost signal could, among other reasons, be due to patient movement. But, no matter what the reason, [Cr] and all other metabolites are likely to be similarly affected, and reassuringly, metabolite ratios will not be affected.

The fraction of CSF may be listed, in particular for HMRI's occipital grey matter location, because we use it as an atrophy index. The CSF fraction is given as the CSF volume in percentage of the total VOI volume. A typical value for grey matter in healthy adults is 5%.

Dependency on age for all the metabolite values, and total brain water and CSF fractions, is a major issue in the first year of life, and a less dramatic, but still important issue in the elderly. Age dependence is not included in any of the mentioned calculations; instead graphs are given for reference to correct normal values (see Chapter 2).

E. Location 2 (Grey Matter)

1. Peak Identification and Quality Assurance: Reading the Spectrum from Right to Left

The procedure for reading the grey matter spectrum (Fig. 4.3) is the same as for white matter. Starting with an evaluation of quality, this grey matter spectrum is excellent with good S/N, well resolved resonances, and no artifacts.

Reading the spectrum systematically from right to left, the region between 0.8 and 1.5 ppm (lipid/lactate) is unremarkable. Continuing, the following peaks are identified: NAA, β,γ-Glx, the second NAA, Cr, Cho, sI, mI, α-Glx, the

*The values listed in Tables 1 or 2 were used as normal values in most of the case studies in this book, but not all. The exceptions include spectra acquired with a different scanner, or by another MRS center, where different measurement parameters were used, and thus a relevant set of normal data were used when available.

Tables 1 and 2 present normal ratios obtained on different occasions, and although the numbers are not exactly identical, the values agree with each other, in itself reassuring considering that they were obtained by different operators at different times. In the data analysis of the patient cases we state how much a ratio is increased or decreased based on these normal ratios, in some cases these percentage values have been rounded of to a multiple of 5%. In other cases a "precise" number has been given, understanding however that this "precision" ± 1–2%, depending on which set of normative ratios have been used.

second Cr, and the second mI. The sizes of all secondary peaks are appropriate.

2. Interpretation and Analysis

The metabolite ratios are: NAA/Cr = 1.4 (normal); Cho/Cr = 0.6 (normal); and mI/Cr = 0.6 (normal). The concentration of Cr (given in Institutional Units) is equal to one ([Cr] = 1), which is normal. β,γ-Glx/Cr = 0.35, and α-Glx/Cr = 0.48, therefore Glx/Cr is normal.

F. Conclusion

This section contains the conclusion based on the one or more VOIs measured. The major findings are summarized and a diagnosis, prognosis, or other clinical comments are stated. Possible differential diagnoses are also mentioned along with suggestions of how to rule these out.

G. Summary of the Patient's History Post MRS

If available, we have given a summary of the patient's history after the MRS examination, in some cases illustrating how MRS influenced patient care.

H. Other Comments

In this section we discuss various aspects of or related to the case.

5

Clinical Cases

5.1 Healthy Subjects: Normal Proton Spectra of All Ages

5.1.1 NORMAL PREMATURE, GESTATIONAL AGE 35 WEEKS

A. Purpose of MRS

(1) Rule out perinatal hypoxic encephalopathy. (2) Compare long and short TRs.

B. Summary of Patient's History

Female, age 2 weeks postnatal (gestational age 35 weeks), and in incubator at the time of MRS. The infant had frequent apnoeic attacks, but was otherwise normal. Patient was delivered by lower segment Cesarean section at 33 weeks gestation.

C. MRS Method, Volume of Interest, and Comments about Examination

Localized STEAM (TE = 30 ms, TM = 13.7 ms) was performed in HMRI's parietal white matter right using a short TR of 1500 ms, and for comparison was repeated in the same location using a long TR of 5000 ms (Location 1). A third spectrum (Location 2) was acquired in the parietal white matter left (TE = 30 ms, TM = 13.7 ms, TR = 5000 ms).

D. Location 1 Acquired at TR 1500 ms (White Matter Right)

1. Peak Identification and Quality Assurance: Reading the Spectrum from Right to Left

The technical quality of the spectrum acquired at TR 1500 ms (Fig. 5.1A) is good, except that details of the NAA peak are not clear. Nevertheless, with a voxel size of 2.88 cm^3, the signal-to-noise is adequate for interpretation.

Reading from right to left, there is no obvious lipid or lactate at 1.33 ppm. Continuing from right to left, the following resonances are identified: NAA, β,γ-Glx, the second NAA peak (prominent), Cr, Cho, sI and glucose, mI, α-Glx, and the

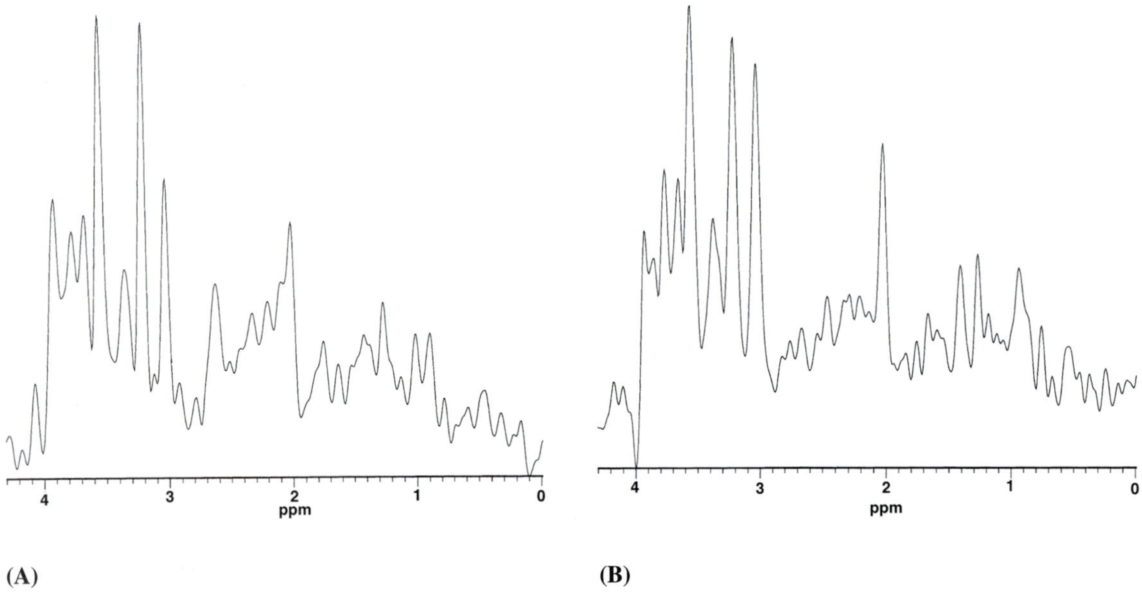

Fig. 5.1 (A) Premature infant spectrum acquired under standard conditions from the right parietal cortex (TE 30 ms/TM 13.7 ms/TR 1500 ms). This proton spectrum, although striking, with low NAA, and high Cho and mI peaks, is appropriate (and therefore normal), for this premature infant of postnatal age 2 weeks, but gestational age only 35 weeks. (B) MRS of normal premature infant, acquired with TR 5000 ms, rather than 1500 ms, to allow for the much prolonged T_1 relaxation times of the principal cerebral metabolites in very young infants. Note that while no new pathology is identified, lactate is better resolved, and both Cr and NAA peaks have become more prominent.

second resonances of Cr and mI. The inverse intensities of mI, Cho, Cr and NAA (as compared to adult) appear generally appropriate for the newborn.

2. Interpretation and Analysis

Reading from right to left, the metabolite ratios are: NAA/Cr = 0.82 (appropriate for 35 weeks gestational age); Cho/Cr = 1.61, and mI/Cr = 1.64 (both fit age-related data). [Cr] is 62%, and is again assumed to be appropriate to this very premature infant. β,γ-Glx/Cr = 0.59, and α-Glx/Cr = 0.92 (both in normal range). Brain water is 91.3%, as expected for neonate. The peak at 2.6 ppm, generally assigned as the second NAA peak, is in this case more likely free aspartate as the proportion 2:3 to the principal NAA peak is entirely inappropriate.

E. Location 1 Acquired at TR 5000 ms (White Matter Right)

1. Peak Identification and Quality Assurance: Reading the Spectrum from Right to Left

A less "T_1-weighted" localized STEAM spectrum acquired from the same location, during the same examination, but with a considerably longer TR (5000 ms), is of excellent technical quality (Fig. 5.1B). After 64 scans, this spectrum has somewhat superior S/N compared with 128 scans acquired at TR 1500 ms (Fig. 5.1A), in part because acquisition time was 5.4 minutes, compared to 3.2 minutes, respectively.

The pattern of the major peaks is subtly different from that observed at TR 1500 ms. New features include the more conclusive identification of lipid (0.9 ppm) and lactate (doublet at 1.33 ppm) peaks, more prominent NAA, but less prominent second NAA, and a more pronounced Cr peak.

Reading systematically from right to left, there is a possible lipid peak at 0.9 ppm, and an unmistakable lactate doublet at 1.33 ppm, followed by NAA at 2.02 ppm, an unusual β,γ-Glx region and a small second NAA peak. Cr is prominent, followed by Cho, sI, mI, a triplet for α-Glx and a small second Cr peak.

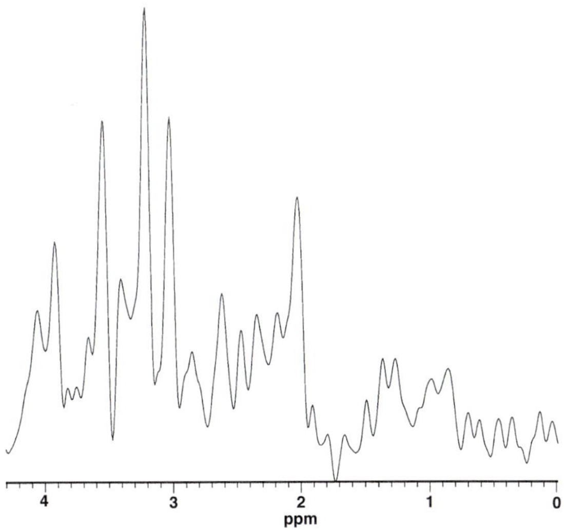

Fig. 5.2 MRS of contralateral hemisphere (parietal white matter left) in premature infant, acquired at TR 5000 ms. There are only minor differences between this and the matching spectrum acquired from the right hand side. The apparent reduction of mI/Cr may be related to the artifacts at 3.5 and 3.75 ppm. An unlabelled peak at ~2.85 ppm is possibly also an artifact.

2. Interpretation and Analysis

The obvious lactate doublet may be an effect of improved S/N, a T_1 effect, or even possibly a physiological effect. By the latter is meant the real accumulation of a small excess of cerebral lactate during the course of this somewhat lengthy MRI/MRS examination conducted with mechanical ventilation and in the presence of sedative doses of barbiturate.

Normative data are not available at this age and TR. Reading from right to left, the calculated peak ratios are: NAA/Cr = 0.74; Cho/Cr = 1.07; mI/Cr = 1.18; β,γ-Glx/Cr = 0.27; α-Glx/Cr = 0.62. Absolute [Cr] moves closer to 100%. This latter value is not corrected for T_1, however. When an appropriate correction is applied, [Cr] still lies between 90–100% of normal. All ratios are substantially different from those determined in the "standard" assay. T_1 for each metabolite is known, and is substantially longer than one second.

F. Location 2 Acquired at TR 5000 ms (White Matter Left)

1. Peak Identification and Quality Assurance: Reading the Spectrum from Right to Left

The less "T_1-weighted" localized STEAM spectrum (Fig. 5.2), acquired from the opposite parietal white matter location, is of excellent technical quality apart from minor artifacts. These include (reading from right to left), an unidentified "peak" at 2.85 ppm, poor resolution of sI from Cho, at 3.2 ppm and loss of definition in the α-Glx region. In contrast to the previous spectrum acquired under identical conditions, β,γ-Glx appears to be well defined, and comparable in its coupling pattern to that seen routinely with TR 1500 ms.

Reading the spectrum systematically from right to left, the lipid peak at 0.9 ppm is dubious as is the presence of a lactate doublet at 1.33 ppm. NAA, β,γ-Glx, a second NAA peak and an unknown (or artifactual peak), are followed by Cr, Cho, sI, mI, artifacted α-Glx, and clearly resolved second peaks for Cr and mI.

2. Interpretation and Analysis

The calculated peak ratios are: NAA/Cr = 0.78; Cho/Cr = 1.31; mI/Cr = 0.99; β,γ-Glx/Cr = 0.42, and α-Glx/Cr = error. Brain water is 91.6%, identical to that determined in the right hemisphere, and appropriate to a normal preterm neonate.

G. Conclusion

A remarkably rare MRS examination in such a premature infant is conceivably quite normal. If so, the onset of NAA synthesis, extrapolated to the end of the second trimester may explain persistently elevated free aspartate. Another remarkable negative (normal) finding is the absence of lactate (except in one spectrum, most likely explained by temporarily less-sufficient ventilation), and low total Cr, making it unlikely that glycolysis is the principal energy source at this late stage of brain development.

The spectra acquired at longer TR exploit the known prolongation of metabolite T_1s in early infancy, to provide better signal to noise. Metabolite ratios are slightly different from those observed at shorter TR. Notably [Cr] and [NAA] appear to increase more than do Cho and mI.

In practice, automated PROBE acquisitions consume excessive amounts of time when run with long TR intervals. (This is because automated prescans occupy more than threefold additional time, in proportion to TR.) This is not repaid by additional clinical information. A possible exception in this case, was the clear demonstration of lactate, not previously recognized. Lactate is almost certainly a normal constituent of the newborn brain spectrum, acquired at longer TR.

For economy of time, we do not advocate longer TR examinations for routine use.

H. Summary of Patient's History Post-MRS

The infant made an uneventful graduation from the neonatal intensive care unit, and showed normal development at age 5 years.

I. Other Comments

The best location for ruling out hypoxic encephalopathy in adults (> 2 years) is grey matter, but in babies the choice of grey versus white matter is not quite so important.

Although we rarely see MR spectra acquired at this gestational age, it seems to fit on a smooth developmental curve with the "commencement" of NAA synthesis at the end of the second trimester, perhaps.

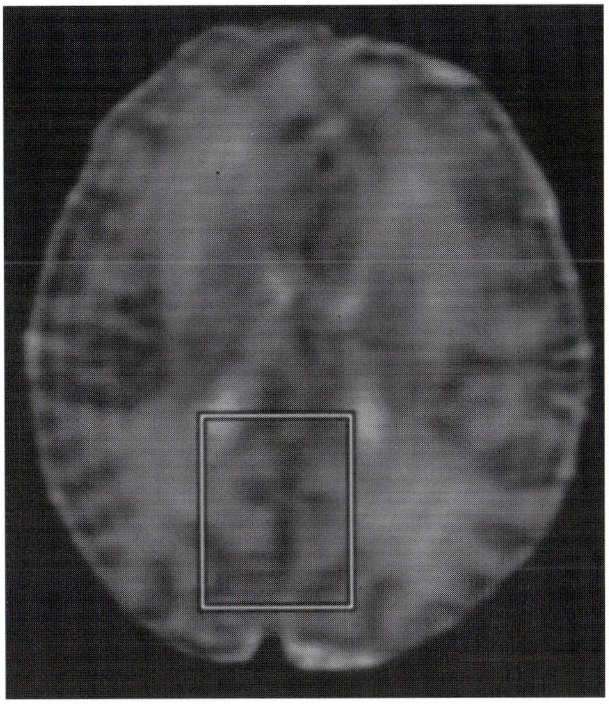

Fig. 5.3 MRI showing the occipital location chosen for MRS. The standard size VOI appears large because the head is so small.

5.1.2 NORMAL PREMATURE, GESTATIONAL AGE 36 WEEKS

A. Purpose of MRS

Rule out perinatal asphyxia.

B. Summary of Patient's History

Male, 3 days postnatal, delivered at 36 weeks gestational age by emergency Cesarean section for fetal distress. Apgar scores were satisfactory.

C. MRS Method, Volume of Interest, and Comments about Examination

Localized STEAM (TE = 30 ms, TM = 13.7 ms, TR = 1500 ms) was performed in HMRI's occipital grey matter (see Fig. 5.3) only, three days after delivery. In an infant the term "grey matter" is used to signify the standard occipital location rather than to describe the composition of the VOI, accurately.

D. Grey Matter Location

1. Peak Identification and Quality Assurance: Reading the Spectrum from Right to Left

Good quality spectrum from grey matter (Fig. 5.4) showed typical, early postnatal result. Reading systematically from right to left, there is a little lipid at 0.9 ppm, but no lactate at 1.33 ppm. Identity of peak(s) at 1.44 ppm is uncertain, but consistent with second lipid peak. Continuing right to left, the following resonances are identified: NAA, Cr, Cho, mI at 3.58 ppm, α-Glx, and the second Cr and mI peaks. The reverse "slope" from mI to NAA is normal for a premature infant.

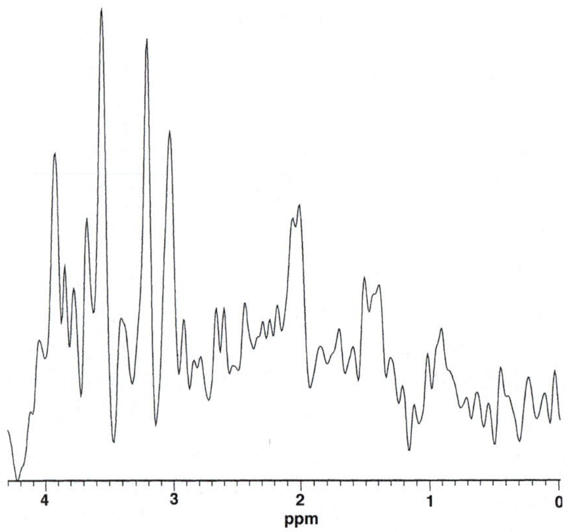

Fig. 5.4 Normal grey matter spectrum in premature infant. The "reversed angle" of the peak heights corresponds to gestational age of 36.5 weeks seen in chapter 2. (An imaginary line through the tops of the peaks slopes down to the right, which is reversed compared to normal, where mI is usually the smallest and NAA the tallest peak, and the imaginary line slopes down to the left. The angle of this imaginary line and the horizontal axis is called "Hunter's angle," after Dr. Hunter Shelden, who initially suggested this straight forward method of pattern recognition in the evaluation of MRS).

2. Interpretation and Analysis

The metabolite ratios are: NAA/Cr = 0.74; Cho/Cr = 1.29; and mI/Cr = 1.45. By comparison with age-adjusted values, these ratios are consistent with a normal infant of gestational age 36.5 weeks.

E. Conclusion

Normal maturation and myelination probably dictates the MRS pattern of early postnatal life. Hypoxic injury, either *in utero* ("fetal distress"), or perinatal, is entirely excluded by this MRS examination.

F. Summary of the Patient's History Post-MRS

Good progress until discharge from neonatal intensive care unit. No long-term follow-up available, but presumed normal.

G. Other Comments

MRS performed in the perinatal period must be matched to gestational age within ± 1 week. The power of this method for assessing developmental delay has not been fully explored.

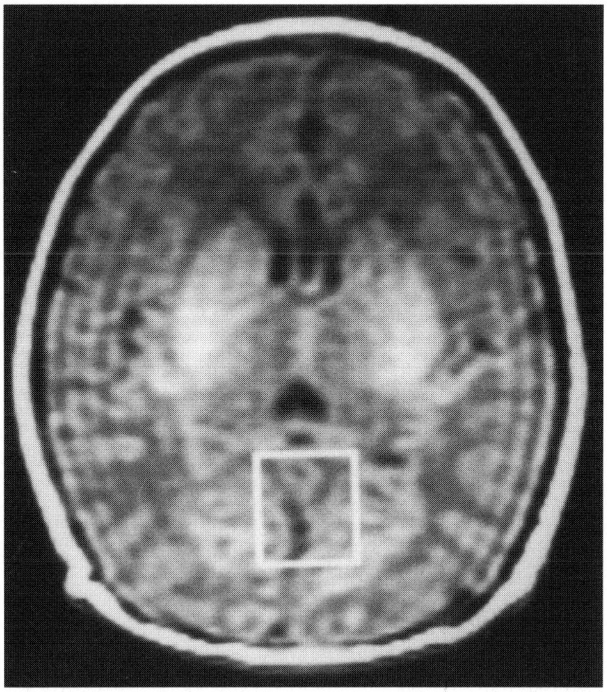

Fig. 5.5 The MRI was acceptable for localizing a VOI and prescribing the spectroscopy sequence despite the movement artifacts. The noisy MRI sequence disturbed the baby, but the much quieter MRS sequence did not, so there is no reason to expect major MRS movement artifacts. However the water suppression artifact seen in the spectrum may be caused by a slight shift in frequency (<10 Hz) occurring after the optimization of the water suppression and before the start of acquisition. Such a shift could for example be caused by the baby moving an arm or a leg.

5.1.3 NORMAL BABY

A. Purpose of MRS

To acquire data from an age matched healthy volunteer.

B. Summary of Patient's History

The healthy baby was 1 day old, and born at term.

C. MRS Method, Volume of Interest, and Comments about Examination

The automated MRS method PROBE-STEAM (TE = 30 ms, TM = 13.7 ms, TR = 1500 ms) was used to measure a VOI in HMRI's occipital location shown on Fig. 5.5.

D. Occipital "Grey Matter" location

1. Peak Identification and Quality Assurance: Reading the Spectrum from Right to Left

The first impression of the spectrum, Fig. 5.6 might be that it is of very poor quality; the peaks appear small and there is a large "wiggle" at 4–4.2 ppm. This "wiggle" is an effect (artifact) of nonideal water suppression and it is not a serious

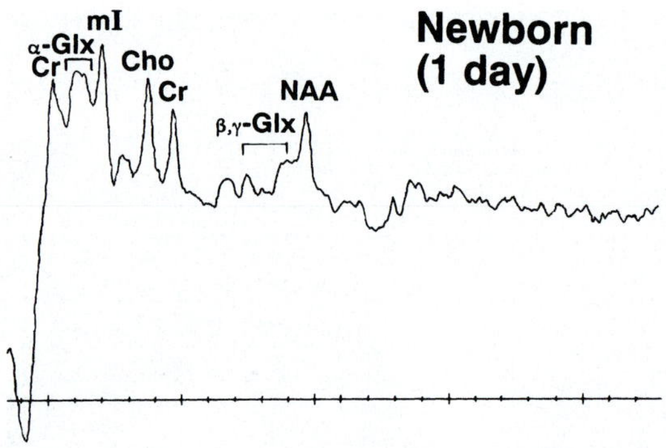

Fig. 5.6 The grey matter spectrum from a 1 day old healthy term baby. The spectrum was acquired by the automated MRS method PROBE-S, and is here displayed as it appeared on the console.

problem, certainly not compromising the diagnostic value of a spectrum like this. The S/N and the resolution are good, and the only reason the peaks appear small is that the signals have been automatically scaled down so that the water suppression artifact would still fit inside the range shown. This is simply how the automated STEAM MRS sequence PROBE-S presents the result.

Reading the spectrum systematically from right to left the following resonances are identified: NAA, β,γ-Glx, the second NAA, Cr, Cho, mI, α-Glx, and the second Cr. The second Cr is somewhat affected by the water suppression. One also notices that the peak heights of NAA, Cr, Cho and mI are reversed compared to normal adult spectra.

2. Interpretation and Analysis

When the peaks are analyzed the elevated baseline should be taken into account. The ratios are estimated by visual inspection to be NAA/Cr = 1.0, Cho/Cr = 1.4 and mI/Cr = 1.6, and [Cr] is not available. These values are all within the normal range for the occipital location in a 1-day-old, term baby (see Chapter 2).

E. Conclusion

Normal.

F. Summary of Patient's History Post-MRS

Not relevant.

G. Other Comments

When evaluating an MR spectrum, it is important to know the gestational age, GA. However be aware that GA for a term baby is not defined equally in all countries. In this text we have used GA = 40 weeks for a term baby.

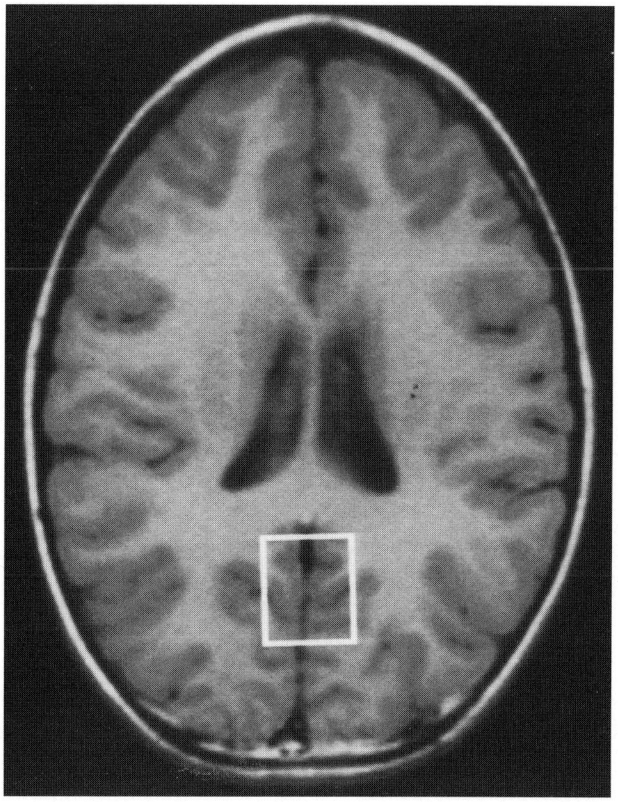

Fig. 5.7 The MRI showing the VOI location.

5.1.4 NORMAL CHILD

A. Purpose of MRS

To acquire data from an age matched healthy volunteer.

B. Summary of Patient's History

The healthy boy was 8 years old.

C. MRS Method, Volume of Interest, and Comments about Examination

The automated MRS method PROBE-STEAM (TE = 30 ms, TM = 13.7 ms, TR = 1500 ms) was used to measure a VOI in HMRI's grey matter as location shown (Fig. 5.7).

D. Grey Matter Location

1. Peak Identification and Quality Assurance: Reading the Spectrum from Right to Left

The spectrum is well resolved, the S/N is good, but the baseline near water is "rolling" a little. Reading the spectrum, Fig. 5.8, systematically from right to left first we see the typical pattern of lipids in the form of long chain fatty acids: Two broad peaks, a smaller at 0.9–1.0 ppm, and a larger at 1.3–1.4 ppm. In some cases lipids are not seen at all, in others some lipid is seen; the amount seen here is still within the normal range, and should normally not warrant mentioning. See further comments on the lipid region below. Continuing reading, the following resonances are identified: NAA, β,γ-Glx, the second

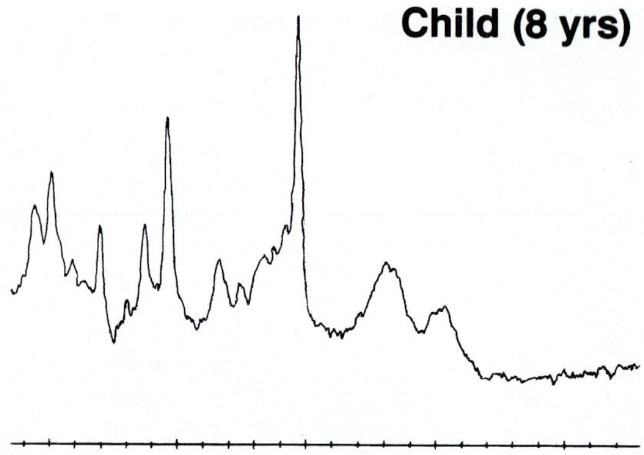

Fig. 5.8 The grey matter spectrum from a 8 year old healthy boy. The spectrum was acquired by the automated MRS method PROBE-S, and is here displayed as it appeared on the screen after acquisition.

NAA, Cr, Cho, mI, α-Glx, and the second Cr and mI. The second Cr and second mI peaks are somewhat affected by the baseline being lifted up and rolling a little here, close to water.

2. Interpretation and Analysis

The ratios are estimated by visual inspection accounting for the somewhat rolling baseline: NAA/Cr = 1.5, Cho/Cr = 0.45 and mI/Cr = 0.6. [Cr] is not available. These values are within the normal range for the age and location except for the Cho/Cr = 0.45, which is low. The lower Cho/Cr may be caused by inadequate compensention for the rolling baseline, or it may be real. The interindividual variation of Cho is larger than for the other metabolites. Furthermore, Cho/Cr is systematically lower within visual cortex, inferior but adjacent to the "standard" grey matter VOI.

E. Conclusion

The boy is known to be normal. Despite a somewhat low Cho/Cr, which is seen occasionally, there is no reason to call the MRS abnormal.

F. Summary of Patient's History Post-MRS

Not relevant.

G. Other Comments

The NAA/Cr ratio is about 10% above the normal adult value, which is exactly what is expected in such a young boy.

The region from 0.7–1.6 ppm needs a little further discussion. First of all, this region of the spectrum is variable even in normal volunteers. Lipids in the scalp may contribute to the signal if the VOI is placed too close to the scalp, but in some cases large lipid signals have been found to truly originate from inside the VOI.

Precise assignments of signals in this region should only be done with great care, and diagnoses based on minor broad signals in the region should be avoided due to the variability of the region. We have made it a habit not to mention insignificant "lipid" resonances when reading a spectrum. The amount seen in this spectrum are however getting close to being worth mentioning. For simplicity we have also made it a habit to call the broad resonances in this region "lipids," well knowing that precise assignments are troubled by the question of signal from the inside or the outside of the VOI, and that these resonances can be caused by many types of lipids and proteins.

A last word of caution concerns the amount of lipid seen in the spectrum. The lipids may appear increased relative to the NAA, Cr, Cho and mI, but you may be misled. Thus in pathologies in which NAA, Cr, Cho and mI are all decreased, lipids may not be truly increased.

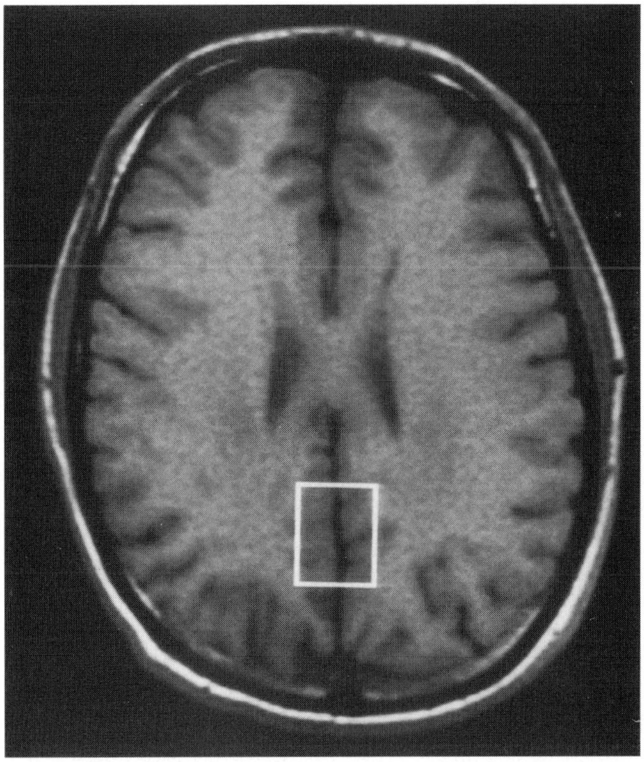

Fig. 5.9 The MRI showing the VOI location, selected to maximize grey matter contribution.

5.1.5 NORMAL ADULT

A. Purpose of MRS

To acquire data from an adult healthy volunteer.

B. Summary of Patient's History

The healthy volunteer was 31 years old.

C. MRS Method, Volume of Interest, and Comments about Examination

The automated MRS method PROBE-STEAM (TE = 30 ms, TM = 13.7 ms, TR = 1500 ms) was used to measure a VOI in HMRI's grey matter at location shown, Fig. 5.9.

D. Grey Matter Location

1. Peak Identification and Quality Assurance: Reading the Spectrum from Right to Left

The spectrum, Fig. 5.10, is of high quality with good resolution, good S/N, and no artifacts. Reading the spectrum, systematically from right to left we see the following resonances: NAA, β,γ-Glx, the second NAA, Cr, Cho, mI, α-Glx, the second Cr, and the second mI.

2. Interpretation and Analysis

The ratios are estimated by visual inspection, and accounting for the baseline to be NAA/Cr = 1.4, Cho/Cr = 0.55 and mI/Cr = 0.5. [Cr] is not available. These values are well within the normal range for the age and location except for the mI/Cr = 0.5, which is at the low end of normal. However by visual inspection we can allow for a little more deviation.

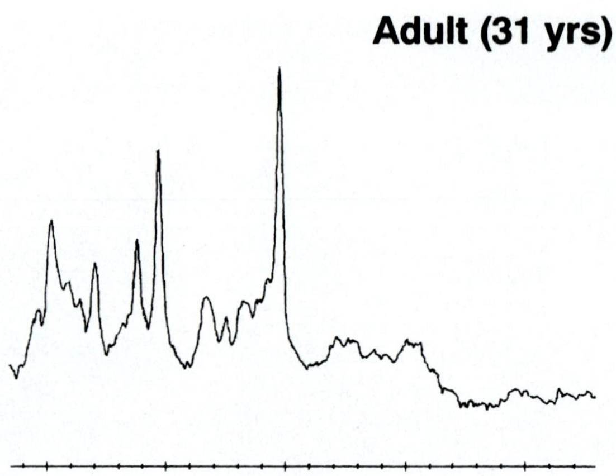

Fig. 5.10 The grey matter spectrum from a 31 year old volunteer. The spectrum was acquired by the automated MRS method PROBE-S, and is here displayed as it appeared on the screen.

E. Conclusion

Normal.

F. Summary of Patient's History Post-MRS

Not relevant.

G. Other Comments

Notice that the lipids in this spectrum are insignificant, indicating that the VOI was placed correctly away from the skull.

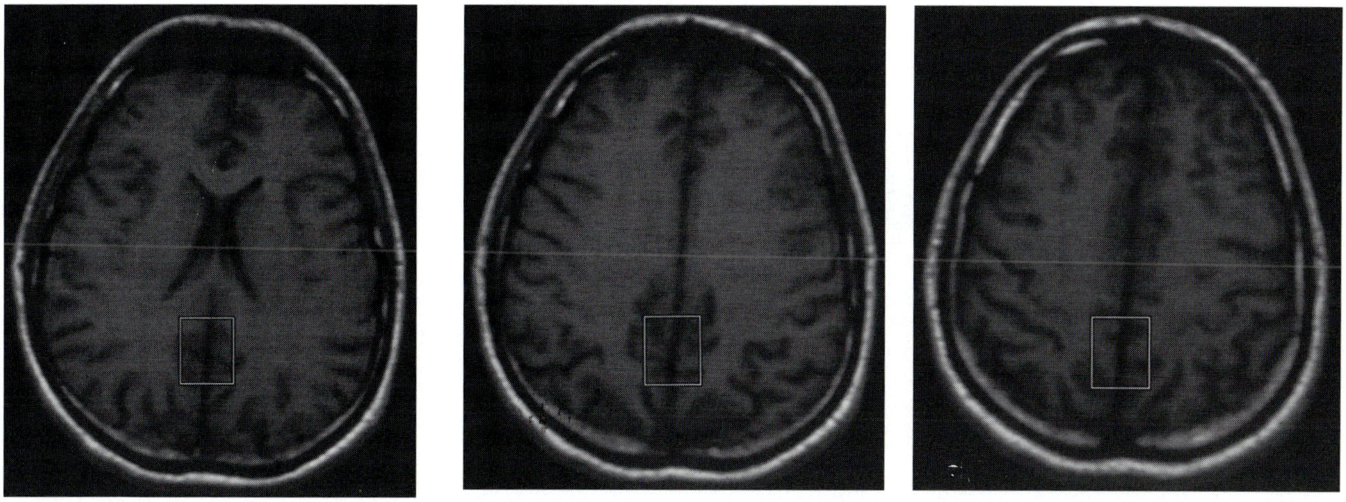

(A) (B) (C)

Fig. 5.11 Axial T_1-weighted MRI establishing all three planes (A–C) included in HMRI's grey matter location for PROBE examination. Landmarks are those normally employed and modified slightly in respect to dimensions of the prescribed VOI and its upper and lower limits, to suit brain size of the patient.

5.1.6 NORMAL PROBE: ADULT

A. Purpose of MRS

Detailed description of VOI, and normal grey matter in a 30-year-old adult male.

B. Summary of Patient's History

Normal, healthy male, age 30 years.

C. MRS Method, Volume of Interest, and Comments about Examination

HMRI's occipital grey matter (Location 1) was measured with automated PROBE-STEAM (TE = 30 ms, TM = 13.7 ms, TR = 1500 ms) as displayed in three axial views (Fig. 5.11). The VOI was adjusted to head size: 21 mm × 26 mm and 20 mm slice thickness, providing a volume of 10.9 cm^3. Ideally the axial view above and below the three selected would also be reviewed to avoid lipid from the skull, scalp, or meninges. It should be remembered that slice selection in three orthogonal planes is intrinsic to the STEAM sequence, and that imperfections in radio frequency pulses result in the inclusion in the measured volume anything up to 20% more brain than that calculated.

Landmarks for routine selection of HMRI's grey matter VOI are displayed in the three images. The VOI is centered across the midline to maximize "grey" and minimize "white" matter included. The falx which not infrequently includes visible "fat" is within the VOI. Altering the center slice by 5 or 10 mm can be a useful device to minimize this lipid contamination of the spectrum. In the example shown, the source of "lipid" is not obvious and would have been difficult to avoid. In older subjects, infants with a small head, or in cases with marked atrophy, the lowest axial slice might safely include a view of the posterior commissures of the corpus callosum (white matter) to serve as landmarks, but not to be included in the VOI.

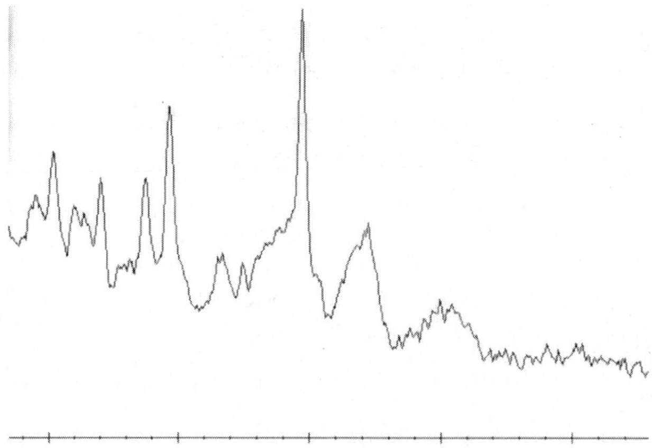

Fig. 5.12 Result of automated PROBE examination obtained in normal adult grey matter (STEAM; TE = 30 ms, TM = 13.7 ms, TR = 1500 ms).

D. Grey Matter Location

1. Peak Identification and Quality Assurance: Reading the Spectrum from Right to Left

The technical quality of the PROBE spectrum (Fig. 5.12) is good, with some residual water which elevated baseline, with a sharp and possibly artifactual "lipid" resonance centered at 1.6 ppm, and with a further possibly artifact at 4.1 ppm. None of these features impairs our ability to read and interpret this PROBE spectrum.

Reading the spectrum from right to left, the intrinsic protein-lipid peak at 0.9 ppm appears normal. The peak at 1.6 ppm is asymmetric and excessively large, possibly lipid within the VOI. NAA, β,γ-Glx, second (and third) NAA peaks, Cr, Cho, no obvious sI, mI, somewhat asymmetric α-Glx, and a second Cr peak are all clearly identified. The second mI peak 4.1 ppm is likely elevated by an artifact.

2. Interpretation and Analysis

The ratios determined by inspection are: NAA/Cr = 1.74; Cho/Cr = 0.57; mI/Cr = 0.48; β,γ-Glx/Cr is appropriate, and α-Glx/Cr not elevated, despite unusual appearance.

E. Conclusion

Normal.

F. Other Comments

Visual inspection of PROBE establishes normal values for MRS examinations. The normal ranges of the ratios from inspection of PROBE may differ in detail from those achieved with automated PROBE-Q or HMRI's quantitative MRS method. The two latter apply identical computerized data processing to achieve metabolite ratios.

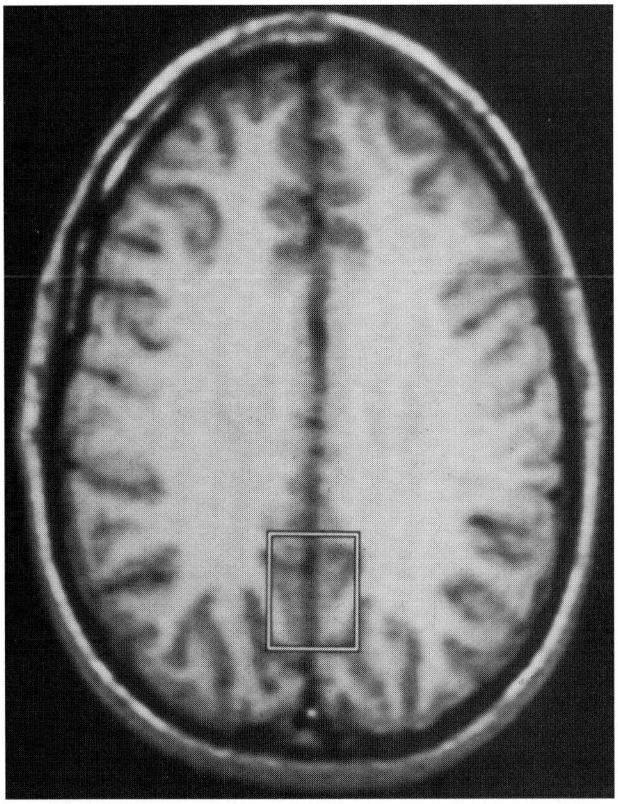

(A)

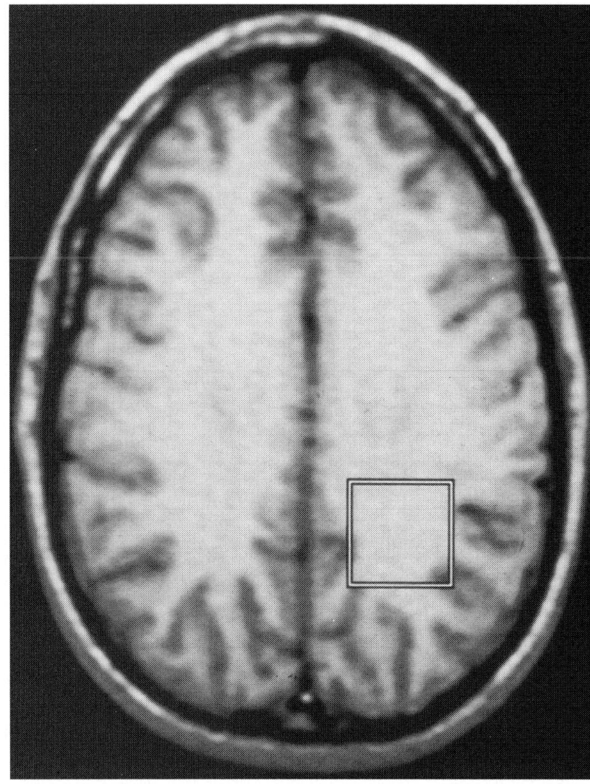

(B)

Fig. 5.13 The grey matter (A) and the white matter (B) locations in a young normal volunteer.

5.1.7 NORMAL YOUNG ADULT

A. Purpose of MRS

To obtain a normal occipital grey and parietal white matter PROBE spectrum in a healthy young adult.

B. Summary of Subject's History

Male, age 23; healthy.

C. MRS Method, Volume of Interest, and Comments about Examination

Automated PROBE using STEAM (TE = 30 ms, TM = 13.7 ms, TR = 1500 ms) was performed in HMRI's occipital grey (Location 1) and parietal white (Location 2) matter as shown on the image (Fig. 5.13).

D. Location 1 (Grey Matter)

1. Peak Identification and Quality Assurance: Reading the Spectrum from Right to Left

The technical quality of this PROBE spectrum is excellent (Fig. 5.14), with no artifacts. Water suppression is also very good, with a residual slope of some 5–10° in the baseline.

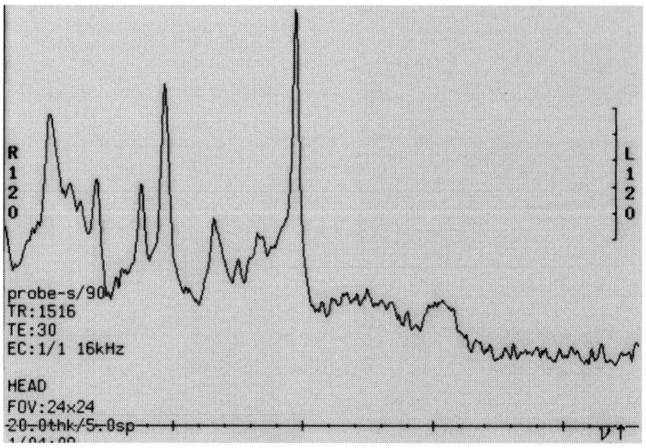

Fig. 5.14 Normal automated PROBE spectrum in the occipital grey matter of a 23 year old male subject.

Reading this normal PROBE spectrum systematically from right to left, there is a small protein-lipid peak at 0.9 ppm, and no lactate doublet at 1.33 ppm. The following metabolite resonances are identified: NAA, β,γ-Glx, a second NAA, Cr, Cho, barely detectable sI, an obvious mI, α-Glx, and a broadened second Cr peak.

2. Interpretation and Analysis

Reading from right to left, the metabolite ratios determined by "eye" are: NAA/Cr = 1.43; Cho/Cr = 0.49; mI/Cr = 0.53, with a normal pattern and intensity for β,γ-Glx/Cr and α-Glx/Cr.

E. Location 2 (White Matter)

1. Peak Identification and Quality Assurance: Reading the Spectrum from Right to Left

The quality of the PROBE spectrum (Fig. 5.15) is excellent. Note the residual water signal establishes a steeper slope for the baseline than in the previous spectrum, but does not impede the calculation of metabolite ratios or limit diagnostic utility.

Reading the spectrum from right to left, there is the characteristic broad spectrum of intrinsic protein-lipid centered at 0.9 ppm and ~1.4 ppm. The resonances identified are NAA, β,γ-Glx, the third and second NAA peaks (see comment below), Cr, Cho, sI, mI, the triplet of α-Glx, and second peaks of Cr and mI. Because of unsuppressed water, a difference in the baseline lifts the origins of those portions of the spectrum which include mI and beyond, to between 0.3 and 0.5 of the Cr peak. This must be accounted for in deriving metabolite ratios to Cr in uncorrected PROBE spectra. The residual water correction is performed automatically by the postprocessing routines before calculating metabolite ratios in PROBE-Q and in HMRI's MRS examinations.

2. Interpretation and Analysis

Reading from right to left, the metabolite ratios by eye are: NAA/Cr = 1.62; Cho/Cr = 0.79; mI/Cr = 0.57, all of which are normal for a young adult. Inspection of β,γ-Glx/Cr and α-Glx is also normal; the ratio of the third NAA peak/second NAA peak being well below 0.5. Since β,γ-Glx shares the resonance at 2.5 ppm, with that of the third NAA peak, the most sensitive indicator of increasing β,γ-Glx is to be sought in this ratio.

F. Conclusion

Differences between HMRI's grey matter and white matter are reliably reproduced in PROBE-S, PROBE-Q, and MRS examinations performed rigorously in terms of TE/TM/TR and selection of VOI.

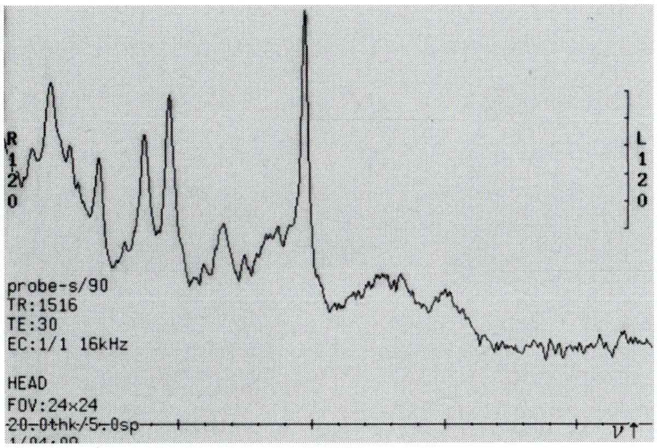

Fig. 5.15 Normal automated PROBE spectrum of parietal white matter.

G. Summary of Patient's History Post-MRS

Not relevant.

H. Other Comments

Notice *Hunter's angle*—the angle between the horizontal axis and the imaginary line drawn through the tops of the NAA, Cr, Cho, and mI peaks in the white matter spectrum. Compare to "Hunter's angle" in a normal baby spectrum, where it is reversed (see also Chapter 4).

Hunter's angle

→ reversed between "preemies" + adults

→ Seems to apply only to short TE, not long TE studies

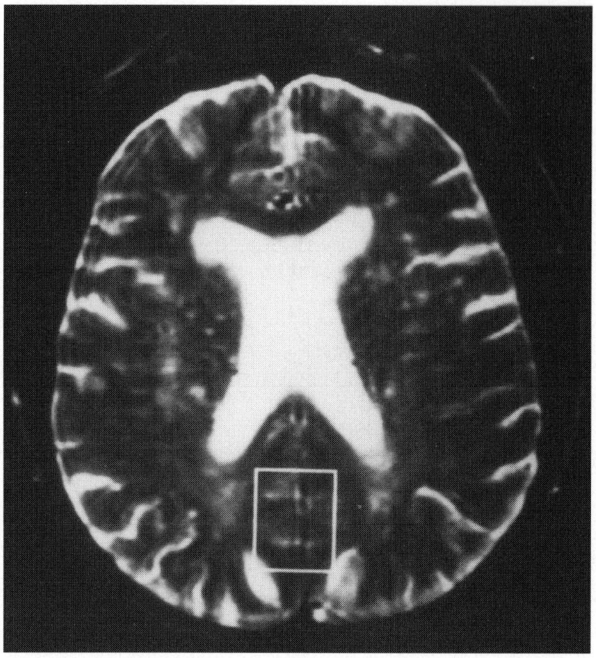

Fig. 5.16 The MRI showing the VOI location in a 83 year old normal volunteer.

5.1.8 NORMAL ELDERLY

A. Purpose of MRS

To acquire data from an elderly healthy volunteer.

B. Summary of Patient's History

The healthy volunteer was 83 years old.

C. MRS Method, Volume of Interest, and Comments about Examination

The automated MRS method PROBE-STEAM (TE = 30 ms, TM = 13.7 ms, TR = 1500 ms) was used to measure a VOI in HMRI's grey matter in location shown (Fig. 5.16).

D. Grey Matter Location

1. Peak Identification and Quality Assurance: Reading the Spectrum from Right to Left

The spectrum, Fig. 5.17, is of excellent quality. The resolution is good. The S/N is good despite the lower available signal due to atrophy normal for age. No artifacts are seen. Reading the spectrum first lipid is seen at about 0.9 and 1.3–1.4 ppm.

Continuing reading from right to left we see the following resonances: NAA, β,γ-Glx, the second NAA, Cr, Cho, sI, mI, α-Glx, the second Cr, and the second mI.

2. Interpretation and Analysis

The ratios are estimated by visual inspection accounting for the baseline to be NAA/Cr = 1.3, Cho/Cr = 0.6, and mI/Cr = 0.6. [Cr] is not available. These values are within the normal range. NAA/Cr down to 1.25 would have been acceptable considering the age.

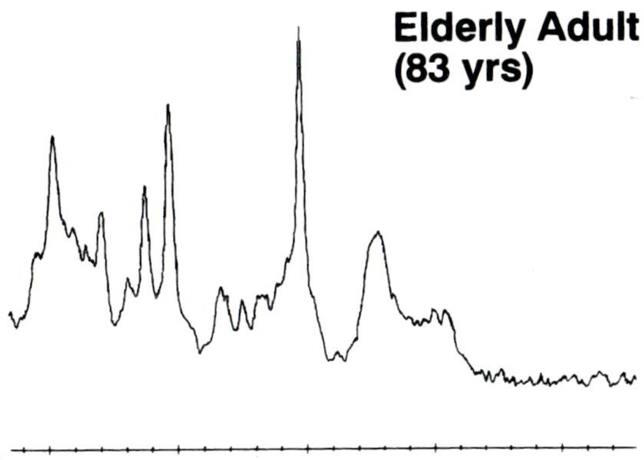

Fig. 5.17 The grey matter spectrum from a 83 year old volunteer. The spectrum was acquired by the automated MRS method PROBE-S, and is here displayed as it appeared on the screen.

E. Conclusion

Normal.

F. Summary of Patient's History Post-MRS

Not relevant.

G. Other Comments

The atrophy will cause a larger fraction of the VOI to be cerebrospinal fluid (CSF) which contains glucose and lactate, but hardly any of the metabolites normally seen in a spectrum; therefore the general signal level, and the S/N is, everything else being equal, lower than in a younger volunteer.

Notice that the apparently large lipid is not unusual considering that we see it in relation to the metabolite peaks, which are smaller than usual due to atrophy.

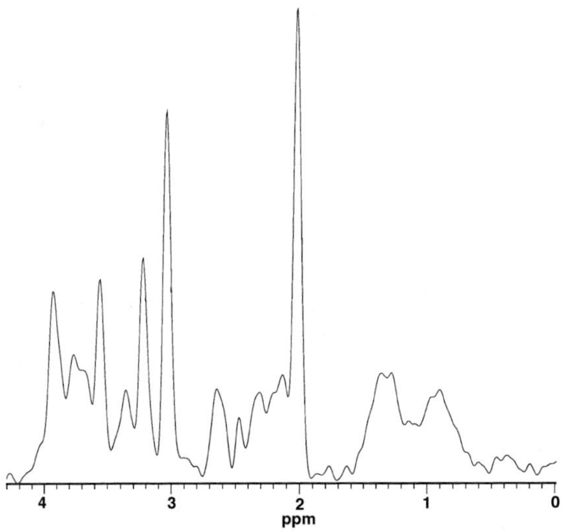

Fig. 5.18 STEAM grey matter spectrum in a normal subject, age 83 years, (TR = 3000 ms, 64 averages).

5.1.9 NORMAL MRS: ELDERLY ADULT

A. Purpose of MRS

To establish normal peak ratios by inspection and direct measurement.

B. Summary of Patient's History

Male, age 83 years, with normal MRI, normal Mini Mental State on neurological examination, no history of diabetes mellitus, renal failure, or of other systemic or neurological disorders.

C. MRS Method, Volume of Interest, and Comments about Examination

STEAM (TE = 30 ms, TM = 13.7 ms, TR = 3000 ms) was performed in HMRI's occipital grey matter location (Fig. 5.18). Acquisition time was 3.5 minutes (64 averages).

D. White Matter Location

1. Peak Identification and Quality Assurance: Reading the Spectrum from Right to Left

The technical quality of the MR spectrum is excellent and obviously shows no pathology.

Reading the spectrum systematically from right to left, the region 0.9 ppm and 1.3 ppm shows the normal intensity of poorly defined brain protein or lipid described for white matter, and no lactate doublet at 1.33 ppm. Continuing reading, the following resonances are identified: NAA, β,γ-Glx, second and third peaks of NAA (2.65 and 2.5 ppm, respectively), Cr, Cho, sI, glucose, mI, two of the three peaks of the α-Glx triplet, and the second peak of Cr.

2. Interpretation and Analysis

There are two approaches to interpretation. First, simply reading by eye, peak ratios can be estimated fairly accurately. The second method involves using a ruler and choosing a point for the baseline of each peak to measure directly and calculate the ratios to the Cr peak.

a. Visual Inspection Method

Metabolite peak height ratios are estimated to be: NAA/Cr = 1.25; Cho/Cr = 0.6; mI/Cr = 0.55; β,γ-Glx/Cr is not elevated as NAA_3/NAA_2 ($\leqslant$ 0.5).

b. Direct Measurement Method

Metabolite peak height ratios are calculated to be: NAA/Cr = 1.27; Cho/Cr = 0.60; mI/Cr = 0.55.

F. Conclusion

While there is some variation, and no one method of inspection is superior, it is suggested that a rigorous adherence to one reading method is likely to provide the radiologist with the necessary confidence to read spectra reproducibly and accurately, with errors of less than 10%.

G. Summary of Patient's History Post-MRS

The normal elderly subject continued healthy.

H. Other Comments

Notice that relevant clinical information was obtained before the examination. It is a good practice always to exclude systemic disorders before an MRS examination, so that a systemic disorder does not have to be included in a differential diagnosis for the patient. An example is hyponatremia, which influences the MRS even several weeks after it has been corrected.

5.2 Degenerative Disease: Alzheimer's Disease, Dementia, Central Pontine Myelinolysis

5.2.1 ALZHEIMER'S DISEASE

A. Purpose of MRS

The patient was referred for MRS to rule out Alzheimer's disease.

B. Summary of Patient's History

Male, age 82 years, gradual deterioration of memory over the preceding three years. Mini Mental State score was 22 (perfect is 30), and Blessed Memory score was 11½ (perfect is 0).

C. MRS Method, Volume of Interest, and Comments about Examination

STEAM (TE = 30 ms, TM = 13.7 ms, TR = 3000) was performed in HMRI's occipital grey matter VOI (Fig. 5.19).

D. Grey Matter Location

1. Peak Identification and Quality Assurance: Reading the Spectrum from Right to Left

The quality of the spectrum is good, and there are no artifacts, Fig. 5.20.

Reading from right to left, there are no significant lipid resonances, and no lactate. Continuing right to left, NAA, β-γ-Glx, the second NAA peak, Cr, Cho, mI, α-Glx, and the second resonance of Cr are present, along with significant resonances of sI and glucose. No other unusual resonances are observed.

2. Interpretation and Analysis

The metabolite ratios are: NAA/Cr = 1.06 (decreased); Cho/Cr = 0.68 (slightly increased); and mI/Cr = 0.65 (increased). The marked reduction of NAA and increased mI is confirmed by the absolute values: [NAA] = 75%, [Cr] = 95%, and [mI] = 120% of the normal values. The elevated glucose appears to be intracerebral. The CSF fraction in the VOI is 11%. The MR images are very unusual with considerable ventricular dilatation, and pronounced white matter changes.

E. Conclusion

Clearly within the Alzheimer's disease range (Fig. 5.21).

F. Summary of Patient's History Post-MRS

No further MRS examinations were performed.

G. Other Comments

This MRS examination is an example of how a short MRS examination can diagnose Alzheimer's disease and aid the decision making for the geriatrician and the psychiatrist.

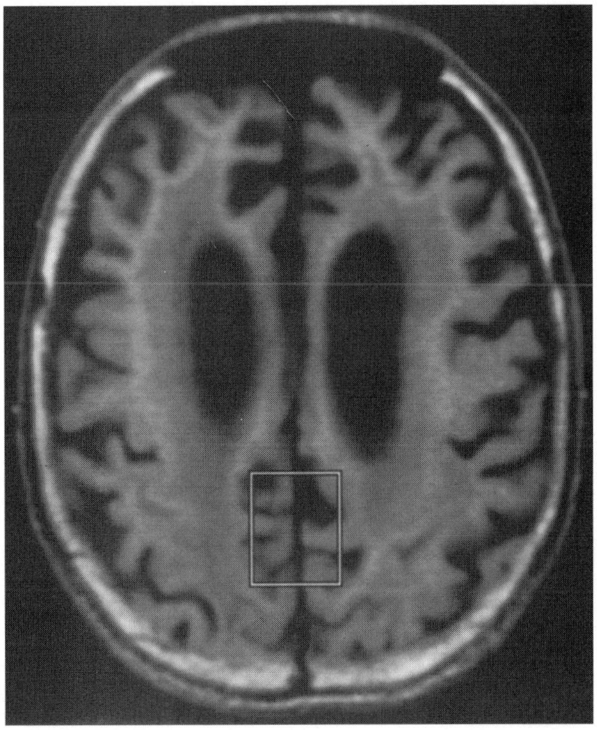

Fig. 5.19 The MRS location, HMRI's normal grey matter location, used in dementia and rule out Alzheimer cases.

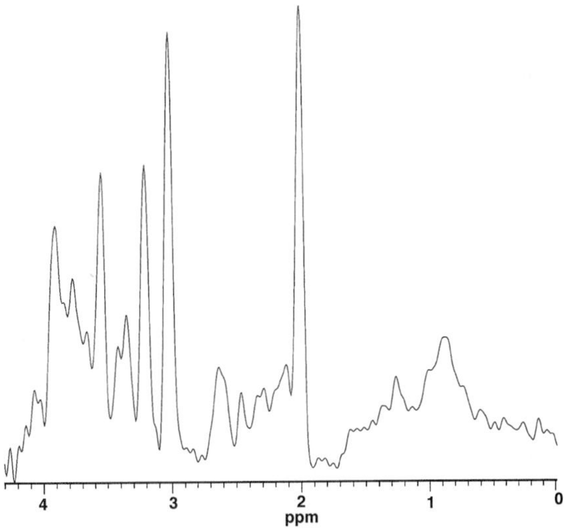

Fig. 5.20 The grey matter MR spectrum shows decreased NAA/Cr and increased mI/Cr, thus consistent with Alzheimer's disease.

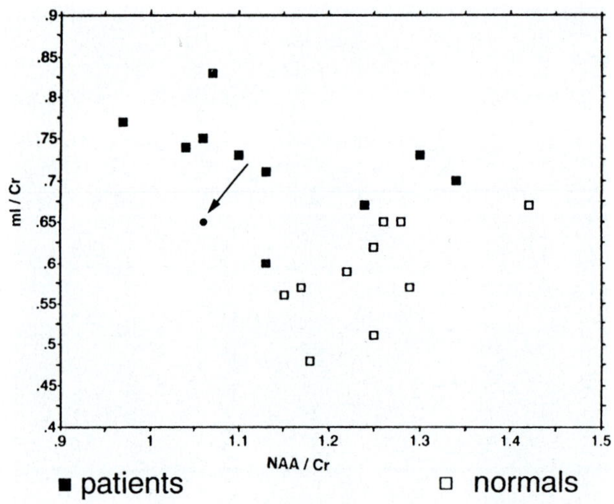

Fig. 5.21 Nomogram plot of NAA/Cr on the horizontal axis and mI/Cr on the vertical axis. The values of NAA/Cr and mI/Cr for this patient when plotted are in the Alzheimer range.

∴ mI/Cr > 0.55 (TE30) is almost certainly c/w AD

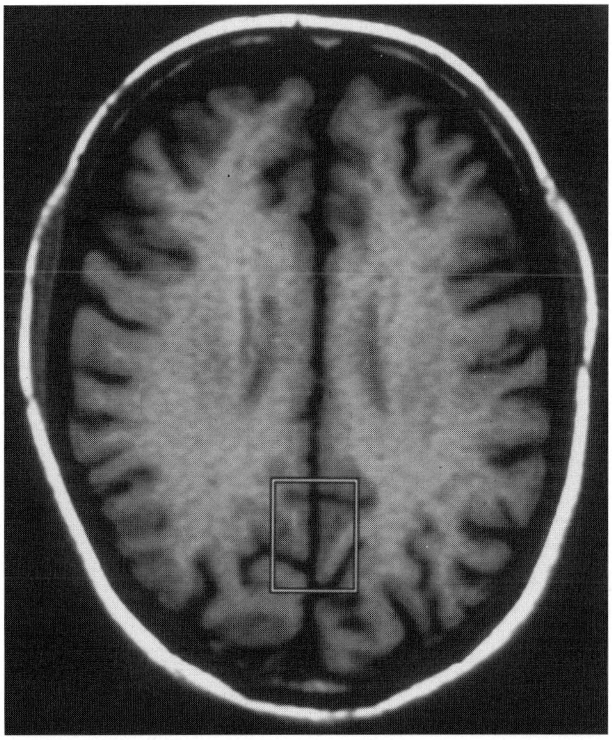

Fig. 5.22 The grey matter location which is the standard MRS location for dementia cases. When the VOI is placed carefully exactly as described in Chapter 4, the MRS is very reproducible, so that metabolite ratio deviations of ±5–10% of normal can be detected.

5.2.2 RULE OUT ALZHEIMER'S DISEASE: NEGATIVE

A. Purpose of MRS

The patient was referred for MRS to rule out Alzheimer's disease.

B. Summary of Patient's History

Female, age 65 years, admitted for depression to an in-patient psychiatric care facility.

C. MRS Method, Volume of Interest, and Comments about Examination

STEAM (TE = 30 ms, TM = 13.7 ms, TR = 3000 ms) was performed in HMRI's occipital grey matter VOI (Fig. 5.22).

D. Grey Matter Location

1. Peak Identification and Quality Assurance: Reading the Spectrum from Right to Left

The technical quality of the spectrum is good (Fig. 5.23) with no artifacts. Reading from right to left, there is no significant lipid resonances, and no lactate. Continuing on from right to left, the following resonances are identified: NAA, β,γ-Glx, the second NAA peak, Cr, Cho, mI, α-Glx, and the second resonance of Cr. Smaller resonances of sI and glucose are seen, and no unusual resonances are observed.

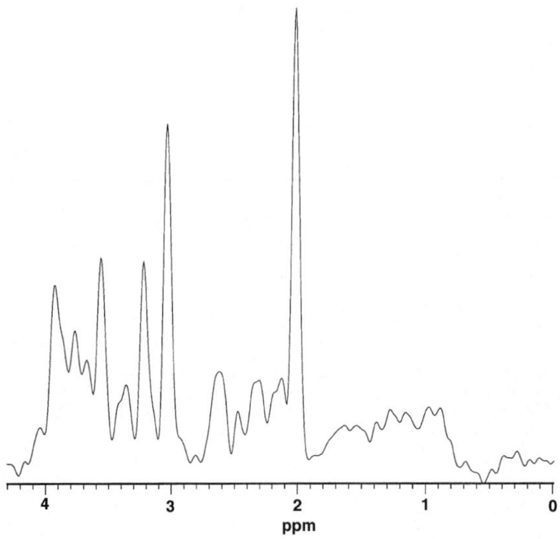

Fig. 5.23 The MR Spectrum from grey matter is normal, NAA is not decreased, and mI not increased. The patient does not have dementia; a diagnosis of depression is more likely.

2. Interpretation and Analysis

The metabolite ratios are: NAA/Cr = 1.33 (normal for age); Cho/Cr = 0.60 (normal for age); mI/Cr = 0.61 (normal for age); and [Cr] is normal. The CSF fraction in the VOI is 9%; images did not show any evidence of deep white matter infarcts, which is also consistent with the MRS result of normal Cho.

E. Conclusion

No evidence of Alzheimer's disease (Fig. 5.24).

F. Summary of Patient's History Post-MRS

The patient was treated for depression and improved. No further MRS examinations were performed.

G. Other Comments

This MRS examination is an example of how a short MRS examination can rule out Alzheimer's disease and dementia, and aid clinical decision making for the gerontologist and the psychiatrist.

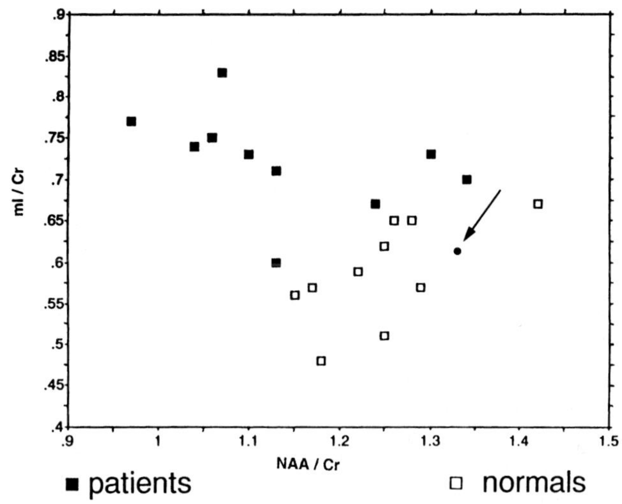

Fig. 5.24 Nomogram plot of NAA/Cr on the horizontal axis and mI/Cr on the vertical axis. Normal elderly are found in the lower right hand corner, Alzheimer patients in the upper left hand corner. This patient falls in the normal range.

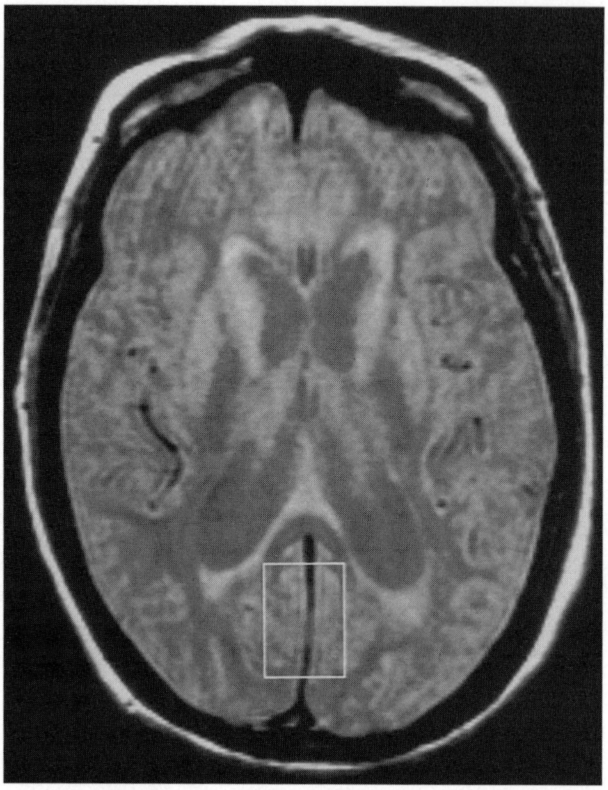

Fig. 5.25 The MRS location, HMRI's normal grey matter location, used in dementia and rule out Alzheimer cases.

5.2.3 CENTRAL PONTINE MYELINOLYSIS

A. Purpose of MRS

The patient was referred for MRS to rule out Alzheimer's disease, but also had hyponatremia.

B. Summary of Patient's History

Female, age 72 years, with psychological disturbances. She was frail, depressed, and on oxygen (3 liters/minute) at the time of the MRS examination.

C. MRS Method, Volume of Interest, and Comments about Examination

STEAM (TE = 30 ms, TM = 13.7 ms, TR = 3000 ms) was performed in HMRI's occipital grey matter (Fig. 5.25).

D. Grey Matter Location

1. Peak Identification and Quality Assurance: Reading the Spectrum from Right to Left

The technical quality is good. There are no unusual resonances observed, and there are no artifacts (Fig. 5.26).

Reading from right to left, there are minor lipid resonances, but no lactate, NAA, β,γ-Glx, the second NAA peak, Cr, and Cho, *myo*-inositol is completely absent, but both the α-Glx and the second resonance of Cr are present.

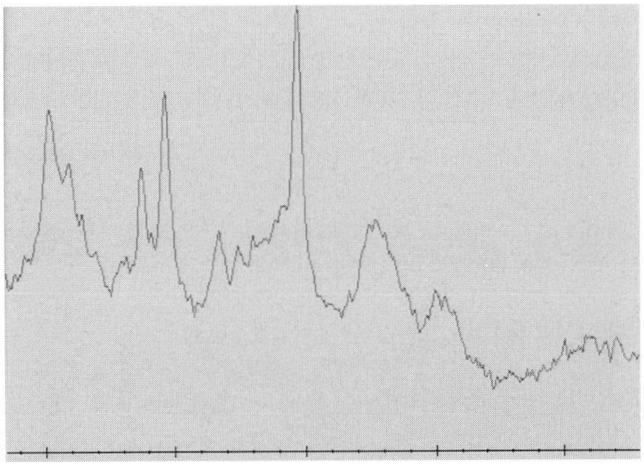

Fig. 5.26 The MR Spectrum in grey matter is not consistent with dementia as NAA/Cr is normal for age. But the main finding is that mI is absent. Absent mI is consistent with hepatic encephalopathy or hyponatremia. The patient was found to have significant hyponatremia, which probably accounts for reduced mI/Cr with normal Glx.

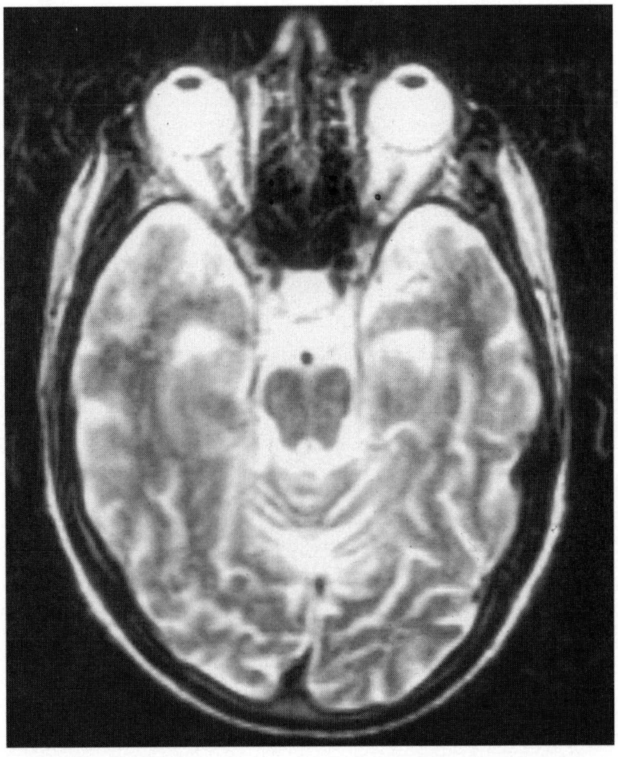

Fig. 5.27 MRI consistent with (but not specific for) central pontine myelinolysis.

2. Interpretation and Analysis

The estimated metabolite ratios are: NAA/Cr = 1.25 (normal for age); Cho/Cr = 0.56 (reduced about 10%); and mI/Cr is close to zero. Absolute metabolite values, and the CSF fraction, are not available for this examination.

E. Conclusion

The mI resonance is absent, which is not consistent with Alzheimer's disease. The major differential diagnoses that can explain the severely decreased mI are liver disease and hyponatremia.

F. Summary of Patient's History Post-MRS

The patient's sodium level was 127, thus the absent mI was due to hyponatremia. The patient was treated and improved partially.

G. Other Comments

This MRS examination is an example of how systemic diseases can influence the brain chemistry. Systemic disease, such as liver disease and hyponatremia, that affect the mI/Cr ratio should be ruled out (or treated) before performing MRS to rule out Alzheimer's disease.

A closer examination of the MRIs (Fig. 5.27) reveals a picture consistent with central pontine myelinolysis, see also Case 5.3.10 and Fig. 5.60.

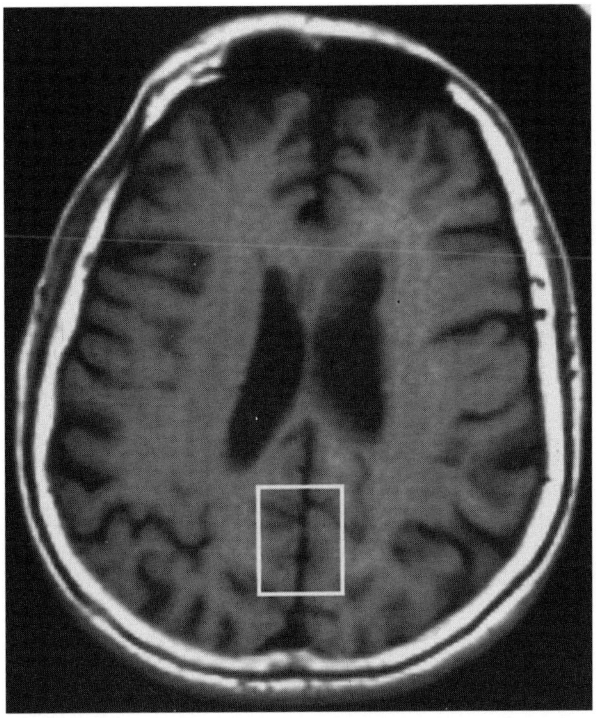

Fig. 5.28 T_1 weighted MRI showing the MRS location in the occipital grey matter.

5.2.4 ALZHEIMER'S DISEASE VERSUS OTHER DEMENTIA

A. Purpose of MRS

MRS was requested to rule out Alzheimer's disease.

B. Summary of Patient's History

A 76-year-old female who is suffering from memory loss.

C. MRS Method, Volume of Interest, and Comments about Examination

STEAM, TE/TM/TR 30/13.7/1500. HMRI's normal occipital grey matter VOI (Fig. 5.28). The patient tolerated MRS well.

D. Grey Matter Location

1. Peak Identification and Quality Assurance: Reading the Spectrum from Right to Left

This is a well resolved high quality spectrum with good S/N. No artifacts are seen apart from a minor "dip" on the right hand side of mI, which does not compromise the interpretation or the clinical value of the spectrum, Fig. 5.29. NAA, β,γ-Glx, the second NAA, Cr, Cho, sI, mI, α-Glx, and the second Cr resonances are noted. No other significant resonances are observed.

2. Interpretation and Analysis

The ratios are NAA/Cr = 1.09 (decreased 13% as 1.25 is normal NAA/Cr ratio for a 76-year-old), Cho/Cr = 0.64 (normal for age), and mI/Cr = 0.53 (reduced about 10% probably due the small "dip" on the right hand side of mI). [Cr] is normal, so the metabolite ratios can be interpreted directly. The CSF fraction in the VOI is 13% in agreement with the ventricular dilatation seen on the MRI.

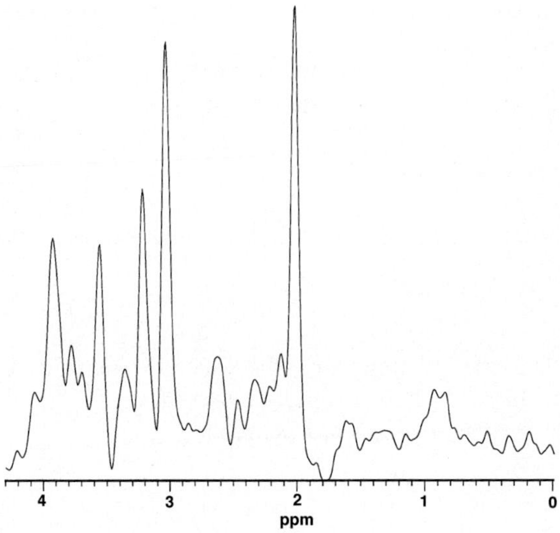

Fig. 5.29 The MRS spectrum consistent with dementia, but not with Alzheimer's disease.

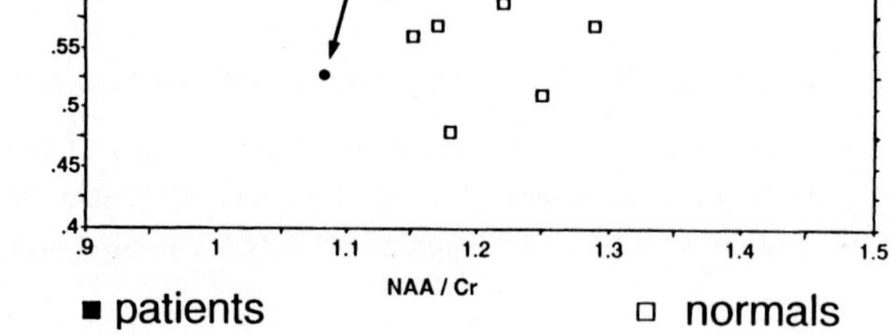

Fig. 5.30 Nomogram plot of NAA/Cr on the horizontal axis and mI/Cr on the vertical axis. Normal elderly are found in the lower right hand corner, Alzheimer patients in the upper left hand corner. Upper right hand corner (high mI/Cr and normal to high NAA/Cr) is not consistent with Alzheimer's disease, but is not normal; a hyperosmolar state can be considered. The lower left hand corner, (low to normal mI/Cr and low NAA/Cr) is consistent with dementia of a non-Alzheimer type. This patient falls in the latter range.

E. Conclusion

NAA is severely decreased, consistent with a dementia, but the MRS is not of the Alzheimer type because the mI is not increased (see plot Fig. 5.30). Cho is not increased, so MRS is therefore not consistent with multi-infarct dementia.

F. Summary of Patient's History Post-MRS

Other causes of dementia were considered, but a long-term history following MRS is not available.

G. Other Comments

The MRS examination, that can be performed in 15 minutes including a routine MRI, has a sensitivity of 82% versus other dementias. Furthermore, MRS has a negative predictive value better than 90% for ruling out AD. MRS is therefore an effective clinical tool that can help in evaluating dementia patients.

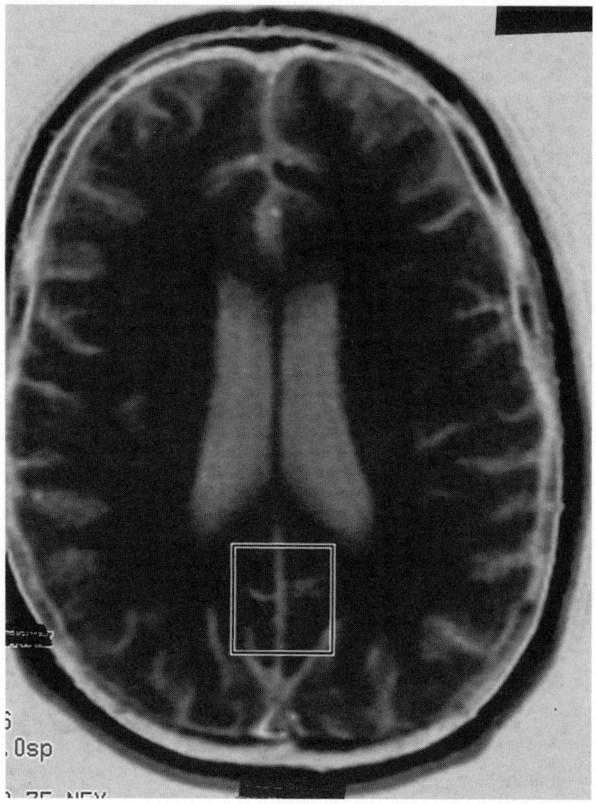

(A)

Fig. 5.31 (A) Axial T_2-weighted MRI showing location for diagnosis of Alzheimer disease (grey matter), and its differentiation from other dementias. Note that there was no diagnostic information from MRI in this patient, even when the remaining series of images were evaluated. (B) PROBE spectrum of grey matter acquired at TR 1800 ms. This small change of TR does not influence metabolite ratios significantly. The diagnostic quality of PROBE-STEAM is excellent, demonstrating elevated mI/Cr. The numerical display refers to automated data processing available in PROBE-Q. Not discussed in the text: (C) and (E) Images illustrating the location of standard white matter automated PROBE acquisitions, symmetrically placed on the left (E) and right (C). (D) and (F) Two automated PROBE-STEAM spectra acquired from the left and right white matter locations. It should be noted that although not necessary in establishing a diagnosis in this patient, some of the features discussed, notably increased mI/Cr, and a smaller increase in Cho/Cr are also seen in these PROBE spectra. Without reading in detail, the numerical output demonstrates the excellent spectral reproducibility and robustness of this automated method. (Courtesy of Dr. D. Hungerford, Mobile Infirmary, Alabama.)

5.2.5 RULE OUT ALZHEIMER'S DISEASE

A. Purpose of MRS

Differential diagnosis of dementia and depression.

B. Summary of Patient's History

Male, age 61 years, who reported memory loss and depression. Mini Mental State and clinical testing was consistent with probable Alzheimer's disease; was not on treatment at the time of the MRS examination.

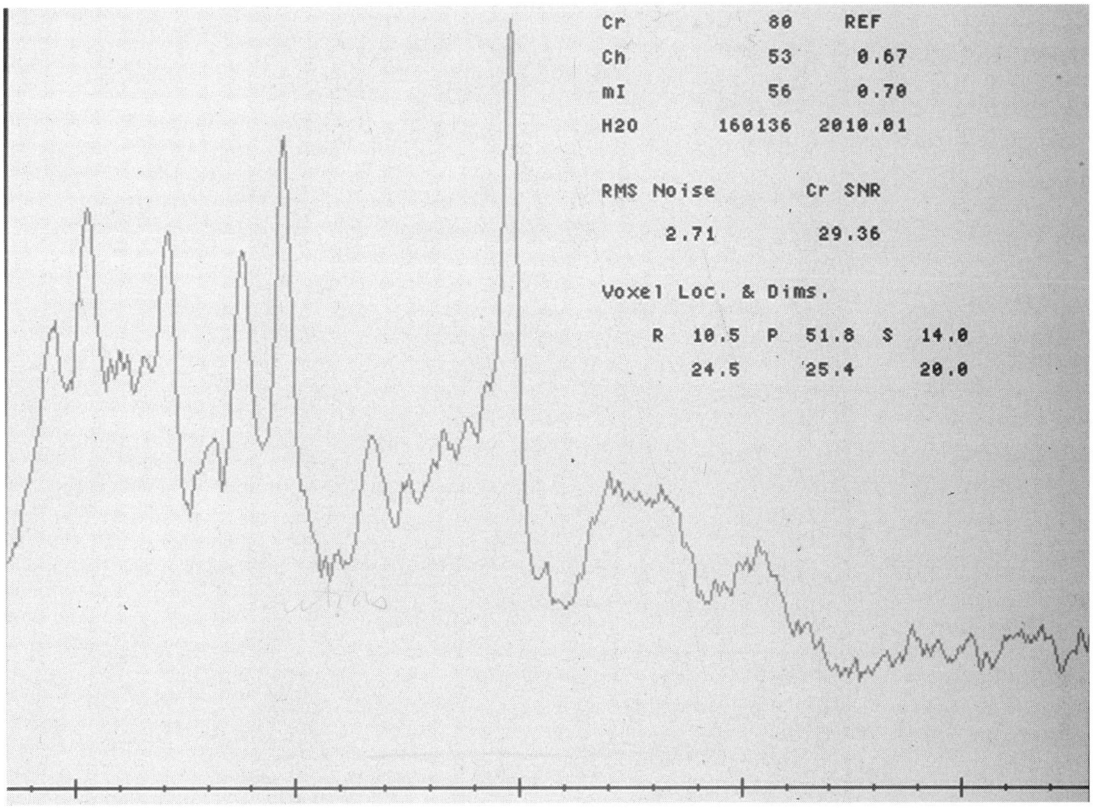

(B)

C. MRS Method, Volume of Interest, and Comments about Examination

PROBE-Q STEAM (TE = 30 ms, TM = 13.7 ms, TR = 1800 ms; note 1800 ms *not* 1500 ms), which includes automatic calculation and display of the principal metabolite ratios, was performed in HMRI's occipital grey matter location (Fig. 5.31A). An additional spectrum was acquired from the same location with PROBE without quantitation (not shown).

D. Grey Matter Location (PROBE-Q)

1. Peak Identification and Quality Assurance: Reading the Spectrum from Right to Left

The PROBE-Q spectrum (Fig. 5.31B) is well-resolved, and has no artifacts. Water suppression is perfect, giving an ideal baseline.

Reading the spectrum from right to left, the lipid-protein regions, 0.9 ppm and 1.4 ppm, are normal, without a lactate doublet at 1.33 ppm. The small peak at 1.9 ppm is normal, but unidentified (possibly GABA). NAA, β,γ-Glx, second and third NAA peaks, Cr, Cho, sI, glucose shoulder, mI (at 3.56 ppm), α-Glx, and well-resolved second peaks of both Cr and mI, are readily identified.

2. Interpretation and Analysis

Peak ratios are printed on the screen, as they appear in Fig. 5.31. Peak ratios are: NAA/Cr = 1.36 (normal); Cr is assumed to be 1.00 (reference); Cho/Cr = 0.67 (elevated for the grey matter location); mI/Cr = 0.70 (much elevated). Scaling of metabolite concentrations may be attempted using [Cr] signal-to-noise ratio, which is 29 and very approximately normal.

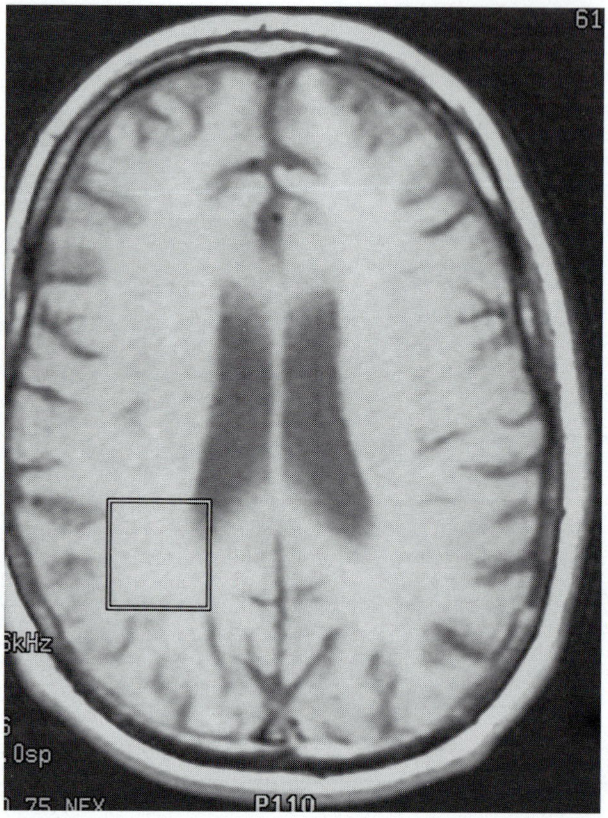

(C)

Fig. 5.31 Continued.

While NAA/Cr is normal, and would make dementia unlikely, mI/Cr is sufficiently elevated to bring the result into the range of probable Alzheimer's disease, when plotted on the nomogram (Fig. 5.32).

An additional abnormality is the excess Cho/Cr, a finding in multi-infarct dementia, rather than probable AD. The upper limit of Cho/Cr increases with age, but at age 61 years this is not appropriate.

Unrelated causes of increased mI/Cr, which include diabetes mellitus, and chronic renal failure (even if on hemo or peritoneal dialysis) must be excluded.

E. Conclusion

The diagnosis is probably Alzheimer's disease, until proved otherwise. Appropriate tests for the latter purpose now include serum ApoE and CSF tau-protein determinations.

F. Summary of Patient's History Post-MRS

No further history is available.

G. Other Comments

This is an example of a more difficult case. Note however, that it has been demonstrated in a large trial, with many borderline case like this one, that Alzheimer's disease can be separated from normal with a specificity of 98% and sensitivity of 83%. For Alzheimer's disease versus other dementias the negative predictive rate is 80% and the sensitivity is 82% (see Chapter 6).

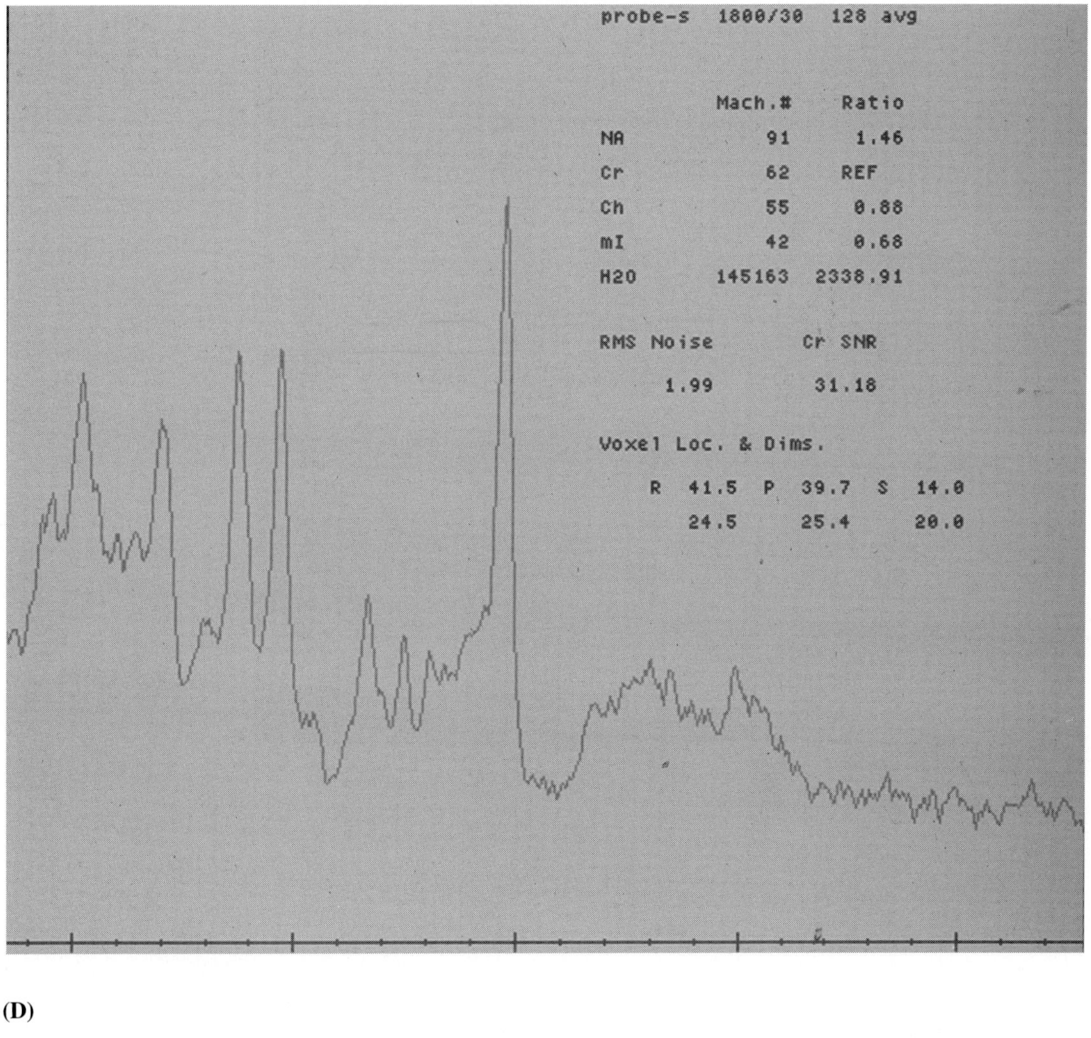
probe-s 1800/30 128 avg
Mach.# Ratio
NA 91 1.46
Cr 62 REF
Ch 55 0.88
mI 42 0.68
H2O 145163 2338.91
RMS Noise Cr SNR
1.99 31.18
Voxel Loc. & Dims.
R 41.5 P 39.7 S 14.0
24.5 25.4 20.0

(D)

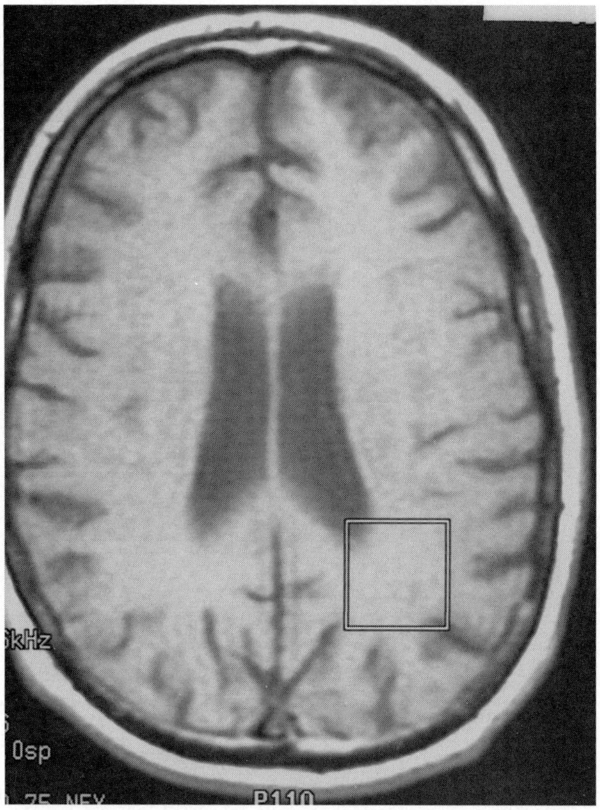

(E)

Fig 5.31 Continued.

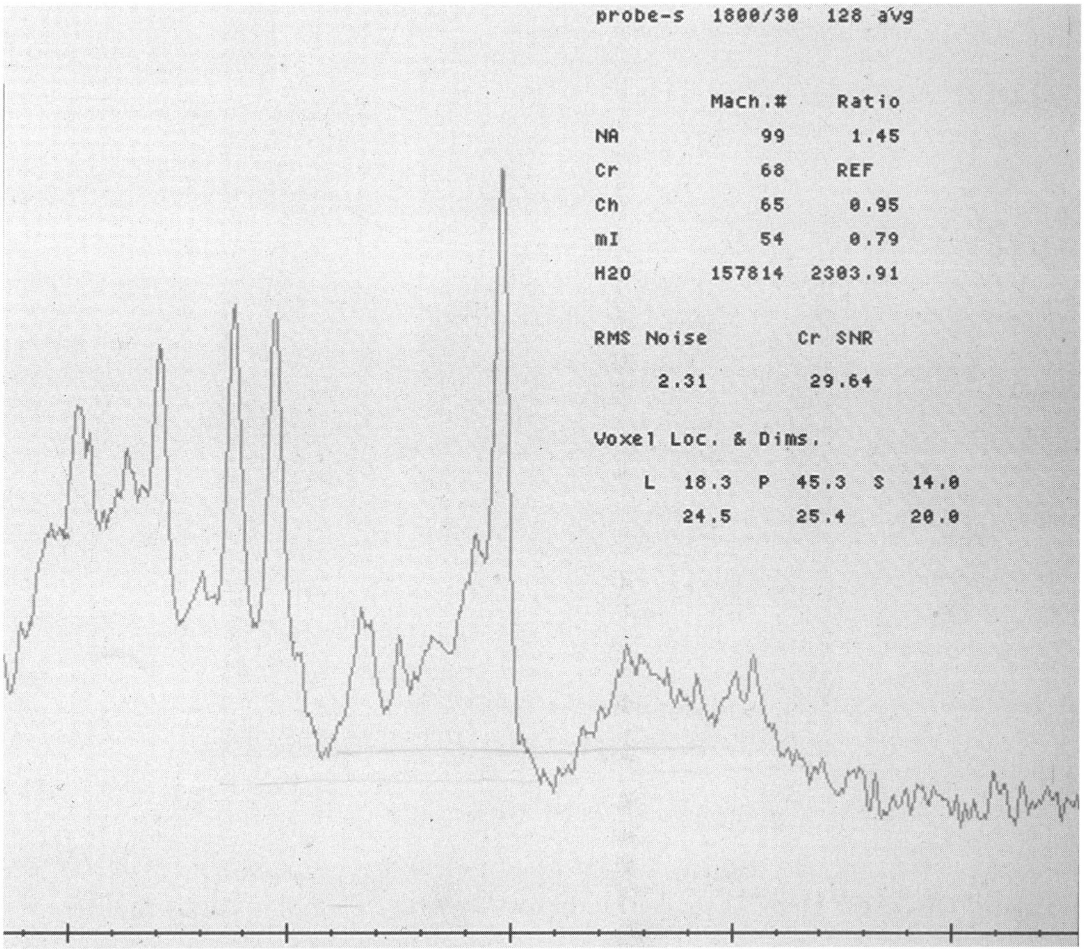

(F)

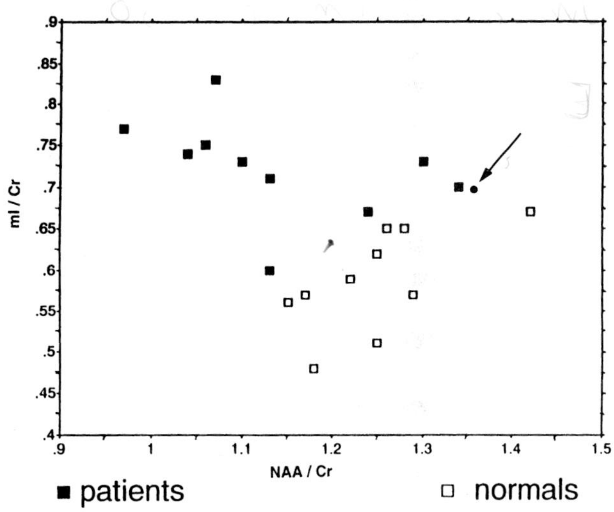

Fig. 5.32 Nomogram plot of NAA/Cr on the horizontal axis and mI/Cr on the vertical axis. The result of the plot in this patient is consistent with Alzheimer's disease.

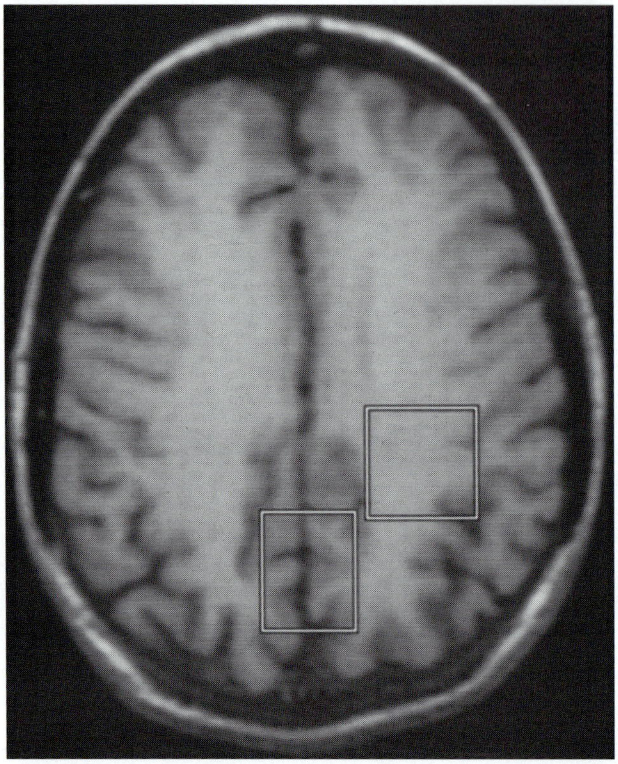

Fig. 5.33 T_1 weighted MRI showing the MRS locations, HMRIs normal grey and white matter locations.

5.3 Systemic Diseases: Liver Disease, Diabetes Mellitus, Osmolar Abnormalities

5.3.1 LIVER DISEASE, SEVERE HEPATIC ENCEPHALOPATHY

A. Purpose of MRS

MRS was requested to confirm presence of and to evaluate the severity of hepatic encephalopathy in a patient awaiting liver transplantation.

B. Summary of Patient's History

Male, age 65 years, who appears ill with ascites and has deteriorated over the past 6 months, he has had one episode of hemorrhage but no coma.

C. MRS Method, Volume of Interest, and Comments about Examination

STEAM (TE = 30 ms, TM = 13.7 ms, TR = 1500 ms). HMRI's normal white matter VOI, location 1, and HMRI's normal grey matter VOI location 2, Fig. 5.33.

D. Location 1 (White Matter)

1. Peak Identification and Quality Assurance: Reading the Spectrum from Right to Left

This is a well resolved good quality spectrum with good S/N and no artifacts, Fig. 5.34. The first impression is that something is not right; a peak is missing from this spectrum of good technical quality. Reading systematically from right to left NAA,

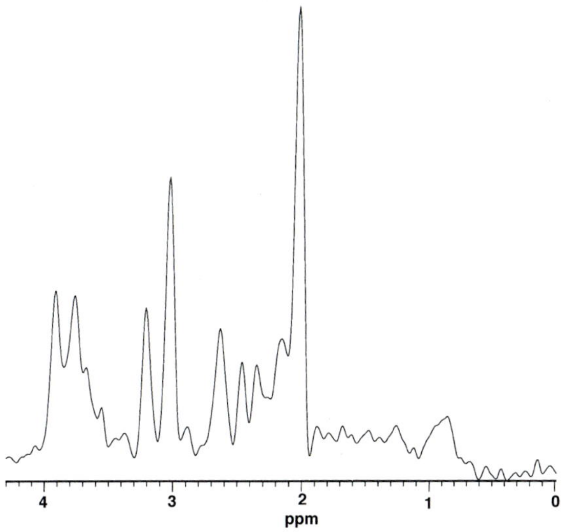

Fig. 5.34 The white matter MRS spectrum, Location 1, consistent with severe hepatic encephalopathy, missing mI, reduced Cho and increased glutamine.

β,γ-Glx, the second NAA, Cr, Cho, α-Glx, and the second Cr are seen. But mI at 3.56 ppm is missing. No unusual resonances are present.

2. Interpretation and Analysis

The ratios are NAA/Cr = 1.56 (increased 10–15%), Cho/Cr = 0.59 (decreased about 25%), and [Cr] is reduced about 10%, possibly explaining the increased NAA/Cr ratio. The glutamine is severely elevated as estimated by α-Glx/Cr = 0.64 and β,γ-Glx/Cr = 0.43, which are 40–60% above normal.

E. Location 2 (Grey Matter)

1. Peak Identification and Quality Assurance: Reading the Spectrum from Right to Left

This is a similarly high quality spectrum with the same features as the white matter spectrum, Fig. 5.35. Reading from right to left NAA, β,γ-Glx, the second NAA, Cr, Cho, α-Glx, and the second Cr are seen, and the mI is missing. No unusual resonances are present.

2. Interpretation and Analysis

The ratios are NAA/Cr = 1.29 (normal for age), Cho/Cr = 0.60 (normal). The α-Glx/Cr = 0.70 and β,γ-Glx/Cr = 0.57 being more than 50% above normal values are again severely elevated.

F. Conclusion

The three marked changes of severe chronic hepatic encephalpathy are present with severe elevation of glutamine, total depletion of mI and reduction of Cho. The Cho reduction is found in the white matter location only. MRS is therefore consistent with severe chronic hepatic encephalpathy. The patient is likely to recover following liver transplantation.

G. Summary of Patient's History Post-MRS

The patient was put on a waiting list for liver transplantation, but died before a suitable liver was found. No further MRS studies were performed.

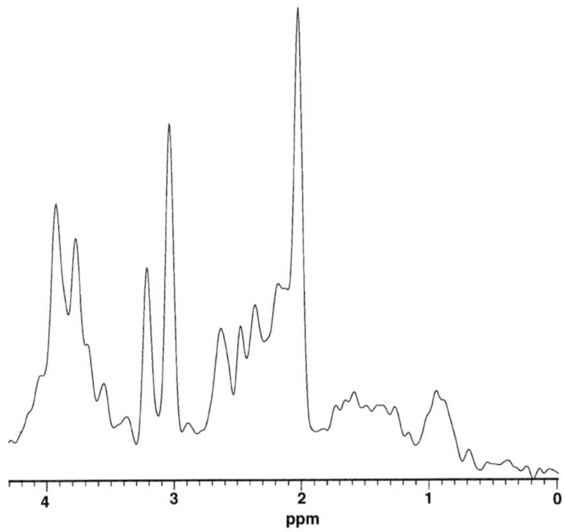

Fig. 5.35 The grey matter MRS spectrum, Location 2, confirming the result from white matter.

H. Other Comments

In a liver disease patient the first priority is to measure the white matter location, because this is the location where the three abnormalities of hepatic encephalopathy can be expected: Missing mI, reduced Cho, and increased glutamine. The grey matter location is less informative as Cho is not necessarily reduced. But an additional spectrum from occipital grey matter can be helpful in an older patient where the clinician wishes to rule out the presence of a dementia, prior to a possible liver transplantation.

measure WM c̄ STEAM TE = 30 to exclude or confirm HE

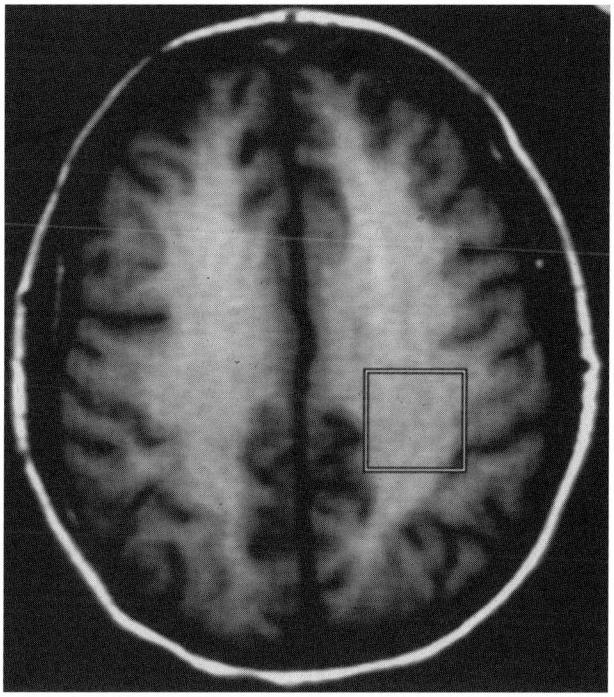

Fig. 5.36 T_1-weighted MRI showing the MRS location.

5.3.2 LIVER DISEASE

A. Purpose of MRS

MRS was performed to evaluate for chronic hepatic encephalopathy (HE), *before* a transjugular intrahepatic portal systemic shunt (TIPS) procedure.

B. Summary of Patient's History

Male, age 57 years, with a clinical diagnosis of viral hepatitis, and a history of alcohol abuse.

C. MRS Method, Volume of Interest, and Comments about Examination

STEAM (TE = 30 ms, TM = 13.7 ms, TR = 1500 ms) was performed in HMRI's parietal white matter (Fig. 5.36). A spectrum from a typical grey matter location was also acquired (not shown).

D. White Matter Location

1. Peak Identification and Quality Assurance: Reading the Spectrum from Right to Left

This is a well-resolved, high quality spectrum, with good S/N, and no artifacts (Fig. 5.37). The trained observer should, however, immediately notice a remarkable abnormality—a missing peak.

Reading the spectrum from right to left, NAA, β,γ-Glx, the second NAA, Cr, and Cho are observed. The mI peak at 3.56 ppm is missing, but α-Glx, and the second Cr are clearly seen, hereby reassuring that the spectral quality is good, and confirming that the absence of mI is real. No unusual resonances are observed.

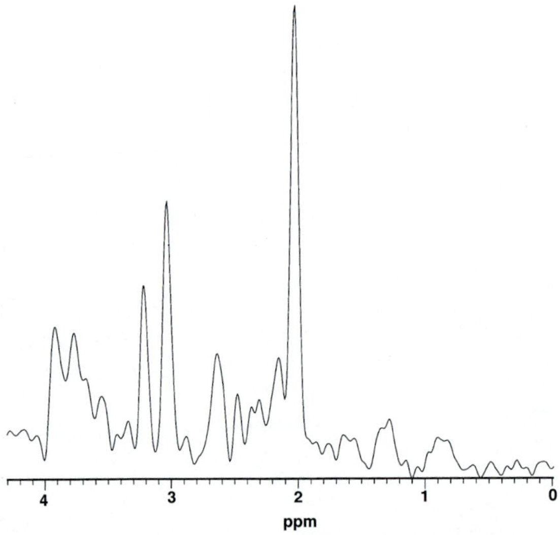

Fig. 5.37 The MRS spectrum of white matter is consistent with hepatic encephalopathy.

2. Interpretation and Analysis

The metabolite ratios are: NAA/Cr = 1.72 (increased about 20%); Cho/Cr = 0.72 (decreased about 15%); mI is virtually absent (a computer fit resulted in the ratio mI/Cr = 0.33). The β,γ-Glx/Cr and the α-Glx/Cr were found to be increased 30% and 28%, respectively. [Cr] is normal. The CSF fraction in the VOI is 15%. The apparent excess of NAA is unexplained; it cannot be attributed alone to the increased Glx, which has a minor resonance underlying the NAA peak.

E. Conclusion

The absence of mI, combined with decreased Cho and increase Glx is clearly consistent with chronic hepatic encephalopathy. TIPS is expected to further worsen the condition.

F. Summary of Patient's History Post-MRS

Not available.

G. Other Comments

The grey matter MRS showed a similar pattern and confirmed the diagnosis of HE (not shown). The absence of mI, decreased Cho, and increased Glx is a very certain MRS fingerprint of chronic HE. Excluding focal lesions such as stroke, tumor, or some AIDS brain masses, which all have a totally different appearance, the only other known cause of decreased mI is hyponatremia, which rarely depletes mI totally, and which also causes the other organic osmolytes including glutamate and glutamine to be depleted, not increased as in HE.

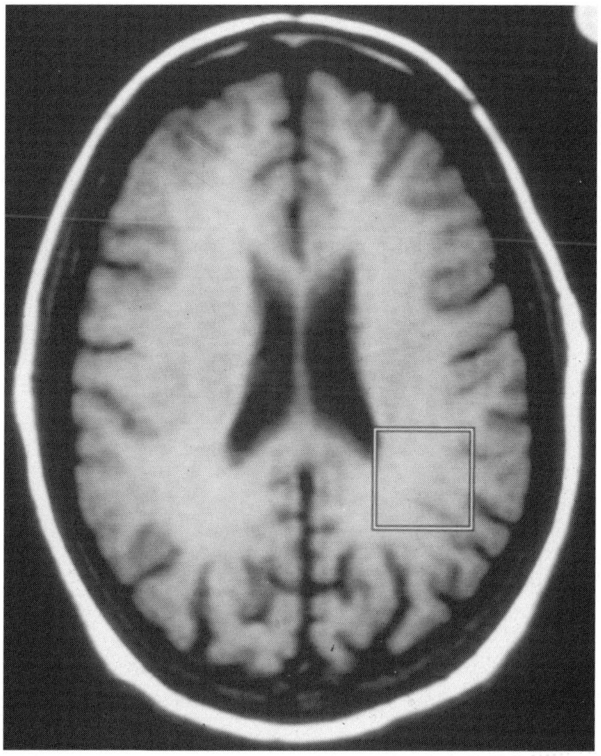

Fig. 5.38 T_1 weighted MRI showing the MRS location.

5.3.3 RULE OUT SUBCLINICAL HEPATIC ENCEPHALOPATHY

A. Purpose of MRS

MRS was performed to evaluate for chronic or subclinical hepatic encephalopathy.

B. Summary of Patient's History

Female, age 41, with liver disease, hematemesis, and a history of ethanol abuse. She does not have a clinical diagnosis of hepatic encephalopathy. The main feature of an MRS examination five months earlier was very low choline, which alone was not sufficient evidence of hepatic encephalopathy.

C. MRS Method, Volume of Interest, and Comments about Examination

Localized STEAM (TE = 30 ms, TM = 13.7 ms, TR 1500 ms) was performed in HMRI's parietal white matter (Fig. 5.38). A spectrum from occipital grey matter was also acquired, but not shown. The examination was limited due to time constraints, yielding metabolite ratios, but no absolute numbers.

D. White Matter Location

1. Peak Identification and Quality Assurance: Reading the Spectrum from Right to Left

This is a well resolved spectrum, with good S/N, and no artifacts apart from some effects of the water suppression on the second Cr peak. The spectrum is of diagnostic quality (Fig. 5.39).

Reading the spectrum from right to left, there is some "in" and "out-of-phase" lipid signals seen; possibly from scalp

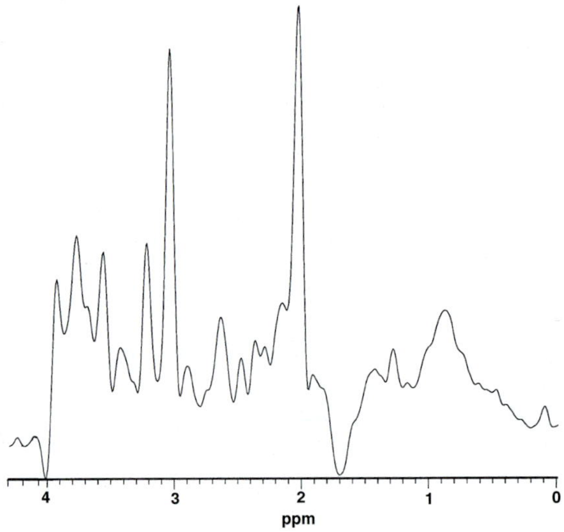

Fig. 5.39 The MRS spectrum consistent with subclinical hepatic encephalopathy.

outside the VOI. The out-of-phase signal may cause an artificial decrease of the neighboring NAA peak. Continuing left to right, β,γ-Glx, the second NAA, Cr, Cho, mI and α-Glx are observed. A trace of glucose at 3.43 and 3.8 ppm may be present. No other unusual resonances are seen.

2. Interpretation and Analysis

The metabolite ratios are: NAA/Cr = 1.16 (decreased 17%); Cho/Cr = 0.53 (decreased 36%); mI/Cr = 0.53 (decreased 13%). [Cr] is not available. α-Glx is elevated (possibly due to glucose) but β,γ-Glx is not (rules out elevated glutamine). Parallel findings are present in the grey matter location (not shown).

E. Conclusion

The striking abnormalities are very low Cho, as seen in the previous MRS examination, and the reduction in mI. However, mI is not within the normal range, nor is it clearly in the chronic HE range (definition: $\leqslant$ 2 standard deviations below normal; $\leqslant$ 0.49). The MRS results of low mI and Cho are consistent with subclinical HE. The only differential diagnosis is hyponatremia which, however unlikely, should be ruled out.

F. Summary of Patient's History Post-MRS

A clinical verification that the patient did in fact have chronic hepatic encephalopathy was not achieved. However, the occurrence of bleeding varices later in the course of her illness was very suggestive of chronic hepatic encephalopathy. Medical treatment was initiated, and after she was discharged the condition was monitored in an out-patient facility. The patient continued to have several medical problems related to abuse of alcohol, and other substances. Her subsequent history included several admissions to the hospital for upper gastrointestinal bleeding, anemia, and bleeding varices. The patient died within four years of the MRS examination.

G. Other Comments

The early detection by MRS of subclinical HE can assist in improving the management of patients with liver disease. Possible treatments can be initiated earlier in the course of the disease, and before the HE has caused major brain damage. Liver transplantation is the only current treatment which restores the brain biochemistry to normal, as illustrated in Cases 5.3.4 and 5.3.5, but in this case liver transplantation was not an option due to the continued substance abuse.

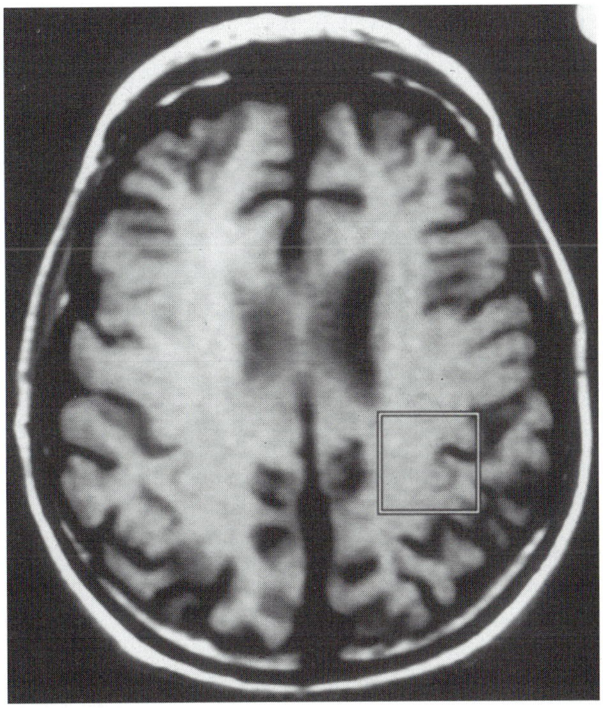

Fig. 5.40 T_1-weighted MRI showing the white matter MRS location, in this case from the second examination. MRIs from the first and third examinations were similar (not shown).

5.3.4 LIVER DISEASE: TRANSPLANTATION

A. Purpose of MRS

MRS was performed to evaluate for chronic hepatic encephalopathy, and for the severity thereof, prior to liver transplantation. The MRS examinations, postliver transplantation, were performed to verify the recovery of normal brain metabolism.

B. Summary of Patient's History

Male, age 65 years, with a clinical diagnosis of sclerosing cholangitis, and a possible diagnosis of chronic hepatic encephalopathy, who underwent a liver transplantation.

C. MRS Method, Volume of Interest, and Comments about Examination

STEAM (TE = 30 ms, TM = 13.7 ms, TR = 1500 ms) was performed in HMRI's parietal white matter VOI (Fig. 5.40). The first examination was performed before the liver transplantation, the second 2 months after, and the third 5½ months postsurgery. In each examination care was taken in positioning the VOIs precisely at the same location. Spectra from a typical grey matter location were also acquired (not shown).

D. Examination 1: White Matter Location (Before Liver Transplantation)

1. Peak Identification and Quality Assurance: Reading the Spectrum from Right to Left

This is a well-resolved, high quality spectrum, with good S/N, and no artifacts (Fig. 5.41). This is clearly an abnormal white matter spectrum, with a peak missing, and a general pattern that is not normal. Reading the spectrum from right to left, NAA, followed by a remarkable β,γ-Glx region, the second NAA, Cr, and Cho are observed, Cho appearing decreased. The mI peak

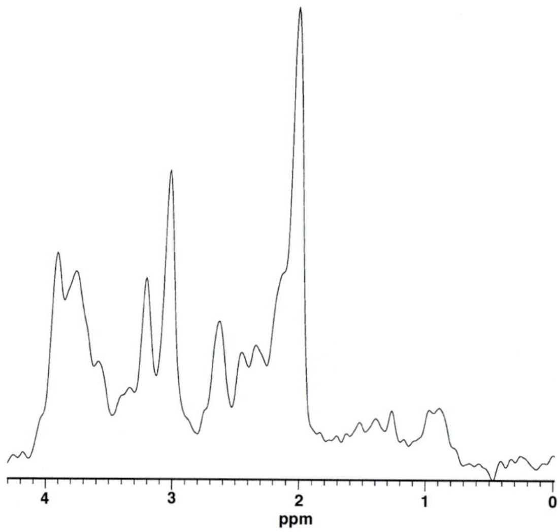

Fig. 5.41 The first examination was performed before liver transplantation. The MR spectrum is consistent with chronic hepatic encephalopathy.

at 3.56 ppm is missing, but the α-Glx is increased in agreement with the large β,γ-Glx. The second Cr is clearly seen in the correct proportion 2:3 compared to the first Cr, hereby assuring that the spectral quality is good, and confirming that the absence of mI is real. No unusual resonances are observed.

2. Interpretation and Analysis

The metabolite ratios are: NAA/Cr = 1.54 (about 10% increased); Cho/Cr = 0.66 (about 20% decreased); mI is virtually absent (a computer fit resulted in the ratio mI/Cr = 0.42). The β,γ-Glx/Cr and the α-Glx/Cr were found to be increased at least 50%. Detailed studies have shown that increased Glx in hepatic encephalopathy patients in fact represents increased glutamine. [Cr] is 10% below normal, hereby explaining the increased NAA/Cr ratio, and leaving Cho even lower than it appears from the Cho/Cr ratio. Similar but less marked results were found in occipital grey matter.

E. Summary of Examination 1

The absence of mI, combined with decreased Cho and increased glutamine, is clearly consistent with chronic hepatic encephalopathy. Follow-up postliver transplantation is expected to show recovery of mI and Cho along with normalization of the glutamine in the brain.

F. Examination 2: White Matter Location (2 Months after Liver Transplantation)

1. Peak Identification and Quality Assurance: Reading the Spectrum from Right to Left

This is another well-resolved, high quality spectrum, with good S/N, and no artifacts (Fig. 5.42). Reading the spectrum from right to left, NAA, β,γ-Glx, the second NAA, Cr, Cho, mI, α-Glx, the second Cr peak, and even the second mI peak at 4.06 ppm are observed. The mI peak at 3.56 ppm appears smaller than normal, Cho larger than normal, and the Glx resonances appear normal. No unusual resonances are observed.

2. Interpretation and Analysis

The metabolite ratios are: NAA/Cr = 1.49 (increased 5%); Cho/Cr = 1.05 (increased 27%); and mI/Cr = 0.56 (8% below normal value). The β,γ-Glx/Cr = 0.27 (normal = 0.26) and α-Glx/Cr = 0.53 (normal = 0.46), thus the glutamine returned to normal. [Cr] also returned to normal.

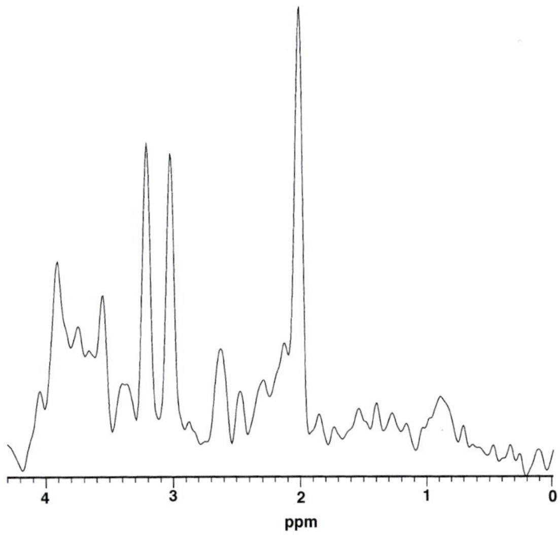

Fig. 5.42 The second examination was performed 2 months after liver transplantation. The MR spectrum shows recovery, though not full recovery from hepatic encephalopathy.

G. Summary of Examination 2

Good result indicating recovery from hepatic encephalopathy after liver transplantation. A repeat examination after another 3–4 months is expected to show further recovery.

H. Examination 3: White Matter Location (5½ Months after Liver Transplantation)

1. Peak Identification and Quality Assurance: Reading the Spectrum from Right to Left

The first impression is a high quality spectrum, which appears normal with the exception of increased Cho (Fig. 5.43). Reading the spectrum from right to left, NAA, β,γ-Glx, the second NAA, Cr, Cho, mI, α-Glx, the second Cr peak, and even the second mI peak at 4.06 ppm are observed. No unusual resonances are observed.

2. Interpretation and Analysis

The metabolite ratios are: NAA/Cr = 1.40 (normal); Cho/Cr = 0.98 (increased 18%); and mI/Cr = 0.69 (increased 13%). The β,γ-Glx and α-Glx are normal. [Cr] is normal.

I. Conclusion

Clear recovery from chronic hepatic encephalopathy. The "missing" metabolite of the first examination, mI, has further recovered from the ratio 0.56 in the second examination to 0.69 in the present.

J. Summary of Patient's History Post-MRS

The patient recovered well. A further history after the third examination was not available.

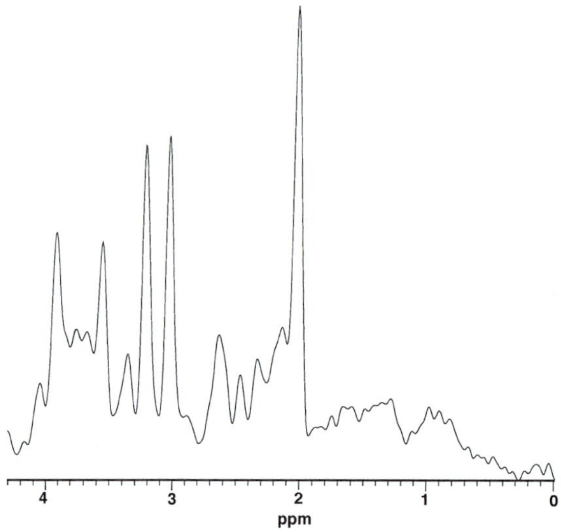

Fig. 5.43 The third examination was performed 5½ months after liver transplantation. Except for a small "overshoot" of Cho, this is a normal spectrum.

K. Other Comments

These three examinations illustrate how liver transplantation totally reverses the brain MRS of the hepatic encephalopathy patient from abnormal to normal. One metabolite, mI, recovers from virtually zero to normal; depleted Cho recovers with a possible overcompensation, and excess glutamine disappears.

Liver transplantation is an expensive but effective treatment for hepatic encephalopathy. The effectiveness of other possible treatment can readily be monitored by MRS, knowing how dramatic recovery of the brain metabolism can be.

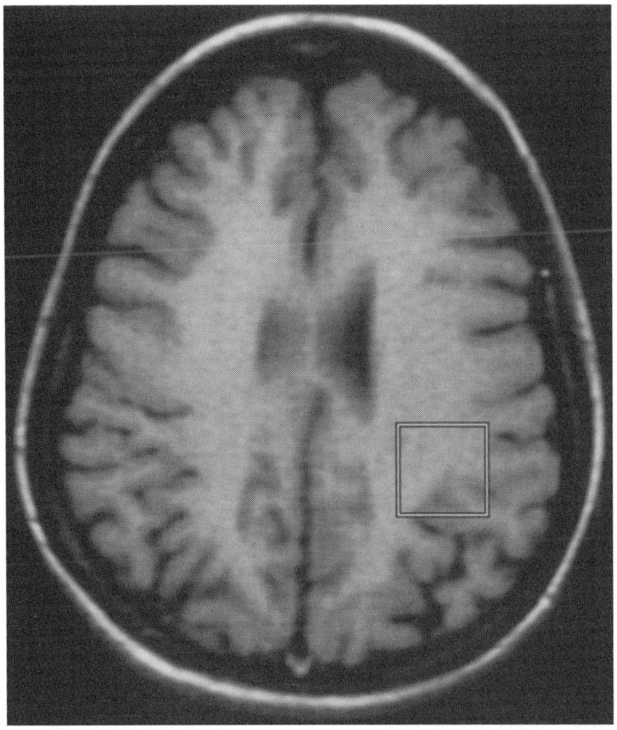

Fig. 5.44 T_1-weighted MRI showing the white matter MRS location, in this case from the first examination, four days before TIPS. MRIs from the second and third examinations were similar (not shown).

5.3.5 LIVER DISEASE, PRE- AND POST-TIPS, AND PRE- AND POSTLIVER TRANSPLANTATION

A. Purpose of MRS

The first MRS was performed to evaluate for chronic hepatic encephalopathy prior to a transjugular intrahepatic portal systemic shunt (TIPS) procedure. The second and third MRS examinations were performed again to evaluate for chronic hepatic encephalopathy after TIPS, and before liver transplantation. The fourth was performed after liver transplantation to confirm full recovery of the brain biochemistry.

B. Summary of Patient's History

Female, age 39 years, initially suffering from cirrhosis-induced hematemesis. She underwent a TIPS procedure, and 1 year and 4 months after TIPS she had a successful liver transplant.

C. MRS Method, Volume of Interest, and Comments about Examination

STEAM (TE = 30 ms, TM = 13.7 ms, TR = 1500 ms) was performed in HMRI's parietal white matter VOI (Fig. 5.44). The first examination was performed 4 days before TIPS, the second 17 days after, and the third was performed 3 months after the TIPS procedure. The fourth MRS examination, 5½ months after liver transplantation (same location), was performed using a different scanner (Siemens SP instead of the GE Signa), and therefore slightly different measurement parameters were used (STEAM, TE = 20 ms, TM = 30 ms, TR = 1500). In each examination care was taken in positioning the VOIs precisely at the same location. Spectra from a typical grey matter location were also acquired, but are not shown because the white matter location is the first priority in hepatic encephalopathy cases.

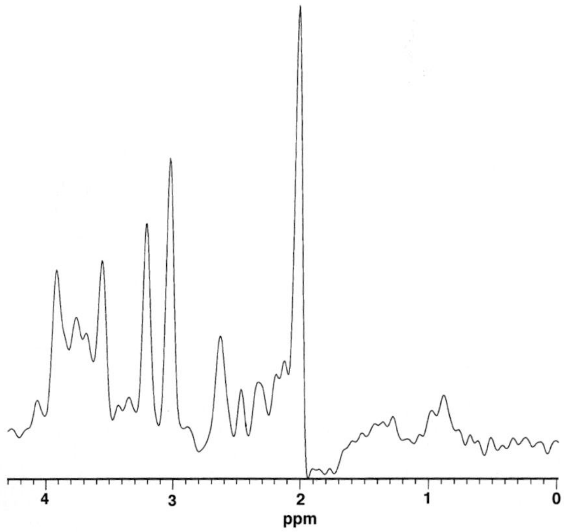

Fig. 5.45 The first examination was performed four days before TIPS. The white matter MR spectrum is broadly normal.

D. Examination 1: White Matter Location (4 days before TIPS)

1. Peak Identification and Quality Assurance: Reading the Spectrum from Right to Left

This is a well-resolved, high quality spectrum, with good S/N, and no artifacts (Fig. 5.45). The first impression is that this is a normal white matter spectrum. Reading the spectrum from right to left, NAA, followed by an unremarkable β,γ-Glx region, the second NAA, Cr, Cho, mI, the α-Glx, the second Cr, and the second mI resonances are observed.

2. Interpretation and Analysis

The metabolite ratios are: NAA/Cr = 1.51 and Cho/Cr = 0.78 (both in the normal range); mI/Cr = 0.67, which is indeed not decreased. The β,γ-Glx/Cr and the α-Glx/Cr were also found to be in the normal range. [Cr] is normal. The MRS result from the occipital grey matter was similarly normal appearing.

E. Summary of Examination 1

The MRS spectra appear normal. In particular, neither mI nor Cho are decreased, and glutamine is not increased; therefore, MRS does not give any evidence of hepatic encephalopathy.

F. Examination 2: White Matter Location (17 days after TIPS)

1. Peak Identification and Quality Assurance: Reading the Spectrum from Right to Left

This is another good quality spectrum (Fig. 5.46). Reading the spectrum from right to left, NAA, β,γ-Glx, the second NAA, Cr, Cho, mI, and α-Glx are identified. The second Cr at 3.94 ppm is clearly seen, and the signal is in the correct proportion, 2:3, compared to the first Cr at 3.03 ppm, hereby reassuring that the spectral quality is good covering the range all the way to at least 4 ppm. Obviously, the spectral quality at 3.56 ppm is therefore good, confirming that the decrease of mI (3.56 ppm) is real.

2. Interpretation and Analysis

The ratios are: NAA/Cr = 1.50 (increased about 5%); Cho/Cr = 0.68 (decreased 18%); and mI/Cr = 0.38 (about 40% reduced). The β,γ-Glx/Cr = 0.47 (normal = 0.26) and α-Glx/Cr = 0.59 (normal = 0.46), thus Glx is severely increased. Because the

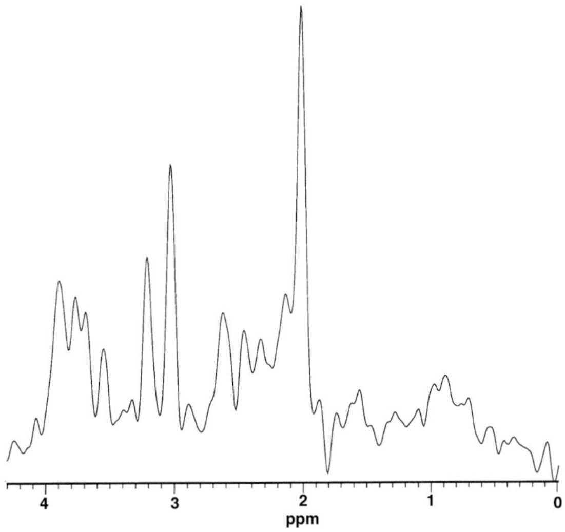

Fig. 5.46 The second examination was performed 17 days after TIPS. The MR spectrum shows changes attributed to sub-clinical hepatic encephalopathy.

relevant peaks are used in defining "Glx," it is readily proven that the increased Glx in patients with liver disease is in fact increased glutamine. [Cr] is 10% below the normal value. The grey matter spectrum showed similar results.

G. Summary of Examination 2

The MRS examination showed decreased mI and Cho, and increased glutamine post-TIPS, which is consistent with subclinical hepatic encephalopathy.

H. Examination 2: White Matter Location (3 Months after TIPS)

1. Peak Identification and Quality Assurance: Reading the Spectrum from Right to Left

The first impression is a high quality spectrum, but a missing peak catches the eye (Fig. 5.47). Reading the spectrum from right to left, NAA is followed by a remarkably increased β,γ-Glx region. The second NAA, Cr, and Cho are observed, Cho appearing decreased. The mI peak at 3.56 ppm is missing, but the α-Glx is increased in agreement with the large β,γ-Glx. The second Cr is clearly seen in the correct proportion compared to the first. No unusual resonances are observed.

2. Interpretation and Analysis

The ratios are: NAA/Cr = 1.45 (normal range); Cho/Cr = 0.66 (decreased 20%); and mI/Cr = 0.29 (about half the normal value as fitted by the computer). The β,γ-Glx/Cr = 0.40 (normal = 0.26) and α-Glx/Cr = 0.61 (normal = 0.46), thus glutamine is markedly increased. [Cr] is normal.

I. Summary of Examination 3

The MRS spectrum is consistent with severe hepatic encephalopathy following TIPS. Marked progression since last MRS is noted. With the very abnormal brain biochemistry reflected in this MRS spectrum, the patient is expected to be clinically symptomatic. Furthermore, all of the neurochemical abnormalities are fully reversible by a successful liver transplant.

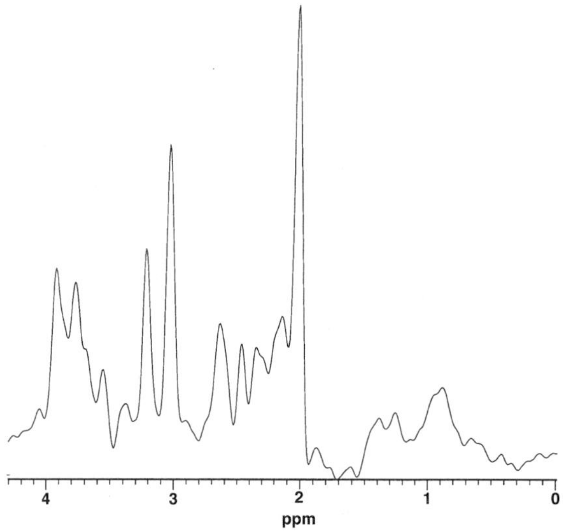

Fig. 5.47 The third examination was performed 3 months post-TIPS. The MR spectrum is consistent with severe symptomatic hepatic encephalopathy following TIPS. Notice the marked progression since the previous MRS examination.

J. Examination 2: White Matter Location (5½ Months after Liver Transplantation)

1. Peak Identification and Quality Assurance: Reading the Spectrum from Right to Left

This is a well-resolved, high quality spectrum, with good S/N, and no artifacts (Fig. 5.48). It can easily be interpreted despite the fact that the MR scanner was not the same as that used in the first three exams, and the MRS measurement parameters are slightly different. The first impression is that this is a near normal white matter spectrum. Reading the spectrum from right to left, NAA, followed by an unremarkable β,γ-Glx region, the second NAA, Cr, Cho, mI, the α-Glx, and the second Cr resonances are observed.

2. Interpretation and Analysis

The metabolite ratios are: NAA/Cr = 1.70 (increased 5%); Cho/Cr = 0.77 (normal); and mI/Cr = 0.68 (increased about 20%). The β,γ-Glx/Cr and the α-Glx/Cr were in the normal range. The [Cr] was also in the normal range. The normal values for the metabolite ratios acquired with STEAM TE = 20 ms and TM = 30 ms are slightly different from the normal ratios related to STEAM TE = 30 ms and TM = 13.7 ms, as seen in most of the MRS cases presented in this book. The occipital grey matter MRS showed a similar result.

K. Conclusion

MRS confirms the full recovery from chronic hepatic encephalopathy. Glutamine returned to normal, Cho and mI have recovered, with an apparent over compensation for mI, and the neuronal marker NAA is still normal, as expected.

L. Summary of Patient's History Post-MRS

The patient recovered after liver transplantation, and no further problems were reported.

M. Other Comments

TIPS is a life-saving procedure in cirrhosis-induced hematemesis, placing a shunt to drain the portal blood into the vena cava. However, not surprisingly, TIPS induces clinical HE in up to 90% of survivors.

Prior to deciding to perform a TIPS procedure, MRS is a worthwhile examination that will identify the patient that al-

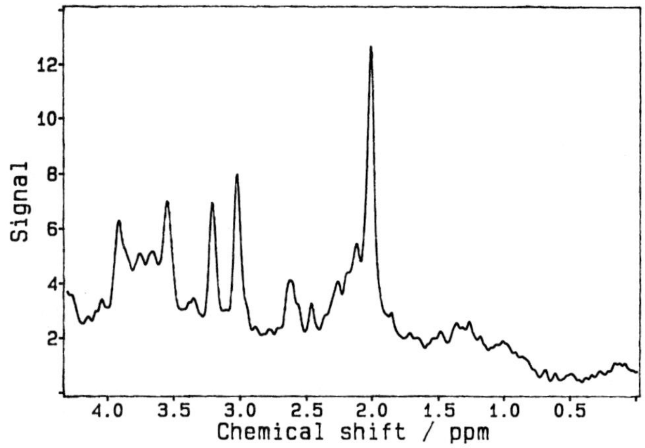

Fig. 5.48 The fourth examination was performed 5½ months after liver transplantation. The near-normal MR spectrum shows the full recovery of brain biochemistry after successful liver transplantation.

ready suffers from chronic hepatic encephalopathy. This case demonstrates that the patient did not suffer from hepatic encephalopathy before TIPS, but that the biochemical changes of hepatic encephalopathy can occur rapidly in the brain because of the TIPS procedure. This case is also an illustration of the full reversibility of HE by liver transplantation.

In selecting patients for liver transplant, ruling out Alzheimer's disease by means of MRS is also potentially valuable, but a theoretical risk that HE may reduce mI/Cr, the marker of AD, has not been totally excluded.

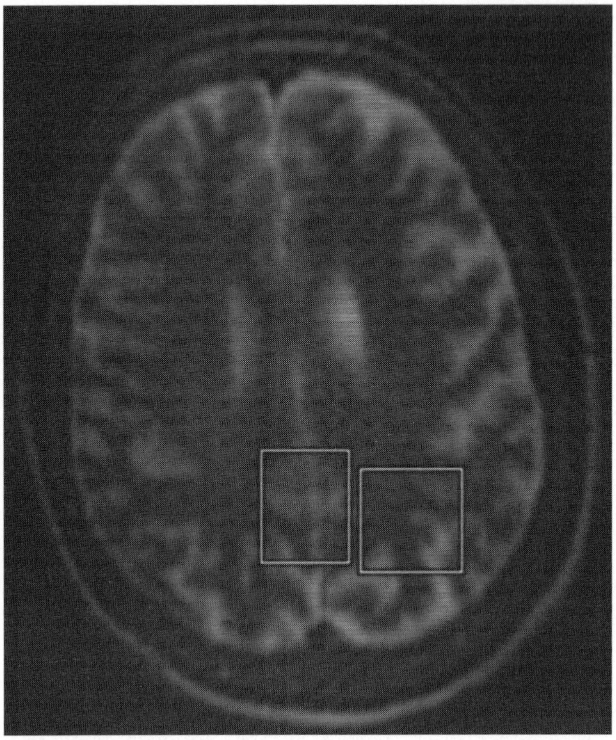

Fig. 5.49 T_2 weighted MRI showing the MRS locations.

5.3.6 HEPATIC ENCEPHALOPATHY VERSUS ALZHEIMER'S DISEASE

A. Purpose of MRS

MRS was performed to rule out Alzheimer's disease or other dementias and to evaluate for chronic or subclinical hepatic encephalopathy in a patient awaiting liver transplantation.

B. Summary of Patient's History

Female, age 62 with a seven year history of biliary cirrhosis. The question of probable Alzheimer's disease has been raised on the basis of neuropsychological testing, and positron emission tomography (PET).

C. MRS Method, Volume of Interest, and Comments about Examination

Localized STEAM (TE = 30 ms, TM = 13.7 ms, TR = 1500 ms) was performed in HMRI's occipital grey matter (Location 1) and occipito-parietal white matter (Location 2) (Fig. 5.49).

D. Location 1 (Grey Matter)

1. Peak Identification and Quality Assurance: Reading the Spectrum from Right to Left

This is a well-resolved spectrum, with good S/N, and no artifacts (Fig. 5.50). Reading the spectrum from right to left, the following resonances are identified: NAA, β,γ-Glx, second NAA, Cr, Cho, mI, α-Glx and the second Cr peak. No other unusual resonances are observed.

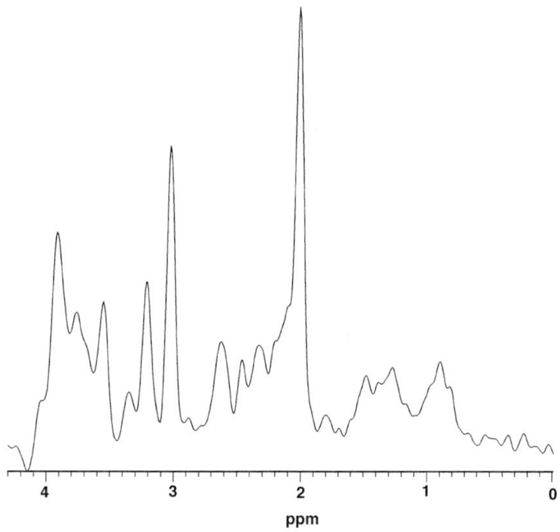

Fig. 5.50 The grey matter MRS spectrum which is not consistent with any form of dementia. A decreased mI and mildly increased Glx is evidence of mild hepatic encephalopathy confirming the result from white matter.

2. Interpretation and Analysis

The metabolite ratios are: NAA/Cr = 1.42 (*above* normal for age), Cho/Cr = 0.57 (normal), mI/Cr = 0.51 (decreased 15%). A computer integration revealed α-Glx/Cr and β,γ-Glx/Cr to be about 30% increased. [Cr] is 85% of the normal value, hereby affecting the interpretation of the ratios to Cr: NAA becomes normal, Glx still mildly increased, Cho becomes decreased, and mI becomes even further decreased than it appears from the mI/Cr ratio. The CSF fraction is 13% which is normal for age in this location.

E. Location 2 (White Matter)

1. Peak Identification and Quality Assurance: Reading the Spectrum from Right to Left

This is also a high quality spectrum (Fig. 5.51).

Reading the spectrum from right to left, the following resonances are identified: NAA, β,γ-Glx, the second NAA, Cr, Cho, mI, α-Glx, the second Cr and the second mI. No other unusual resonances are observed.

2. Interpretation and Analysis

Reading from right to left, the metabolite ratios are: NAA/Cr = 1.58 (above mean for age), Cho/Cr = 0.62 (decreased 27%), mI/Cr = 0.51 (decreased 16%). A small increase of Glx is seen: α-Glx/Cr = 0.57 and β,γ-Glx/Cr = 0.31. [Cr] is normal.

F. Conclusion

NAA is definitely not decreased, MRS is therefore not consistent with dementia of any kind. The MRS is in particular not consistent with Alzheimer's disease as in addition mI is not increased. The mI/Cr ratio is reduced in both locations; also Cho/Cr is reduced in the white matter location, and a small excess of Glx is observed. This is consistent with mild chronic hepatic encephalopathy. The condition is potentially reversible if the patient undergoes liver transplantation.

G. Summary of Patient's History Post-MRS

The patient is still awaiting liver transplantation half a year after the MRS examination.

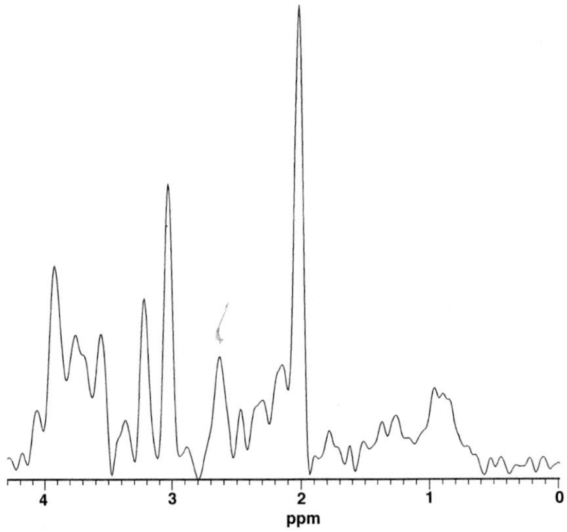

Fig. 5.51 The white matter MRS spectrum which is consistent with mild reversible chronic hepatic encephalopathy. NAA is not reduced, confirming the grey matter result that the patient is not demented.

H. Other Comments

This particular case is a prime example of a liver patient who was suspected of having Alzheimer's disease, or another dementia, but who in fact had hepatic encephalopathy, proved by MRS, and who possibly had a mild depression. Neuropsychological testing was nonspecific. PET for diagnosing AD is still controversial, but in general, hepatologists have little experience with MRS; therefore, more studies are urgently needed.

The evaluation of hepatic encephalopathy versus Alzheimer's disease or other dementias is critical in the management of liver disease in 55–70-year-old patients, and the role of MRS is central. An apparently demented patient could have a depression just due to the pressure of the whole situation, and if chronic hepatic encephalopathy is present the appearance of the patient will be worse. Such a patient is likely to recover fully after a liver transplantation, and can expect 10–20 good years after the transplantation. Most liver centers will on the other hand not perform a transplantation if they find evidence of dementia in 55–70 year old patients. The role of MRS is to rule out that a dementia is present, and to evaluate the severity of hepatic encephalopathy.

The question of mI in hepatic encephalopathy versus Alzheimer's disease is interesting. In one disease mI is increased in the other it is depleted. How will the spectrum look in a patient with both diseases? The answer for mI is not known. But patients with both diseases will have reduced NAA due to the Alzheimer's dementia. Patients with hepatic encephalopathy, except terminal stages, will have a normal NAA. Thus the question of hepatic encephalopathy versus Alzheimer's disease can still be answered by MRS. A question of Alzheimer's disease versus other dementia in a patient with liver disease may be more difficult, but fortunately also will be similarly rare and insignificant for patient care.

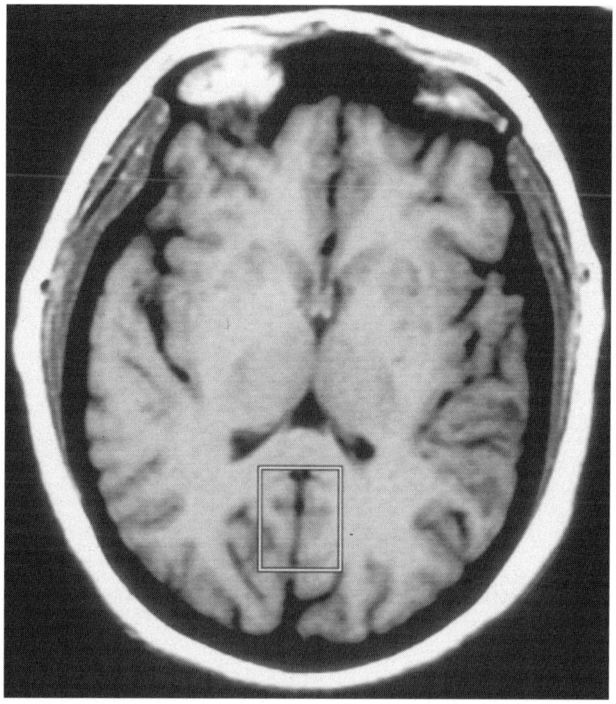

Fig. 5.52 T_1-weighted image showing grey matter MRS location.

5.3.7 LIVER DISEASE: PORTAL HYPERTENSION

A. Purpose of MRS

Exclude subclinical hepatic encephalopathy in patient with portal hypertension.

B. Summary of Patient's History

Female, age 47, with spontaneous portal hypertension with varices, but no evidence of cirrhosis on biopsy of liver.

C. MRS Method, Volume of Interest, and Comments about Examination

Localized STEAM (TE = 30 ms, TM = 13.7 ms, TR = 3000 ms) was performed in HMRI's occipital grey matter (see Fig. 5.52). There were no neurological symptoms or signs at the time of examination.

D. Location (Grey Matter)

1. Peak Identification and Quality Assurance: Reading the Spectrum from Right to Left

The quality of the grey matter spectrum, Fig. 5.53, is excellent, with some irregular signal in the lipid/lactate region, and possible artifacts at 1.5 and 1.7 ppm.

Reading from right to left, the following peaks are identified: NAA, β,γ-Glx, second NAA, Cr, Cho, a little sI and glucose, mI (at 3.56 ppm), α-Glx, and second Cr peak. No obvious abnormalities are identified.

2. Interpretation and Analysis

Reading from right to left, the metabolite ratios are: NAA/Cr = 1.39 (normal for grey matter); Cho/Cr = 0.64 (normal, or even slightly elevated); mI/Cr = 0.63 (normal, not low). [Cr] is 95% of normal. β,γ-Glx/Cr = 0.28, and α-Glx/Cr = 0.44 (both normal).

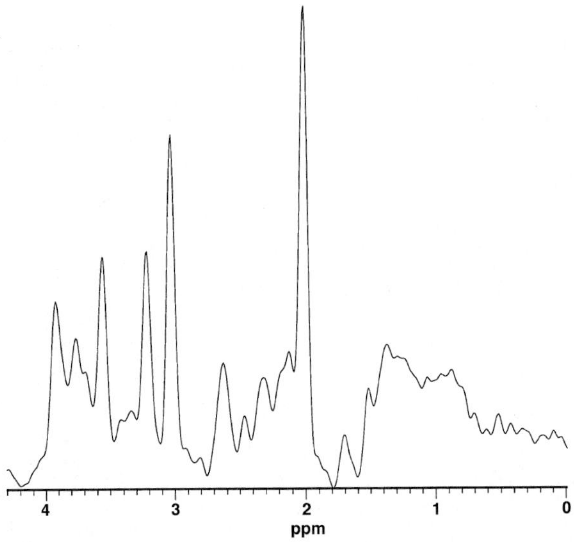

Fig. 5.53 The grey matter spectrum shows entirely normal MRS in this patient with portal hypertension. No evidence of hepatic encephalopathy.

E. Conclusion

This a normal spectrum, excluding, therefore, subclinical hepatic encephalopathy.

F. Summary of the Patient's History Post-MRS

Not available.

G. Other Comments

MRS in patients with *any* liver disease generally show reduced Cho/Cr in HMRI's white matter location, but not systematically in HMRI's grey matter location. Thus, because the white matter location provides more information, the location of choice in ruling out hepatic encephalopathy is parietal white matter. In established cases of cirrhosis, depletion of mI, Cho with or without elevation of Glx makes the diagnosis of hepatic encephalopathy certain. In this context, grey matter and white matter examinations are of equal sensitivity.

Finally, in terms of pathogenesis, this patient helps to explain that portal hypertension alone is *not* responsible for the astrocyte swelling and *myo*-inositol depletion seen in patients with hepatic encephalopathy, even when it is subclinical.

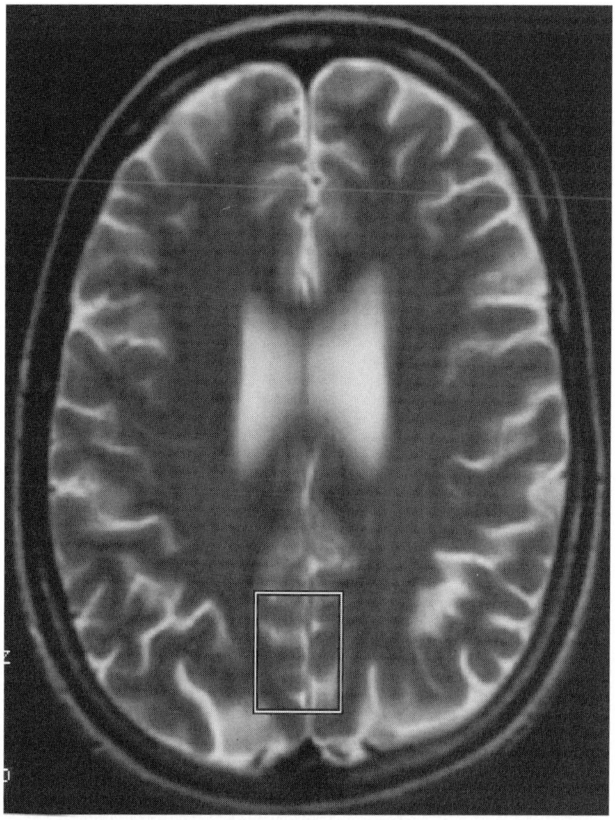

Fig. 5.54 The T_2-weighted MRI shows more atrophy of the cortical sulci than should be expected in a 40 year old. Otherwise, no abnormalities are noted. The position of the grey matter VOI is shown. (Courtesy of Dr. Wang, Rancho Los Amigos Hospital, Downey, California.)

5.3.8 KORSAKOFF-WERNICKE'S PSYCHOSIS, RULE OUT HEPATIC ENCEPHALOPATHY

A. Purpose of MRS

To investigate a patient who had been diagnosed with Korsakoff-Wernicke's psychosis, and to rule out suspected hepatic encephalopathy.

B. Summary of Patient's History

Female, age 40. (A more detailed history is not available.)

C. MRS Method, Volume of Interest, and Comments and Examination

PROBE-STEAM (TE = 30 ms, TM = 13.7 ms, TR = 1500 ms) was performed in HMRI's occipital grey matter as shown on the image (Fig. 5.54).

D. Grey Matter Location

1. Peak Identification and Quality Assurance: Reading the Spectrum from Right to Left

The good S/N, resolution, and the absence of artifacts guarantees the high quality of the spectrum (Fig. 5.55). The slight slope of the baseline is caused by residual water, which is not subtracted out before these spectra are displayed on the screen.

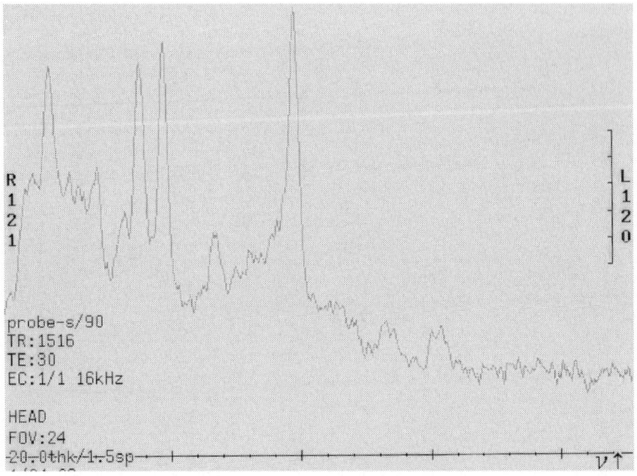

Fig. 5.55 The abnormal grey matter PROBE spectrum. Hepatic encephalopathy can be excluded, but a specific diagnosis cannot be given. (Courtesy of Dr. Wang, Rancho Los Amigos Hospital, Downey, California.)

Reading from right to left, there are no significant signals in the lipid-lactate region before NAA at 2.02 ppm, followed by β,γ-Glx, the second NAA, Cr, Cho, sI, mI, α-Glx, the second Cr, and a very small second mI.

2. Interpretation and Analysis

[Cr] is not available from this automated MRS examination, but based on the clinical picture [Cr] is not likely to be too far from normal; ratios are therefore used. The ratios were obtained by processing the spectrum by computer: NAA/Cr = 1.15 (reduced about 15%); Cho/Cr = 0.85 (increased 44%); and mI/Cr = 0.41 (decreased 28%). Glx is not elevated, however mI/Cr is below normal, and earlier HE now recovered, or hyponatremia must both be considered.

E. Conclusion

The marked increase of Cho/Cr combined with normal, rather than elevated Glx, excludes hepatic encephalopathy. The decreased NAA/Cr is consistent with the findings of atrophic changes in cortical grey matter.

F. Summary of Patient's History Post-MRS

Not available.

G. Other Comments

An increased Cho in grey matter may sometimes be attributed to misplacement of the VOI (into too much white matter), but in this case by re-inspecting the positioning on all relevant axial images misplacement was excluded. The spectrum is abnormal with decreased NAA/Cr and mI/Cr, and increased Cho/Cr, but it is not diagnostic of Korsakoff-Wernicke's psychosis.

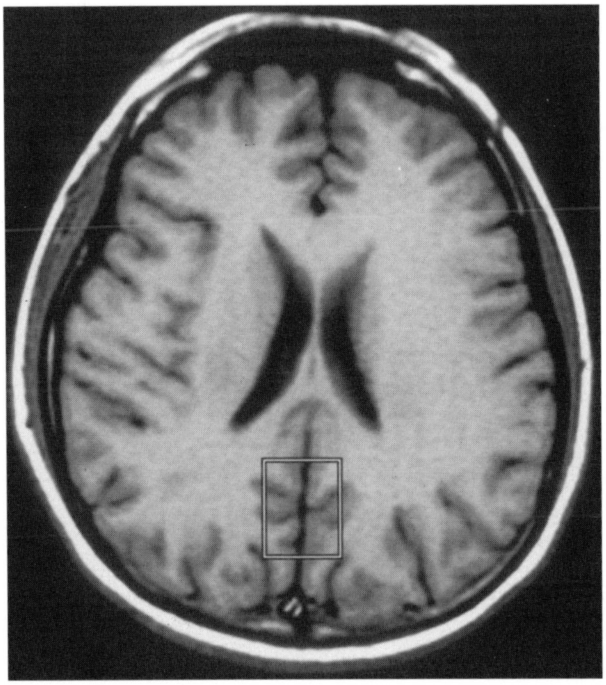

Fig. 5.56 The grey matter location used for MRS in a patient in coma, following diabetes.

5.3.9 DIABETIC: HYPEROSMOLAR COMA

A. Purpose of MRS

To rule out hyperosmolar coma.

B. Summary of Patient's History

Male, age 14.5 years, admitted while in deep diabetic coma (glucose = 1889 mg/dL; Na^+ = 126 mE/L; K^+ = 402; BUN = 46; pH = 7.09; lactate = 2.2mM). Slow recovery from diabetic coma in known diabetic, with keto-acidosis and prolonged coma. At the time of MRS examination glucose had fallen to 269 mg/dL but Na^+ = 179 mE/L, Cl = 142 mE/L, and serum osmolarity of 356 mOsm/L was recorded. Ketones and lactate were no longer elevated. The patient was still in coma, with the question of hyperosmolar versus diabetic acidosis-induced coma?

C. MRS Method, Volume of Interest, and Comments about Examination

Localized STEAM (TE = 30 ms, TM = 13.7 ms, TR = 3000 ms, NAV 64) was performed in HMRI's occipital grey matter (see Fig. 5.56).

D. Location (Grey Matter)

1. Peak Identification and Quality Assurance: Reading the Spectrum from Right to Left

The spectrum is well-resolved, Fig. 5.57. There is a small, but probably significant lactate peak at 1.33 ppm; no abnormal lipid.

Reading from right to left, the following resonances are identified: NAA, β,γ-Glx (but, prominent additional peak at 2.14 ppm), second NAA, Cr, Cho, sI, glucose, mI, α-Glx and glucose, but little evidence of second Cr or mI.

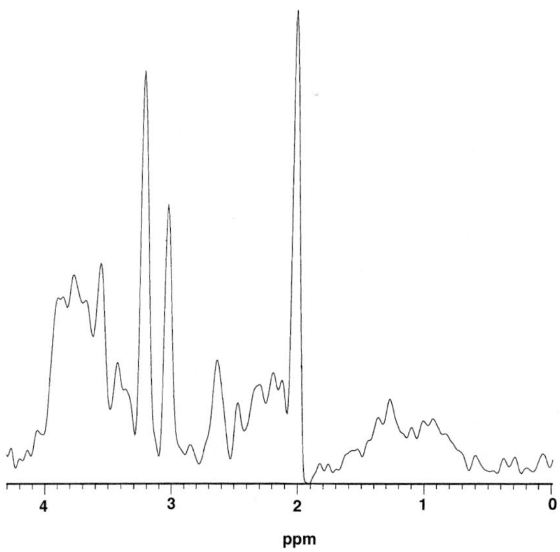

Fig. 5.57 A striking abnormal MRS in hyperosmolar coma, following diabetes. The abnormally high NAA and Cho resonances are partly explained by the 12% reduction in Cr. In fact, all major peaks in the spectrum except NAA are actually present in abnormal amounts.

2. Interpretation and Analysis

Reading from right to left, the presence of lactate is significant in relation to diabetic state. The metabolite ratios are: NAA/Cr = 1.76 (markedly increased); Cho/Cr = 1.52 (three times normal); mI/Cr = 0.77 (25% increased). Other notable abnormalities include a peak at 2.14 ppm (possibly a ketone body), and $glucose_1$/Cr and $glucose_2$/Cr at about three-time normal (integrals are 0.38 vs 0.10, and 1.02 vs 0.40). A crucial finding is [Cr] = 88% of normal, explaining in part the elevated ratios. In absolute terms, [NAA] = 100%; [Cho] = 184%; [mI] = 144%. H_2O is 65% (the normal is 78%).

E. Conclusion

This is a typical "diabetic" spectrum, but includes strong evidence of a (reversible) hyperosmolar state.

F. Summary of the Patient's History Post-MRS

Slow, but uneventful recovery. Slow rehydration introduced, to avoid cerebral edema.

G. Other Comments

This unusual spectrum from a known diabetic may be, as suggested, a manifestation of hyperosmolar state, since Cho is a recognized osmolyte. However, mI, Cr, and NAA have also been implicated as osmolytes. In this patient, elevation of NAA, even though a neuronal osmolyte, is not confirmed by quantitation. Instead, [Cr] appears reduced. A more complex pathogenesis may be involved than that proposed here.

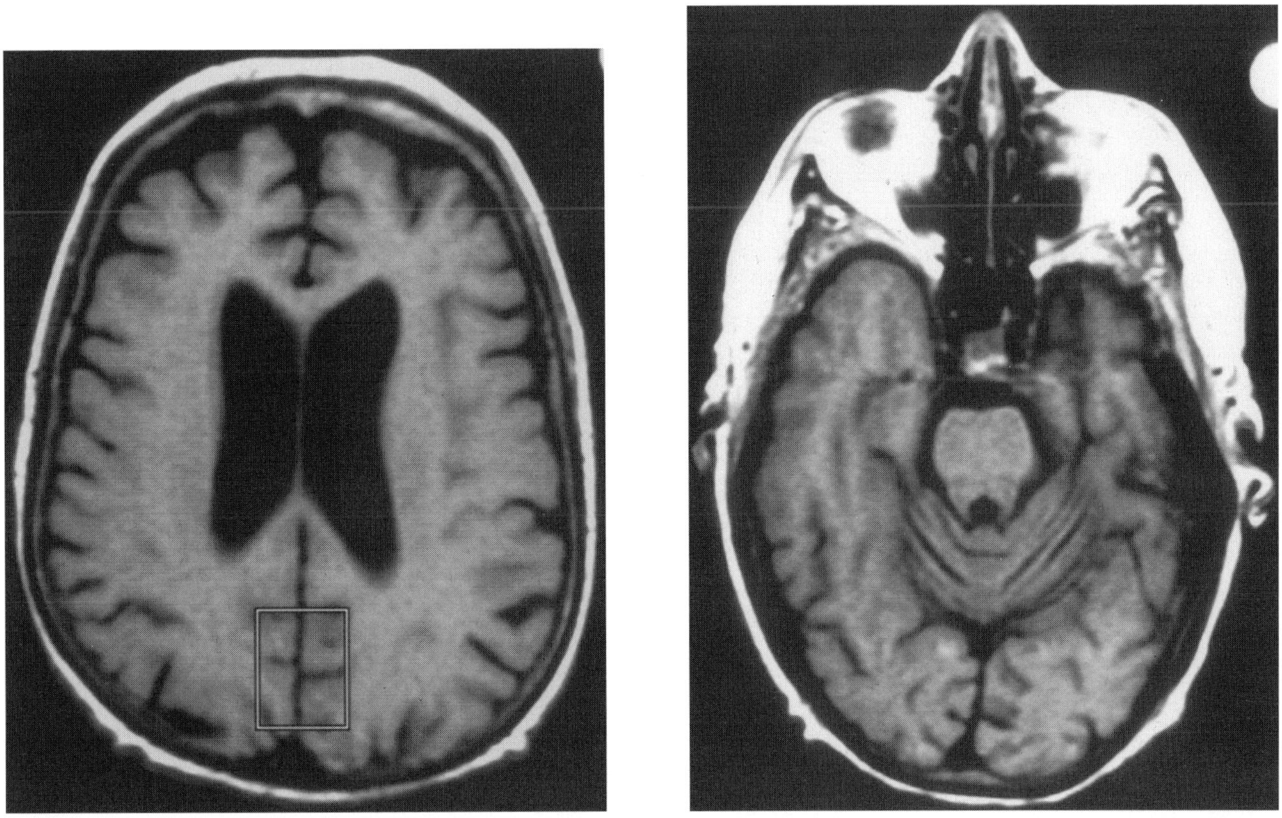

(A) (B)

Fig. 5.58 (A) Axial MRI showing ventricular dilatation and modest cortical atrophy. The location of the grey matter VOI is illustrated. (B) MRI at the level of the pons may show elements of myelinolysis. However, the appearance is not classical and no MRS was performed on the pons in this case (see the figure below of MRS in the pons of a patient with central pontine myelinolysis).

5.3.10 HYPONATREMIA

A. Purpose of MRS

Rule out central pontine myelinolysis.

B. Summary of Patient's History

Female, age 66, admitted with confusion and found to have modest hyponatremia (serum Na^+ was 127 mEq/L).

C. MRS Method, Volume of Interest, and Comments about Examination

Localized STEAM (TE = 30 ms, TM = 13.7 ms, TR = 1500 ms) was performed in HMRI's occipital grey matter (see Fig. 5.58).

D. Grey Matter Location

1. Peak Identification and Quality Assurance: Reading the Spectrum from Right to Left

The technical quality of the spectrum (Fig. 5.59) is excellent and clearly abnormal.

Reading systematically from right to left, there is no excess lipid at 0.9 ppm, and no obvious lactate doublet at 1.33 ppm.

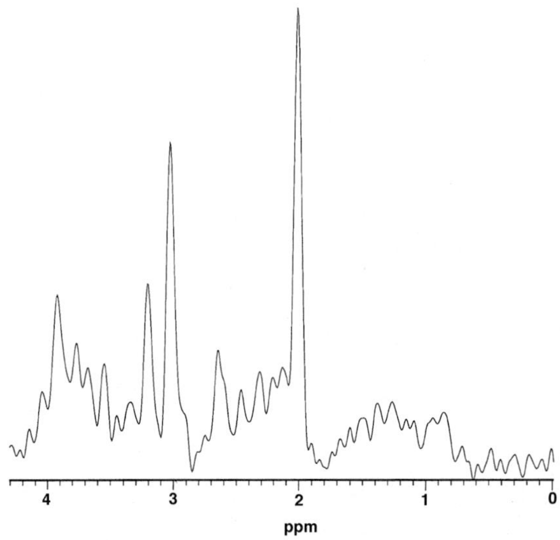

Fig. 5.59 Abnormal MRS of grey matter. NAA/Cr is normal or even elevated, ruling out dementia. *Myo*-inositol is very much reduced, in accordance with a diagnosis of hyponatremia; note β,γ-Glx is well-resolved but not elevated, ruling out hepatic encephalopathy, the other principal diagnosis that results in reduced mI.

Continuing right to left, the following resonances are identified: NAA, β,γ-Glx, the second NAA, Cr, Cho, much reduced mI at 3.56 ppm, α-Glx, and the second Cr. Cho and mI appear reduced, but Glx does not appear increased, and NAA appears normal.

2. Interpretation and Analysis

Reading from right to left, the metabolite ratios are: NAA/Cr = 1.41 (slightly elevated for age); Cho/Cr = 0.55 (normal); and mI/Cr = 0.31 (profoundly reduced). [Cr] is 83% of normal, and brain water is 72% (approximately normal for occipital grey matter). β,γ-Glx/Cr = 0.28; α-Glx/Cr = 0.40 (both are normal or slightly reduced).

Because [Cr] is reduced, all metabolite concentrations are in fact significantly below normal.

E. Conclusion

MRS shows hypo-osmolar encephalopathy due to hyponatremia.

F. Summary of Patient's History Post-MRS

Not available.

G. Other Comments

The results of this examination are typical of hyponatremia, and in strong contrast to the spectrum in hepatic encephalopathy and Alzheimer's disease. *Myo*-inositol, and Cho in particular, are cerebral osmolytes that are reversibly displaced in all hypo-osmolar syndromes, including SIADH in some juvenile trauma patients.

An important exclusion is central pontine myelinolysis (CPM), which can sometimes be identified in MRI, (Fig. 5.60A), but does not necessarily show the MRS features of hyponatremia (Fig. 5.60B). Instead, a strong lipid–triglyceride signal consistent with histological evidence of lipid-filled macrophages at the site of myelinolysis, dominates the CPM-spectrum (Fig. 5.60C).

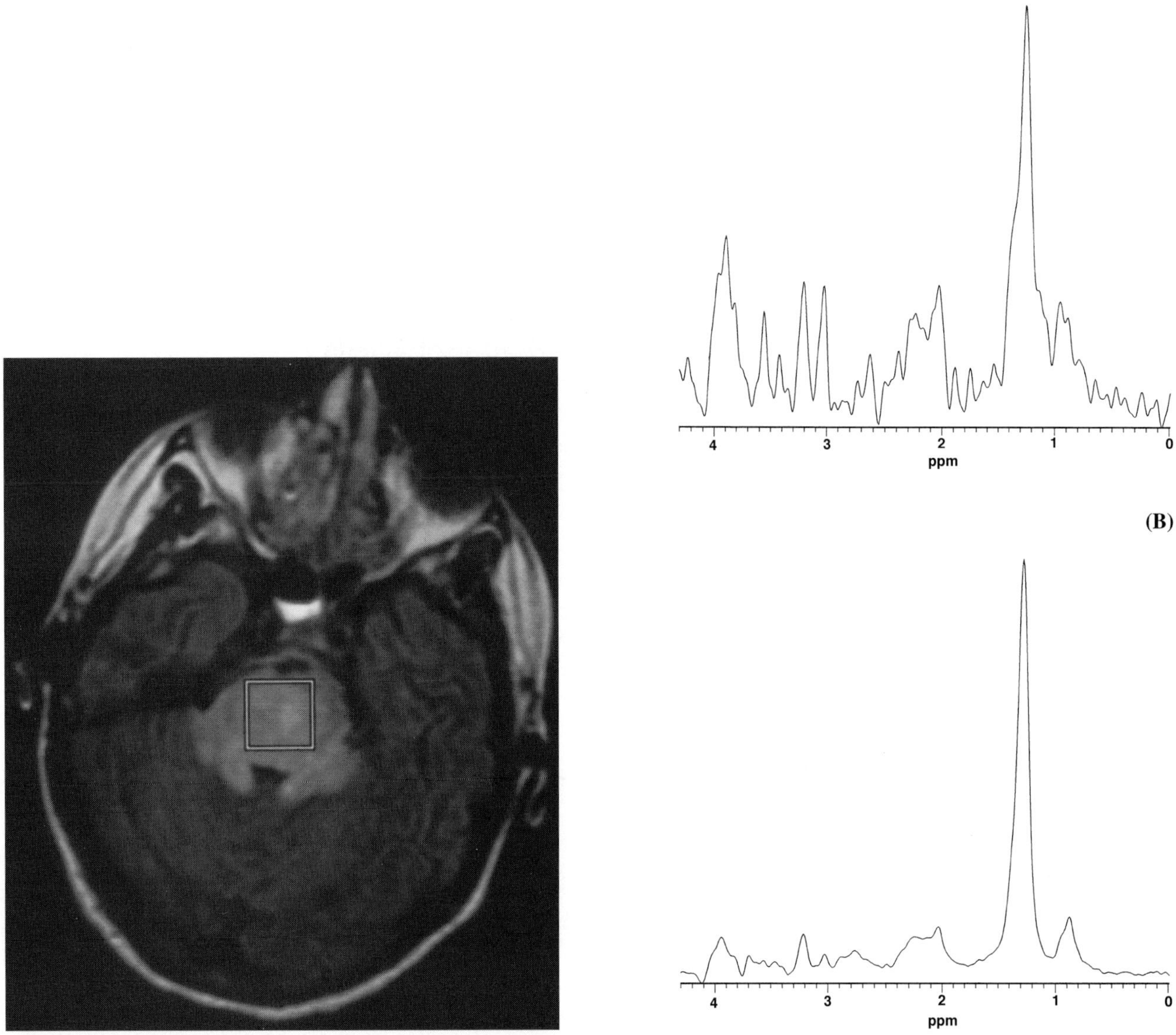

(A) (C)

Fig. 5.60 (A) MRI consistent with central pontine myelinolysis (in another patient). The patient is a 29-year-old female, a week post-partum. (B) MRS of the pons in true central pontine myelinolysis. Dramatic abnormality of cerebral metabolites in central pontine myelinolysis includes two well-resolved peaks, at 0.9 and 1.2 ppm, representing triglycerides or long-chain fatty acids. However, no depletion of mI was noted in this patient, (mI is not decreased in grey and white matter, not shown), in whom hyponatremia was never a feature. (C) MRS one month later shows an approximately four times increase of the triglycerides or long-chain fatty acids resonances from the VOI in the pons.

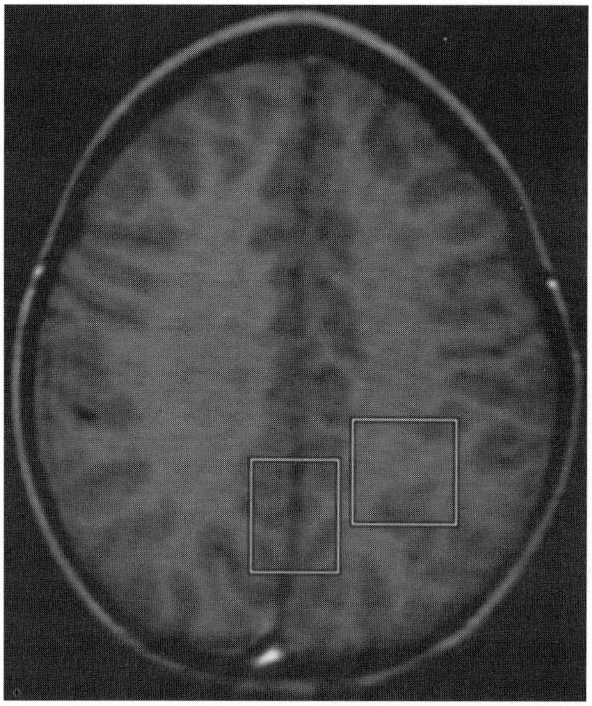

Fig. 5.61 T_1 weighted MRI showing the MRS locations in grey and white matter.

5.4 Trauma: SIADH, Hyperosmolar State, Diffuse Axonal Injury, Hypoxic Injury

5.4.1 HEAD INJURY: SIADH

A. Purpose of MRS

MRS was requested to evaluate outcome after head trauma.

B. Summary of Patient's History

Male, age 6 years, who had been involved in a road traffic accident. The Glasgow Coma Score was 8 when the child was admitted to hospital and had improved to 13 on the fifth day postinjury when the MRS was performed. The child was confused and tearful. The serum sodium had been low at 119 mEq/liter (normal range 135–145) just after the injury, but had been adjusted gradually back to normal (140 mEq/liter) by the day of the MRS examination.

C. MRS Method, Volume of Interest, and Comments About Examination

STEAM (TE = 30 ms, TM = 13.7 ms, TR = 1500 ms) was performed in HMRI's occipital grey matter (Location 1), and occipito-parietal white matter (Location 2) (Fig. 5.61).

D. Location 1 (Grey Matter)

1. Peak Identification and Quality Assurance: Reading the Spectrum from Right to Left

This is a well-resolved, good quality spectrum with good S/N, and no artifacts (Fig. 5.62). There is no lactate or significant lipid resonances. The NAA, β,γ-Glx, the second NAA, Cr, Cho, mI, α-Glx, and the second Cr are identified. No unusual resonances are observed.

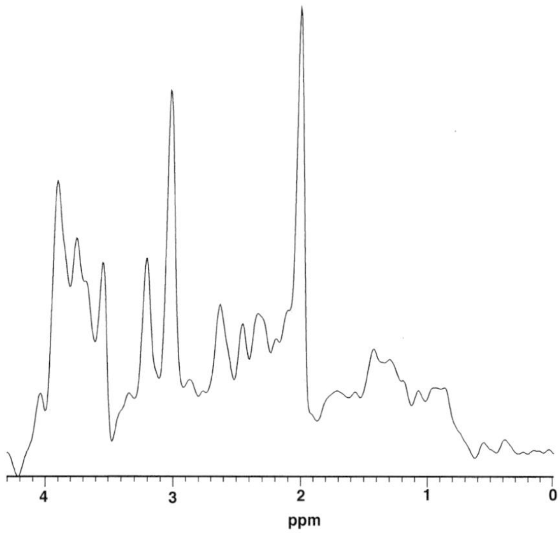

Fig. 5.62 The grey matter MRS spectrum consistent with SIADH in a child with head trauma.

2. Interpretation and Analysis

The metabolite ratios are: NAA/Cr = 1.22 (reduced about 20%), Cho/Cr = 0.53 (reduced about 12%), and mI/Cr = 0.52 (reduced about 13%), and [Cr] is 87% of normal. The absolute metabolite concentrations of [NAA], [Cho] and [mI] are also reduced 20–35%, consistent with the given values for the ratios. The CSF fraction in the VOI is 5%.

E. Location 2 (White Matter)

1. Peak Identification and Quality Assurance: Reading the Spectrum from Right to Left

The white matter spectrum is also of good quality (Fig. 5.63). The peak identification is very similar that of the grey matter spectrum. Again, there is no lactate or significant lipids, and the presence of the same metabolites as in the grey matter location is noted: NAA, Glx, Cr, Cho, and mI.

2. Interpretation and Analysis

The ratios are: NAA/Cr = 1.30 (reduced 10–20%); Cho/Cr = 0.62 (reduced 20–25%); mI/Cr = 0.62 (close to normal); [Cr] is 86% of normal. The remaining metabolite concentrations are also reduced.

F. Conclusion

There is no evidence of severe hypoxic injury because there is no lactate or significant lipids present in the spectra, and no excess of β,γ- or α-Glx. The absolute metabolite concentrations [NAA], [Cr], [Cho] (in particular) and [mI] are reduced by as much as 40% below the normal values. The reduction in NAA is consistent with neuronal/axonal loss as a direct result of head injury, but is only partly due to a minor neuronal loss in this case. The generalized low level of all the metabolite concentrations is consistent with hyponatremia (reduced concentration of organic osmolytes in the brain), and is thus in this case explained by syndrome of inappropriate antidiuretic hormone secretion (SIADH). The identification of SIADH led to a good prognosis; the child was expected to recover as the status of organic osmolytes in the brain returned to normal. Recovery of NAA may be incomplete in proportion to the extent of direct neuronal damage.

G. Summary of Patient's History Post-MRS

The patient recovered fully. At the latest inquiry more than one year after the accident, the mother reports that he attends regular school, and that he is doing very well both at school and in all other aspects of life.

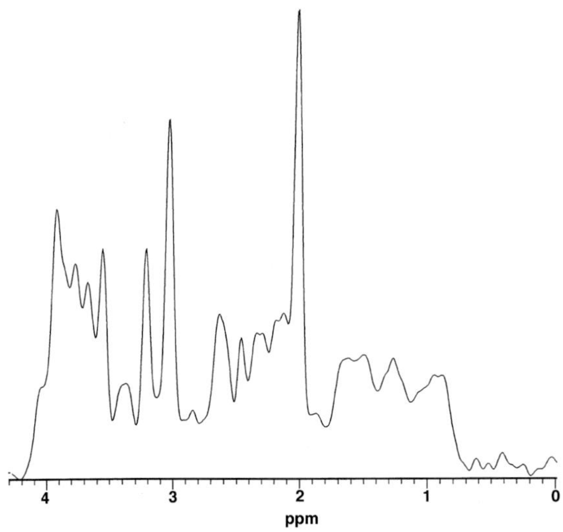

Fig. 5.63 The white matter MRS spectrum consistent with SIADH in a child with head trauma.

H. Other Comments

Endocrine response to trauma, in the form of SIADH, is seen in 40–50% of children with head injury. Effects of the hyponatremia is seen on the MRS in the form of decreased mI, Cho, Cr and NAA, even in cases with less severe reduction of serum sodium than in this case.

The patient's serum sodium level was initially 119 mEq/l, but normal on the day of the MRS examination (140 mEq/l). The MR spectrum was, however, still severely affected because it takes several weeks for the organic osmolytes in the brain to return to normal. A similar, but milder, response can be expected in cases with less severely reduced serum sodium.

The diagnosis of hyponatremia/SIADH is primarily made based on the absolute concentrations of all the main cerebral metabolites, which, in this context, are all better described as osmolytes. The neurospectroscopist who only has metabolite ratios (mI/Cr, Cho/Cr and NAA/Cr) available might be discouraged, because the otherwise assumed constant metabolite Cr in hyponatremia is actually reduced in SIADH. How then can the diagnosis be made from metabolite ratios? The solution is to look at the ratios for Cho/Cr and mI/Cr. The reduction in the concentrations of Cho and mI is usually more than the reduction (if any) in the concentration of Cr. Therefore, the ratios of Cho/Cr and mI/Cr are often also reduced. By combining this observation with the clinical information (i.e., in patients suffering from head trauma, SIADH is a possibility), a correct conclusion can easily be made, even if peak ratios only are available.

The MRS pattern of SIADH is one of the five distinct patterns that can be identified in MRS of trauma patients. Other patterns identify hypoxic injury, diffuse axonal injury, neuronal damage, and a comatose hyperosmolar state (see the other trauma cases). Each individual MRS pattern is believed to correspond to different biochemical responses to the head injury, and as the expected outcome varies, it is important to be able to recognize these varying patterns.

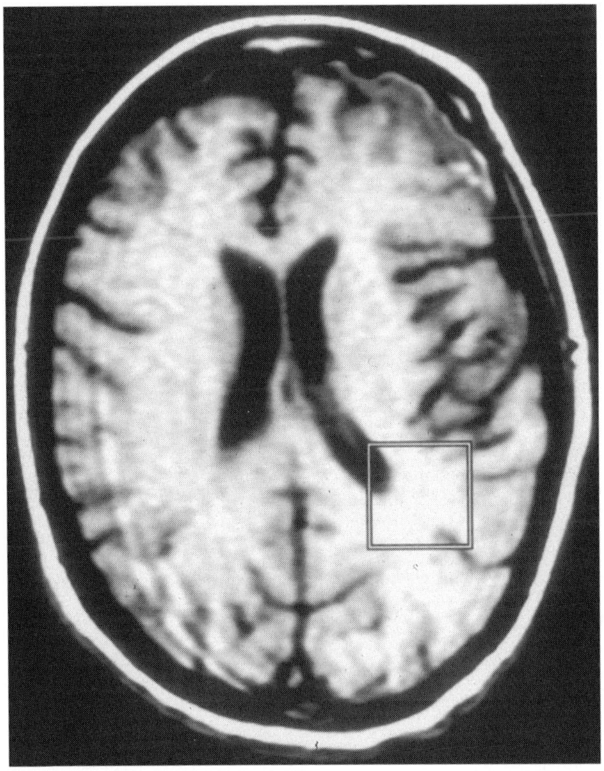

Fig. 5.64 T_1-weighted MRI showing the MRS location in white matter.

5.4.2 HEAD INJURY: HYPEROSMOLAR STATE

A. Purpose of MRS

MRS was requested to evaluate a comatose head injury patient.

B. Summary of Patient's History

Female, age 28, who was found unconscious near a hiking trail in the mountains, and apparently had fallen about 50 feet. The Glasgow coma scale score was 5 when the patient was admitted to hospital, and she was still in a coma at the time of this MRS examination, 2½ months after the accident.

C. MRS Method, Volume of Interest, and Comments and Examination

STEAM (TE = 30 ms, TM = 13.7 ms, TR = 1500 ms) was performed in HMRI's parietal white matter (Fig. 5.64).

D. White Matter Location

1. Peak Identification and Quality Assurance: Reading the Spectrum from Right to Left

This is a well-resolved, good quality spectrum, with good S/N, and no artifacts except a minor "out of phase" lipid signal around 1.4 ppm; this signal is attributed to lipid outside the VOI because it is not in phase with the remaining peaks (Fig. 5.65). There is no lactate, or significant lipid resonances. The NAA, β,γ-Glx, the second NAA, Cr, Cho, a minor sI peak at 3.36 ppm, mI, α-Glx, the second Cr, and the second mI peaks are identified. No unusual resonances are observed.

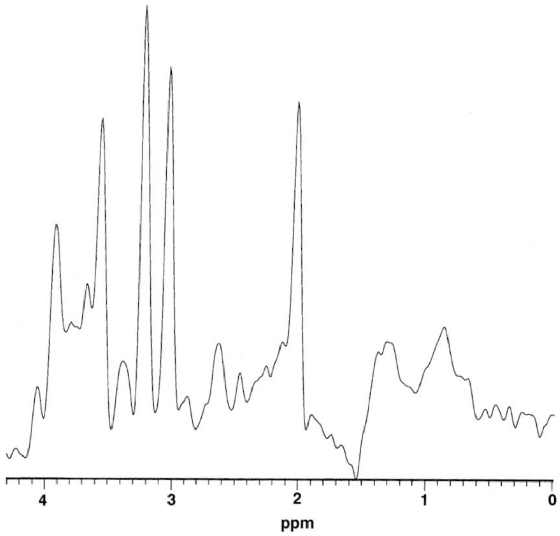

Fig. 5.65 The white matter MR spectrum from a young female with head trauma. Note that using metabolite ratios alone can be misleading in this case.

2. Interpretation and Analysis

The metabolite ratios are: NAA/Cr = 0.90 (reduced about 40%); Cho/Cr = 1.17 (increased about 40%); and mI/Cr = 0.87 (increased about 45%). The absolute concentrations are very striking in this patient: [NAA] is 87%, [Cr] is 120%, [Cho] is 187%, and [mI] is 167% of normal. The water content, which is also measured as an absolute concentration by the same calibration procedure, was normal for white matter, so there is no reason to doubt the absolute metabolite concentrations.

E. Conclusion

There is no evidence of severe hypoxic injury because there is no lactate, or significant lipids present in the spectra. [Cho] is almost double the normal concentration, and [Cr] and [mI] are also markedly increased. The reduction in [NAA] to 87% of normal is consistent with head injury, but entirely consistent with a good prognosis. The severe increase of [Cho], [Cr] and [mI] is consistent with a hyperosmolar state, which is sometimes seen in comatose head injury patients. Our experience is that a slow recovery, but a long-term relatively good outcome, can be expected.

F. Summary of Patient's History Post-MRS

The long term outcome (three years after the accident) was relatively good. The patient was for a long time dependent on various degrees of care and rehabilitation, but was eventually able to live alone and take care of herself (Rancho Outcome Score 14). A number of moderate deficits, however, remained.

Further MRS studies were performed the following years. All four main metabolites, NAA, Cr, Cho, and mI, decreased as the status of the organic osmolytes in the brain slowly normalized. [NAA] was further reduced to 71% after two years, in agreement with other observations that even NAA behaves as an osmolyte. [Cr] was already normalized 4½ months after the injury. Remarkably, [mI] and [Cho] was still about 20–25% increased after two years, but an examination after another 1½ years showed more normal values also for [mI] and [Cho].

G. Other Comments

The hyperosmolar state is one of the distinctly different patterns that can be identified in MRS of trauma patients. Other patterns identify hypoxic injury, diffuse axonal injury, neuronal damage, and a hypo-osmolar state, SIADH (see the other trauma cases). Each different MRS pattern is believed to correspond to different biochemical responses to the head injury, and as the expected outcome varies it is important to be able to recognize these different patterns.

The hyperosmolar state is in particular important to notice because its recognition leads to a much more optimistic prognosis than the first impression that the MRS could give based on the reduced NAA/Cr ratio which is in fact caused by *increased* absolute Cr. These patients may have been comatose for weeks, and may still not wake up for a while, but those patients we have seen with this pattern have all come out of the coma, and had a long-term good outcome with only minor deficits.

The identification of the hyperosmolar state is based on the absolute concentrations of all the main metabolites. The neurospectroscopist who only has metabolite ratios available (mI/Cr, Cho/Cr and NAA/Cr), might be discouraged because the otherwise assumed constant metabolite Cr is actually increased. There is, however, a trick: The effect on mI and Cho is greater than on Cr, and NAA, so the pattern will be of an almost horizontal or downwards "slope" of the peaks from left to right. Combining this observation with information about the patient (in cases of trauma, the hyperosmolar state is a possibility) should even with ratios alone easily lead to the correct conclusion.

A full physiological and biochemical explanation for the hyperosmolar state in some of these comatose trauma patients is currently not available. One suggestion might be that the nature of the comatose state is similar to that of diabetic hyperosmolar coma.

The increased [mI], [Cho], and [Cr] have only been observed in white locations. This is one of the reasons to perform MRS in the white matter location as the first choice in head injury patients, so that the pattern of the hyperosmolar state can be discovered if present.

Because NAA appears to be the neuronal osmolyte, outcome prediction from either NAA/Cr, or [NAA] can be over-optimistic.

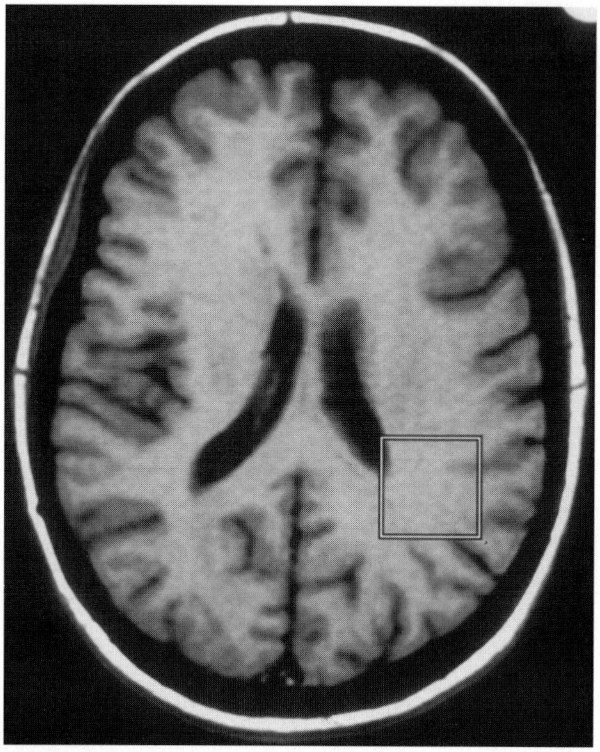

Fig. 5.66 T_1-weighted MRI showing the MRS locations in white matter.

5.4.3 HEAD INJURY: DIFFUSE AXONAL INJURY

A. Purpose of MRS

MRS was requested to evaluate outcome after head trauma.

B. Summary of Patient's History

Female, age 50, who was hit by a car when bicycling. The MRS was performed 20 days after the accident, and the Glasgow Coma Scale (GCS) score was 15.

C. MRS Method, Volume of Interest, and Comments about Examination

STEAM (TE = 30 ms, TM = 13.7 ms, TR = 1500 ms) was performed in HMRI's parietal white matter as shown on the image (Fig. 5.66).

D. White Matter Location

1. Peak Identification and Quality Assurance: Reading the Spectrum from Right to Left

This is a well-resolved, good quality spectrum, with good S/N, and no artifacts (Fig. 5.67). There is no lactate or significant lipids resonances. The NAA, β,γ-Glx, the second NAA, Cr, Cho, sI at 3.36 ppm, mI, α-Glx, and the second Cr are identified. The Cho peak is remarkable in that it is almost as high as Cr. No unusual resonances are observed.

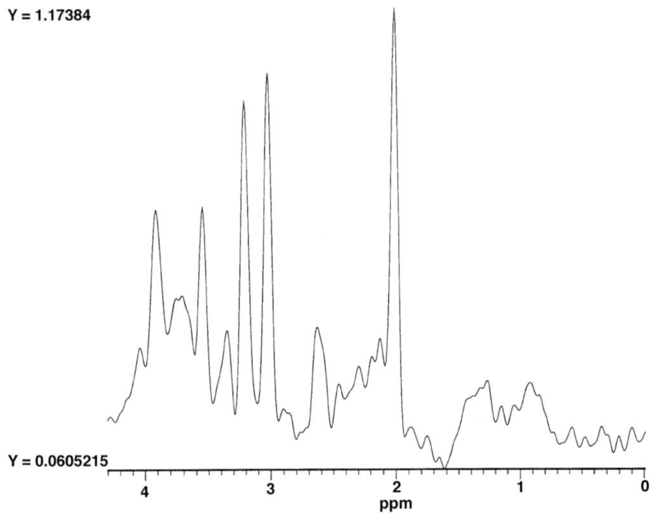

Fig. 5.67 The white matter MR spectrum is consistent with diffuse axonal injury in head trauma. The prognosis in this case is moderate to good outcome.

2. Interpretation and Analysis

The ratios are: NAA/Cr = 1.17 (reduced 15–20%); Cho/Cr = 0.93 (increased 16%); and mI/Cr = 0.63 (close to normal); and [Cr] is normal.

E. Conclusion

The reduced NAA is consistent with some neuronal loss, and the increased Cho is consistent with diffuse axonal injury after the closed head trauma. The prognosis is moderate to good because NAA is not severely reduced, Cr is not reduced, and because there is no evidence of hypoxic injury (no lactate or significant lipids present).

F. Summary of Patient's History Post-MRS

The patient recovered well. She has repossessed full mental capability, but is suffering from a number of motor deficits.

G. Other Comments

In this case low NAA is observed as in the majority of head injury cases, and high Cho alone (which is different for the hyperosmolar state with high mI, Cho and Cr) is observed. Cho is associated with myelin and membrane breakdown, and therefore believed to be consistent with diffuse axonal injury.

Note that Cr and mI are normal in this patient, hereby clearly differentiating the pattern of high Cho from the pattern of hyperosmolar state.

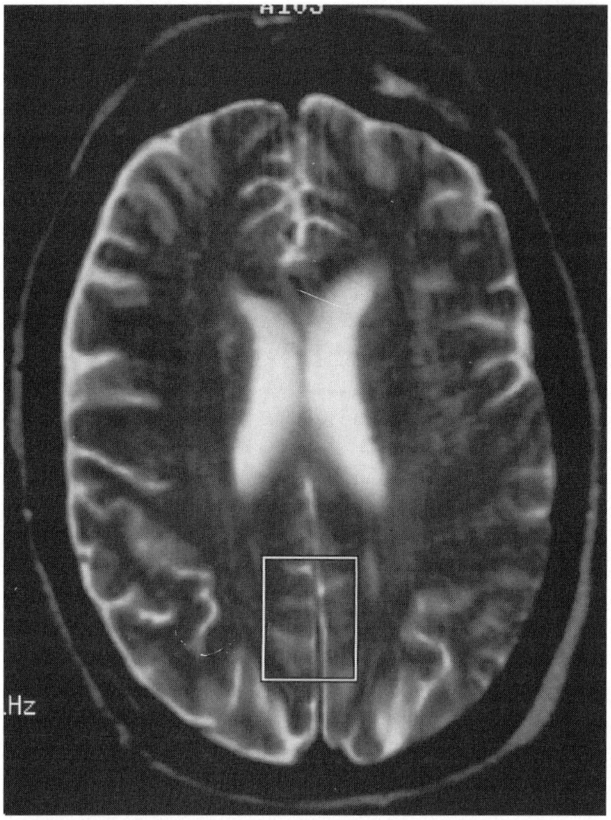

Fig. 5.68 A T_2-weighted MRI showing the location of the VOI. The T_2-weighted MRIs showed a multilobulated $3.5 \times 2.0 \times 2.0$ mass in the left temporal lobe, consistent with a hemorrhagic contusion (not shown). It would have been meaningless to place the MRS VOI in this mass, see discussion of VOI locations elsewhere. (Courtesy of Dr. Wang, Rancho Los Amigos Hospital, Downey, California.)

5.4.4 HEPATIC ENCEPHALOPATHY VERSUS LONG TERM EFFECT OF TRAUMA

A. Purpose of MRS

To rule out hepatic encephalopathy in a recovering trauma patient who also had a history of alcohol abuse.

B. Summary of Patient's History

Female, age 61 years, one month posttrauma (automobile versus pedestrian accident), resulting in left temporal subdural hematoma which was evacuated. She was intubated for a week because of airway obstruction. A history of alcohol abuse, but no hypertension, diabetes, or renal failure, three conditions that influence MRS.

C. MRS Method, Volume of Interest, and Comments about Examination

PROBE-STEAM (TE = 30 ms, TM = 13.7 ms, TR = 1500 ms) was performed in HMRI's occipital grey matter location (Fig. 5.68). The patient received 5 mg i.v. Valium 2× prior to the examination, but was nevertheless not very cooperative, and a second VOI was therefore not measured.

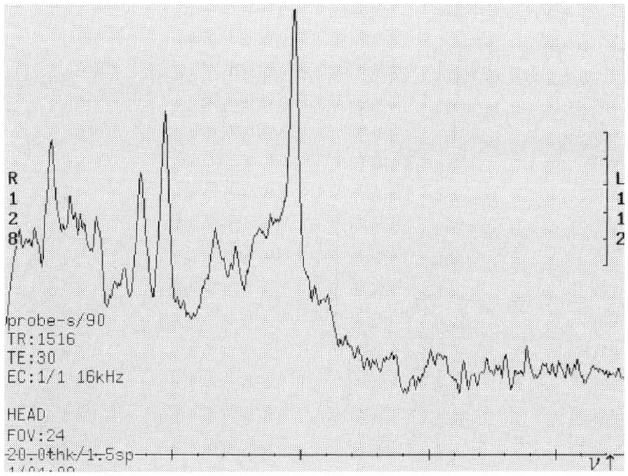

Fig. 5.69 The grey matter spectrum consistent with mild diffuse axonal injury post trauma, but hepatic encephalopathy cannot be completely ruled out. (Courtesy of Dr. Wang, Rancho Los Amigos Hospital, Downey, California.)

D. Grey Matter Location

1. Peak Identification and Quality Assurance: Reading the Spectrum from Right to Left

The quality of the spectrum (Fig. 5.69) is surprisingly good considering how uncooperative the patient was. The S/N is good, the peaks narrow, and there are no artifacts. The baseline slopes smoothly up towards the left because of the residual water signal.

Reading the spectrum from right to left, no lipids or lactate are seen, so the first significant peak is NAA, followed by β,γ-Glx, the second NAA, Cr, Cho, sI, a hint of glucose, a somewhat small mI, α-Glx, and the second Cr.

2. Interpretation and Analysis

The estimated metabolite ratios are: NAA/Cr = 1.5 (not decreased); Cho/Cr = 0.7 (increased about 15% compared to normal grey matter); and mI/Cr = 0.4 (about 35% decreased); note that the baseline has been accounted for. Glx/Cr does not appear increased. The NAA/Cr ratio is high, so it is unlikely that [Cr] is decreased. Furthermore, increased [Cr] is rarely seen unless [mI] and [Cho] are both increased relatively more than [Cr] (hyper osmolar), which is not the case here, so it is likely that [Cr] is close to normal.

The question was whether this was long term effects of trauma or hepatic encephalopathy. Let us review what can be expected for each of the two. In a relatively well functioning patient post trauma, mild versions of two of the trauma patterns can be expected; the "low NAA with no other changes" pattern, and the "low NAA with high Cho" pattern (for details see the previous case). In hepatic encephalopathy, decreased or absent mI is expected, and in some cases accompanied by decreased Cho, especially in white matter, and if the disease is severe, is also accompanied by increased Glx (caused by excess glutamine). Decreased NAA can only be expected in terminal stages of hepatic encephalopathy.

E. Conclusion

The increased Cho/Cr is consistent with diffuse axonal injury posttrauma, but since NAA is surprisingly unaffected, the injury must have been mild. There are no signs of severe post traumatic injuries, or of dementia. Hepatic encephalopathy can on the other hand not be completely ruled out, considering the very low mI/Cr.

F. Other Comments

A white matter VOI in HMRI's standard location would have been more helpful, and should have been done as the first priority.

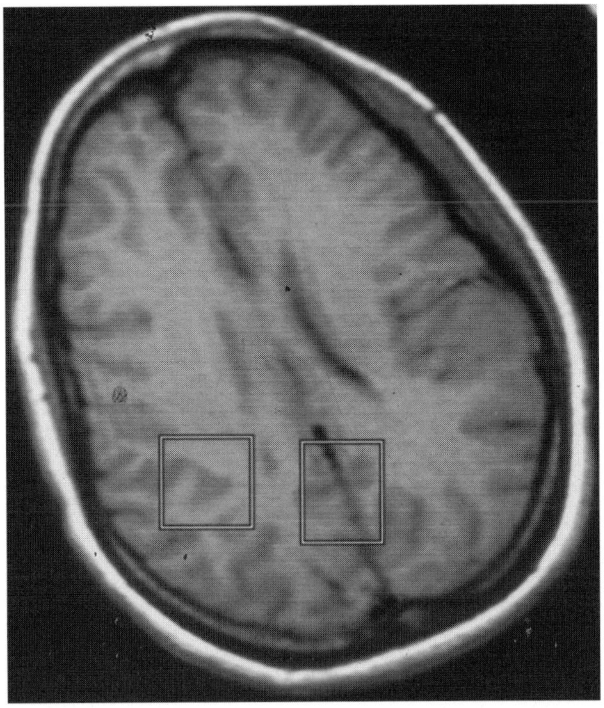

Fig. 5.70 T_1 weighted MRI showing focal hemorrhage and their avoidance in the MRS locations selected in grey and white matter.

5.4.5 HEAD INJURY: DIFFUSE AXONAL INJURY

A. Purpose of MRS

MRS was requested to evaluate outcome after severe, closed head trauma.

B. Summary of Patient's History

Female, age 29 years, who was in an auto accident. The patient was restless but comatose on arrival at the MR Center. One clinical evaluation gave a Glasgow Coma Scale (GCS) of 7 on the day of the MRS examination, 5 days after the accident. The clinical neurosurgical prognosis was described as, "very poor, survival not expected." GCS score was > 6 (not comatose); the doctor scored her 1-5-1 on the day of MRS, five days postinjury. However, stated "comatose for two weeks." Clinical evaluation was therefore somewhat controversial and inconclusive.

C. MRS Method, Volume of Interest, and Comments about Examination

STEAM (TE = 30 ms, TM = 13.7 ms, TR = 1500 ms). HMRI's occipito-parietal white matter VOI (Location 1) and HMRI's grey matter VOI (Location 2) (Fig. 5.70). The patient was difficult to keep perfectly motionless during her exam.

D. Location 1 (White Matter)

1. Peak Identification and Quality Assurance: Reading the Spectrum from Right to Left

This is a well resolved spectrum of acceptable quality (Fig. 5.71). Reading from the right to the left possible lipids or macro molecules are seen at 0.9 ppm and 1.4–1.5 ppm. This signal is likely to be a true lipid signal from the VOI, but an artifact cannot totally be ruled out, due to the restlessness of the patient. There is definitely no excess of lactate.

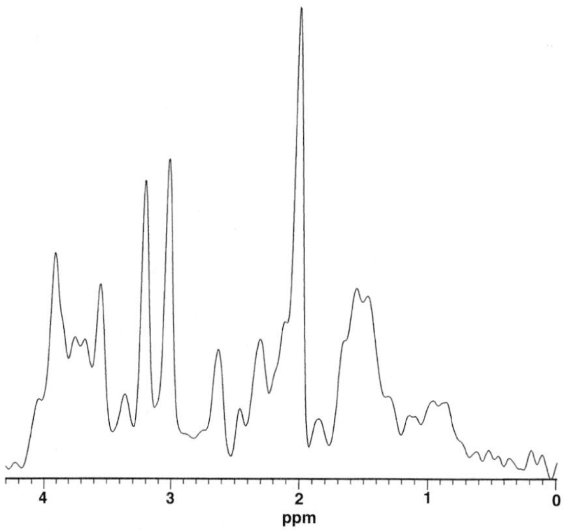

Fig. 5.71 The white matter MRS spectrum consistent with mild diffuse axonal injury in head trauma. The prognosis in this case is good outcome. Notice that despite presence of hemorrhages grey and white matter MRS locations were easily chosen avoiding overlap with any of these focal lesions.

Continuing to read from right to left, the following resonances are identified: NAA, β,γ-Glx, the second NAA, Cr, Cho, mI, α-Glx, and the second Cr.

2. Interpretation and Analysis

The ratios are NAA/Cr = 1.48 (normal range), Cho/Cr = 0.94 (15% above normal) and mI/Cr = 0.60 (normal), and [Cr] is about 95% of the normal value.

E. Location 2 (Grey Matter)

1. Peak Identification and Quality Assurance: Reading the Spectrum from Right to Left

The spectrum is of acceptable diagnostic quality, despite the presence of a small artifact appearing as a dip on the right hand side of NAA around 1.9 ppm (Fig. 5.72). This type of artifact is sometimes seen when lipid from outside the VOI is interfering. This can happen if the VOI is placed too close to the skull or if the patient is not keeping perfectly still. The effect of this small artifact is that it may artificially make the NAA peak smaller, so when analyzing this spectrum an underestimation of NAA can be expected.

Reading from right to left it is again noted that there is no excess of lactate; furthermore the following resonances are identified: NAA, β,γ-Glx, the second NAA, Cr, Cho, mI, α-Glx, and the second Cr.

2. Interpretation and Analysis

The ratios are NAA/Cr = 1.24 (reduced, but not more than can be explained by the "dip" artifact), Cho/Cr = 0.74 (15% above normal) and mI/Cr = 0.57 (normal), and [Cr] is 85% of the normal value. Thus the ratios are probably slightly overestimated, Cho is therefore likely to be closer to normal than the Cho/Cr ratio indicates.

F. Conclusion

Unexpectedly, good MRS result. There is no significant loss of the neuronal marker NAA. There is no lactate or significant lipid and therefore no evidence of hypoxic encephalopathy. The small increase of Cho is consistent with diffuse axonal injury, and the possible lipid signal in white matter could be consistent with the associated membrane breakdown.

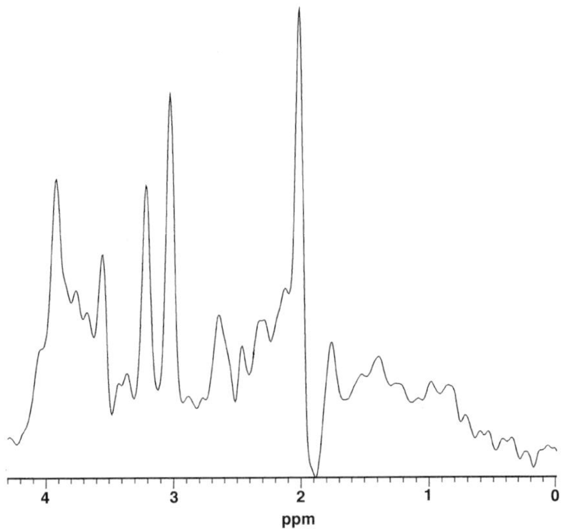

Fig. 5.72 The grey matter MRS spectrum confirming the good MRS results seen in white matter.

Altogether, the spectra are not too abnormal, and the prognosis for outcome should therefore be considerably more optimistic; a good outcome was expected.

G. Summary of Patient's History Post-MRS

The patient recovered relatively well, but minor deficits remain. She lives with family, and is able to manage most daily tasks, but a year after the injury she still has problems with concentration.

H. Other Comments

This is one of the cases that should be of great encouragement for the use of MRS in comatose head injury patients. This patient was virtually not left much hope for survival by the clinical diagnosis, but MRS changed that view: prognosis good! Notice that two VOIs were measured. In this case it was of particular importance to confirm the result which was so surprisingly good.

A mild increase of Cho (high Cho alone, which is different for the hyperosmolar state with high mI, Cho and Cr) was observed which is associated with myelin and membrane breakdown and therefore believed to be consistent with diffuse axonal injury. This pattern of high Cho is just one of the patterns that can be seen in closed head trauma. Others are the hyperosmolar state, the hypo-osmolar state = SIADH, hypoxic injury, and those with neuronal injury but none of the other abnormalities. The different patterns are viewed as different biochemical responses to the trauma.

Common pitfalls include: 1) Prognosis based on MRS results alone, without CT or MRI findings, may underestimate severity of injury. 2) VOI within the visible hemorrhage may fail to provide useful clinical information.

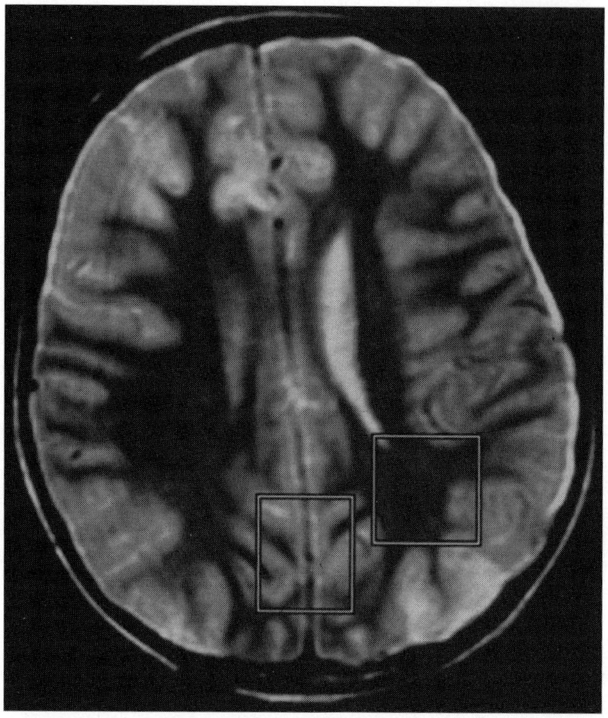

Fig. 5.73 T_2-weighted MRI showing the grey and white matter MRS locations measured 5 days postinjury.

5.4.6 HEAD INJURY: HYPOXIC INJURY

A. Purpose of MRS

Evaluation of outcome prognosis in a comatose head injury patient.

B. Summary of Patient's History

Male, age 8 years, who was struck by an automobile. He was admitted to the hospital with multiple injuries, and a Glasgow Coma Scale score of 4.

C. MRS Method, Volume of Interest, and Comments about Examination

STEAM (TE = 30 ms, TM = 13.7 ms, TR = 1500 ms) was performed in HMRI's parietal white matter (Locations 1 and 3), and grey matter (Locations 2 and 4). VOIs were measured five days (Fig. 5.73), and 4½ months (Fig. 5.74) after the accident. The first examination was performed as soon as it was practically possible, and at that time the Glasgow Coma Scale score was 4. At the time of the second MRS, the GCS score was 6.

D. Examination 1: Location 1 (White Matter)

1. Peak Identification and Quality Assurance: Reading the Spectrum from Right to Left

The technical quality of the spectrum (Fig. 5.75) is very good, but it is immediately clear to the trained eye that the spectrum is abnormal, and that the result is not promising for the patient's outcome.

Reading the spectrum systematically from right to left, a little lipid is seen at 0.9–1.0 ppm, and a minor contribution from lipid is probably also underlying the prominent lactate doublet at 1.33 ppm. Continuing reading, the following resonances are

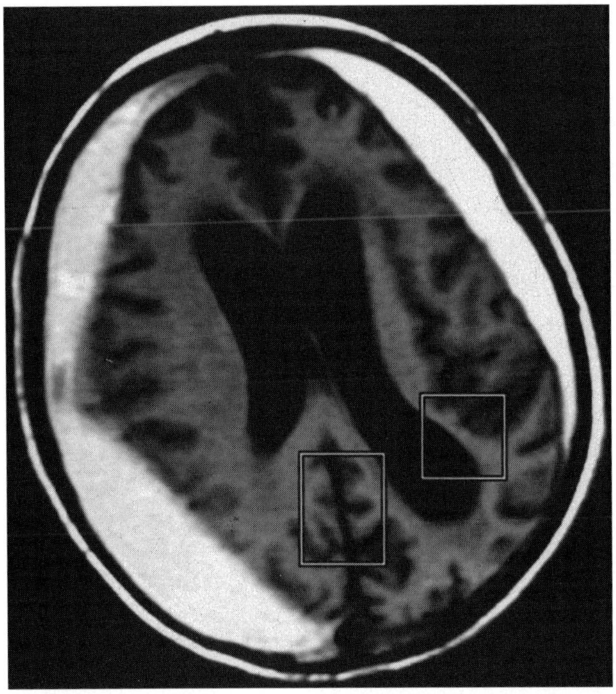

Fig. 5.74 T_1-weighted MRI showing the same MRS locations measured at the second examination 4½ months postinjury. Severe MRI changes are observed.

identified: NAA, β,γ-Glx, the second NAA, Cr, Cho, mI, α-Glx, and the second Cr. NAA appears reduced, and Glx appears increased.

2. Interpretation and Analysis

The metabolite ratios are: NAA/Cr = 1.14 (20% reduced); Cho/Cr = 0.84 (normal); mI/Cr = 0.57 (low end of normal); and [Cr] is about 15% below the normal value. β,γ-Glx/Cr = 0.38, and α-Glx/Cr = 0.68, thus Glx/Cr is about 45% elevated. Accounting for the [Cr] being reduced, [NAA] is reduced more than the ratio indicates, and [Glx] is increased, but probably only about 20%.

E. Examination 1: Location 2 (Grey Matter)

1. Peak Identification and Quality Assurance: Reading the Spectrum from Right to Left

The grey matter spectrum (Fig. 5.76) is also of excellent quality. The abnormalities found in the white matter are immediately recognized, but the grey matter is even more severe.

Reading the spectrum systematically from right to left, more pronounced lactate and lipid are seen, followed by a very reduced NAA, significant appearing β,γ-Glx, the second NAA, Cr, Cho, mI, α-Glx, and the second Cr.

The second Cr peak is supposed to be ⅔ the size of the first Cr peak, due to the number of contributing protons. Comparing the two Cr peaks in this spectrum, the second peak may appear a little too small. In the previous white matter spectrum the second peak appears a little too large. However, looking closely it becomes evident that these small deviations from the ratio of 2/3 between the two Cr peaks are due to minor variations in the baseline this close to the residual water at 4.7 ppm. It is also clear that the technical quality of the spectrum is not compromised even with respect to α-Glx, very close to the second Cr peak.

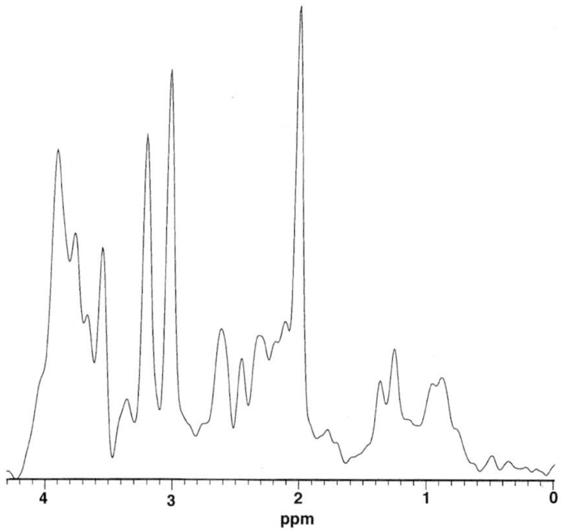

Fig. 5.75 The white matter spectrum measured 5 days after the accident, is consistent with hypoxic injury.

2. Interpretation and Analysis

The ratios are: NAA/Cr = 0.81 (40% reduced); Cho/Cr = 0.68 (15% increased); mI/Cr = 0.51 (10% reduced); and [Cr] is about 15% reduced. β,γ-Glx/Cr = 0.42, and α-Glx/Cr = 0.51, thus Glx/Cr is at the most 20% increased, but accounting for the decreased [Cr], [Glx] is not very much increased in this location. The fraction of CSF in the VOI was measured to be 3%, which is normal.

F. Summary of Examination 1

Excess of lactate and lipid, severely reduced NAA, reduced Cr, and some increase of Glx are all factors consistent with severe hypoxic encephalopathy post trauma. The prognosis is poor.

G. Examination 2: Location 3 (White Matter)

1. Peak Identification and Quality Assurance: Reading the Spectrum from Right to Left

The spectral quality is excellent, except for a small dip and possible artifact at 1.55 ppm (Fig. 5.77). The dip does not compromise the diagnostic value of the spectrum, and it is most likely caused by a very small lipid signal from outside the VOI.

Reading the spectrum from right to left, the lactate doublet is clearly seen at 1.33 ppm. Notice that the splitting of the two peaks of the doublet is almost down to the baseline, which is expected from a well-resolved spectrum as this one. Notice that two peaks are separated by 7 Hz (= 0.11 ppm at 1.5 Tesla). After lactate follows NAA, unremarkable β,γ-Glx, and the second NAA. Cr and Cho are followed by a small somewhat broader peak at 3.4 ppm, which is most likely due to overlapping sI and glucose. The mI is followed by α-Glx, the second Cr, and the second mI.

2. Interpretation and Analysis

The metabolite ratios are: NAA/Cr = 0.96 (28% reduced); Cho/Cr = 1.14 (37% increased); mI/Cr = 1.09 (79% increased); and [Cr] = 0.82 is about 18% reduced. The β,γ-Glx/Cr = 0.25, and the α-Glx/Cr = 0.51, thus Glx is not significantly different from normal. Considering both NAA/Cr = 0.96 and [Cr] = 0.82, thus [NAA] is about half of normal.

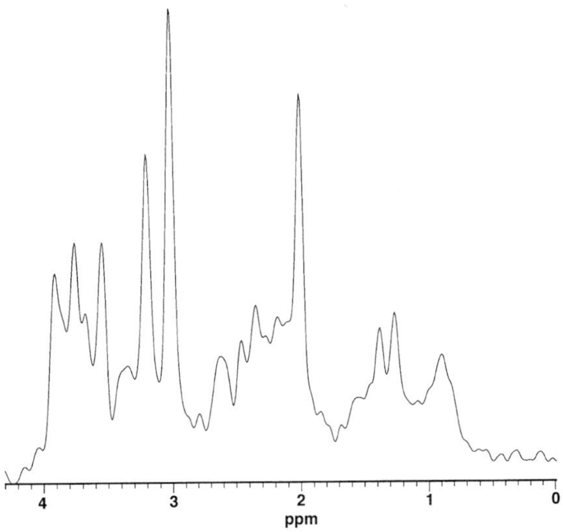

Fig. 5.76 The grey matter spectrum 5 days postinjury, the hypoxic changes are more pronounced in the grey matter spectrum. Both spectra of this first examination signifies a poor prognosis for the patient outcome.

H. Examination 2: Location 4 (Grey Matter)

1. Peak Identification and Quality Assurance: Reading the Spectrum from Right to Left

This is a well-resolved, high quality spectrum (Fig. 5.78). The spectrum is clearly abnormal with the larger peaks to the left side of the spectrum, and with the presence of lactate.

Reading the spectrum from right to left, a broad signal assigned to lipid is found around 0.9 ppm; a small lipid signal also underlies the lactate doublet at 1.33 ppm. The lactate doublet is not fully symmetrical due to underlying signal, but there is no doubt about the identification of lactate. Continuing reading, the following resonances are identified: NAA, β,γ-Glx, the second NAA, Cr, Cho, partially overlapping sI and glucose, mI, α-Glx, and the second Cr.

2. Interpretation and Analysis

The ratios are: NAA/Cr = 0.76 (42% below normal); Cho/Cr = 1.02 (23% increased); mI/Cr = 1.18 (double the normal); [Cr] = 0.74 (74% of the normal value). Considering the low [Cr] and the NAA/Cr ratio, [NAA] is only about 40% of normal. The Glx is not increased: β,γ-Glx/Cr = 0.22 (low) and α-Glx/Cr = 0.48 (normal). The fraction of CSF in the VOI was measured to be 11%, which is increased and reflects the general atrophy.

I. Conclusion

Marked further deterioration and expected long-term effects of severe hypoxic damage posttrauma are observed: severely decreased [NAA] reflecting neuronal loss; increased Cho and mI, perhaps indicative of gliosis; persistent presence of lactate; and increasing atrophy. The neurological outlook is poor.

J. Summary of Patient's History Post-MRS

The outcome is poor, as described. When he was readmitted to the hospital to treat chronic vomiting, dehydration, and gastritis, 1½ years after the accident, the boy was not comatose, but was mentally retarded, suffering frequent seizures, and had to be institutionalized.

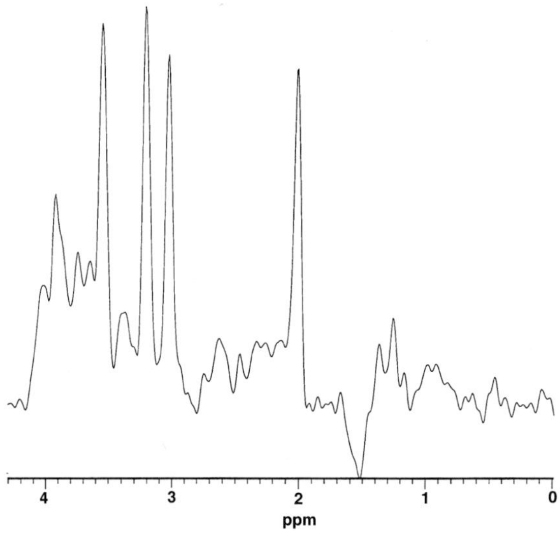

Fig. 5.77 The white matter spectrum 4½ months postinjury. Significant changes since the first MRS are noted: further reduced NAA, normalized Glx, reduced Cr, and increased Cho and mI. Lactate persists.

K. Other Comments

This patient was eight years old. For simplicity, we usually apply age correcting calculations only in infants, and from 1½ years of age we use the normal adult values as a reference. (Keep in mind that in adolescents NAA is about 10% higher than in average adults, and in elderly the NAA normally decreases 5–10%, and Cho may increase a little; see Chapter 2.) In the current patient, this observation will tend to make the picture slightly grimmer.

The presence of lactate, and often also lipid, is one of the MRS patterns seen in closed head trauma; it is consistent with hypoxic encephalopathy, and poor outcome. Other patterns are the hyperosmolar state, the hypo-osmolar state (SIADH), high Cho consistent with diffuse axonal injury, and those with neuronal injury, but none of the other abnormalities. The different patterns are viewed as different biochemical responses to the trauma.

It has been advised (see other trauma cases) to measure white matter as the first choice in trauma cases, and measure grey matter, if feasible. The reason is that most possible features of trauma are readily detected in the white matter location, and some of the above mentioned patterns can better be distinguished in white matter. Hypoxic encephalopathy is the exception. Grey matter is more sensitive to hypoxic encephalopathy, but rarely would it not be recognized in white matter, too. The white matter VOI is therefore still the best first choice unless hypoxic encephalopathy for some reason is already part of the picture.

1. Glutamine Versus Lactate as a Marker of Acute Hypoxic Injury?

In the second MRS examination, the NAA was further reduced, and Cho and mI were further increased, all of which can be understood in the context of neuronal injury (reduced NAA), and membrane breakdown (increased Cho and mI). Glx was mildly increased in the first MRS examination, consistent with hypoxic encephalopathy, but over the long-term Glx recovers towards normal, as seen in near-drowning cases. This contrasts with long persistence of lactate in such cases.

2. Impact of Cortical Atrophy on MRS

Notice that while the atrophy changes on the MRI of the second examination are not surprising, they should be considered as a consequence of the severe neuronal damage, which was already seen on the first MRS examination. The MRS is itself predictive of MRI-visible atrophy, and can be of great clinical value, preceding it by up to two weeks. The severe atrophy

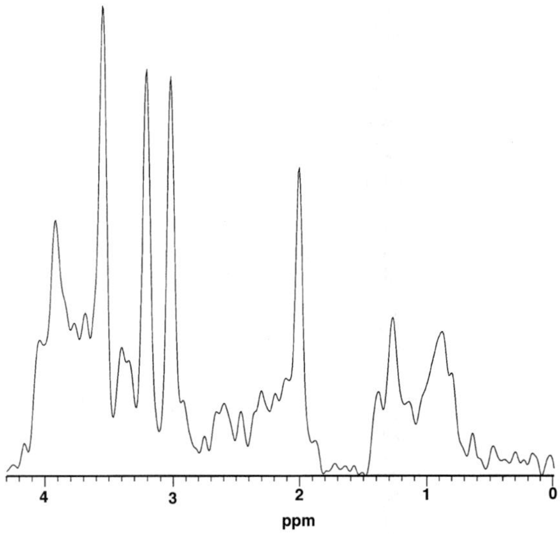

Fig. 5.78 The grey matter spectrum 4½ months postinjury.

made it difficult to totally avoid CSF in the VOIs of the second examination, which is not a problem when it is recognized and corrected for in the measurement of [Cr]. Glucose and lactate are the only metabolites found in CSF, probably explaining the hint of glucose which was found overlapping with sI at 3.4 ppm, but it was so little that the second resonance at 3.8 ppm was not recognized in either spectra of the second examination. The contribution of CSF lactate to the dominant lactate doublets of these spectra are correspondingly small and unimportant.

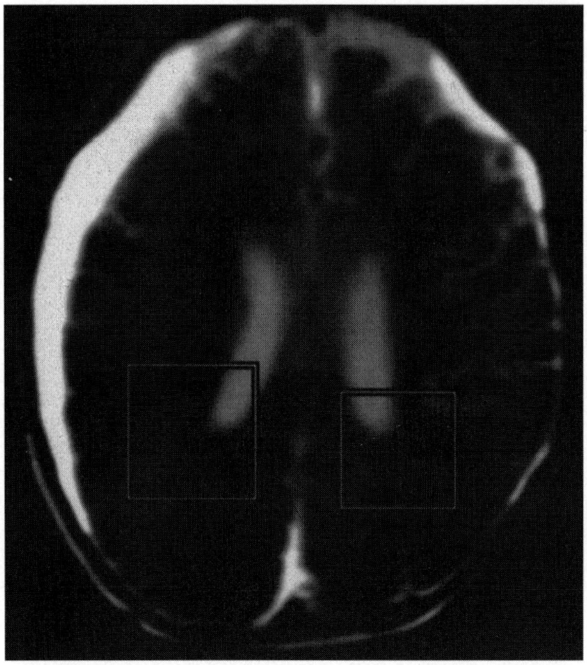

Fig. 5.79 Fast spin echo T_2-weighted image showing subdural hematoma (right), and diffuse changes. Location of white matter MRS was measured 10 days postinjury.

5.4.7 HEAD INJURY, POOR OUTCOME

A. Purpose of MRS

To determine the prognosis in a comatose infant.

B. Summary of Patient's History

Patient was a female, age 4 months. She was struck by an automobile while in a stroller, and admitted to the hospital with Glasgow Coma Scale 3, and increasing Glasgow Coma Scale 5–6; persistent coma at 10 days.

C. MRS Method, Volume of Interest, and Comments about Examination

Localized STEAM (TE = 30 ms, TM = 13.7 ms, TR = 1500 ms) was performed in HMRI's standard parietal white matter left (Location 1), and standard parietal white matter right (Location 2). VOI's were measured 10 days after the accident.

D. Location 1 (White Matter Left)

1. Peak Identification and Quality Assurance: Reading the Spectrum from Right to Left

The spectrum is of poor technical quality (not shown). Excess lipid and lactate are obvious, but no other resolvable peaks are identified.

2. Interpretation and Analysis

Evident loss of normal brain structure has compromised MRS.

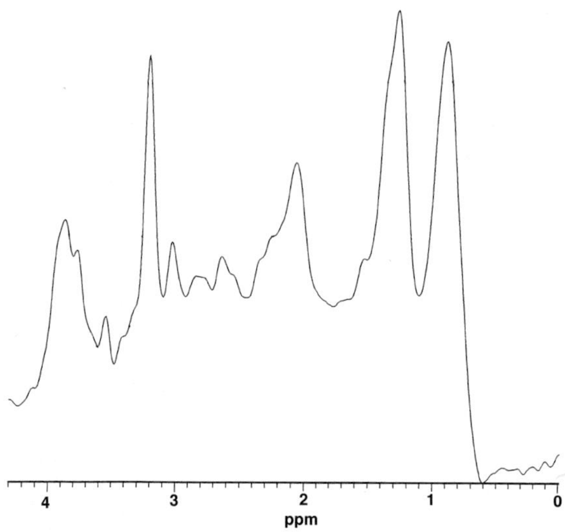

Fig. 5.80 The white matter spectrum from the right hemisphere in the infant 10 days postinjury is typical of irreversible brain trauma.

E. Location 2 (White Matter Right)

1. Peak Identification and Quality Assurance: Reading the Spectrum from Right to Left

The spectrum is from the less severely affected hemisphere shown on the MRI (Fig. 5.79). The technical quality (broad lines well resolved) is affected by the injury, but is obviously very abnormal (Fig. 5.80).

Reading the spectrum from right to left, it is dominated by two huge, and equally intense lipid peaks. The presence of lactate cannot be excluded. Continuing reading from right to left, there is no convincing NAA peak, but β,γ-Glx may be present. A peak at 2.6 ppm may be aspartate; peak at 2.75 ppm is probably insignificant (unidentified). Cr, Cho, mI, α-Glx (dubious), and no second Cr or mI. NAA is absent and total Cr barely remains. Cho is the only convincing resonance other than the lipids.

2. Interpretation and Analysis

Metabolite ratios are meaningless because [Cr] is $< 20\%$. There is a great excess of infantile pattern "lipid" and apparently elevated Cho is probably at or below normal, even though Cho/Cr is 1.9.

F. Conclusion

Excess lipid, complete loss of NAA and almost complete loss of Cr are diagnostic of severe, irreversible brain injury after trauma in very young infants ($<$ 1 year old). The prognosis determined by MRS was grim.

G. Summary of Patient's History Post-MRS

As predicted by MRS, the child died 4 days after MRS examination.

H. Other Comments

MRS performed early in the course of pediatric head injury can be very informative. Metabolite changes occur over the course of hours, and is independent of MRI appearances so that prognosis is usually straightforward.

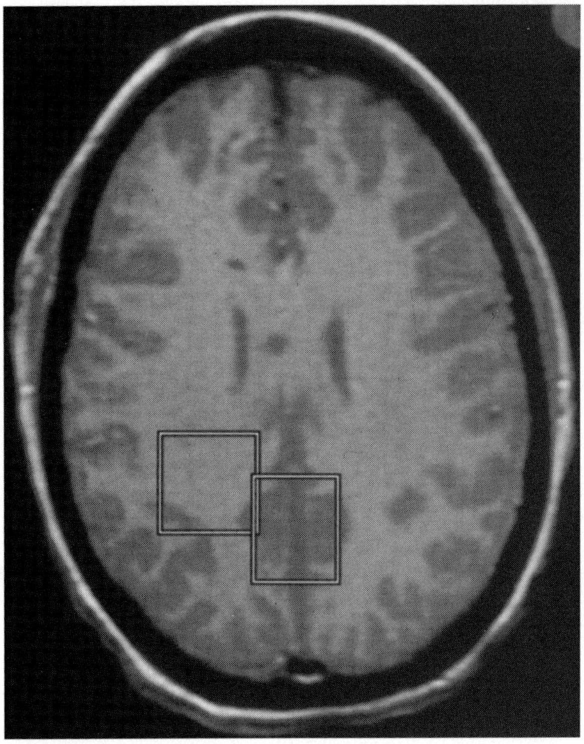

Fig. 5.81 T_1-weighted MRI showing the MRS locations in grey and white matter.

5.4.8 HEAD INJURY, GOOD OUTCOME

A. Purpose of MRS

MRS was requested to evaluate a comatose trauma patient.

B. Summary of Patient's History

Female, age 21 years, who was in an auto accident. The patient was on a ventilator and comatose with a Glasgow Coma Scale score of 3 on the day of the MRS examination, five days after the accident.

C. MRS Method, Volume of Interest, and Comments about Examination

STEAM (TE = 30 ms, TM = 13.7 ms, TR = 1500 ms) was performed in HMRI's parietal white matter VOI (Location 1), and occipital grey matter VOI (Location 2), as seen on the image (Fig. 5.81). The examination was challenging both because a metal staple was hidden in a mass of hair, and because the patient was somewhat restless. Because of these complications, it took more than one attempt to get a reasonable spectrum, and data was not acquired for calculating absolute concentrations. Time constraints and patient movement limited the achieved number of averages to 64 for the white matter, and 96 for the grey matter spectrum.

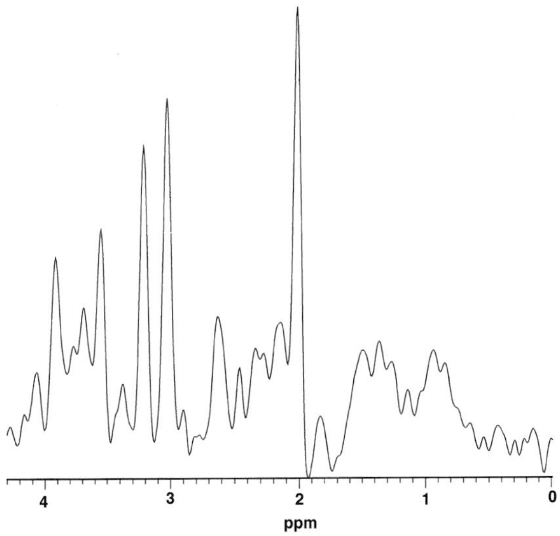

Fig. 5.82 The white matter MRS spectrum from a trauma patient with Glasgow Coma Scale score of 3. The spectrum is consistent with good outcome.

D. Location 1 (White Matter)

1. Peak Identification and Quality Assurance: Reading the Spectrum from Right to Left

This is a spectrum of acceptable prognostic quality despite the fact that only 64 averages were acquired (Fig. 5.82). Reading from right to the left, there is no excess of lactate or lipid. Continuing, the following resonances are identified: NAA, β,γ-Glx, the second NAA, Cr, Cho, sI at 3.36 ppm, mI, α-Glx, and the second Cr, as well as a second mI.

2. Interpretation and Analysis

The metabolites ratios are: NAA/Cr = 1.25 (reduced 15–20%); Cho/Cr = 0.88 (7% increased); mI/Cr = 0.65 (8% increased); [Cr] is not available.

E. Location 2 (Grey Matter)

1. Peak Identification and Quality Assurance: Reading the Spectrum from Right to Left

The grey matter spectrum is also of acceptable diagnostic quality (Fig. 5.83). Reading from right to left it is again noted that there is no excess of lactate or lipid. Furthermore, the following resonances are identified: NAA, β,γ-Glx, the second NAA, Cr, Cho, mI, α-Glx, and the second Cr.

2. Interpretation and Analysis

The ratios are: NAA/Cr = 1.13 (reduced 18%); Cho/Cr = 0.64 (normal range); mI/Cr = 0.57 (normal range); [Cr] is not available.

F. Conclusion

Unexpectedly good MRS result. The only significant abnormality is a 15–20% loss of the neuronal marker NAA. Little evidence of diffuse axonal injury, and no evidence of hypoxic injury. All together, the spectra are surprisingly near normal, and the prognosis for outcome was good.

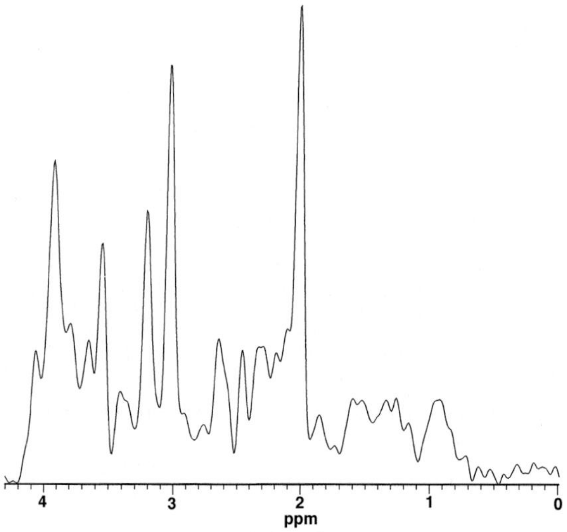

Fig. 5.83 The grey matter MRS spectrum confirms the good MRS result seen in white matter.

G. Summary of Patient's History Post-MRS

The patient showed excellent progress, and recovered well.

H. Other Comments

This case alone should be sufficient motivation to perform MRS in all comatose head injury patients. The patient was in a coma with a Glasgow Coma Scale score as low as 3, and a correspondingly poor prognosis. MRS demonstrated that no severe brain damage had occurred, and good outcome was predicted.

It is noteworthy that a clinically valuable result was obtained despite practical difficulties such as patient movement, and severe time constraints. The availability of ratios alone did not hinder the interpretation of the spectra, nor did the limited number of averages impair the prognostic value of the spectra. The clinical result could, if necessary, have been obtained from the white matter spectrum alone, which took less than two minutes to acquire.

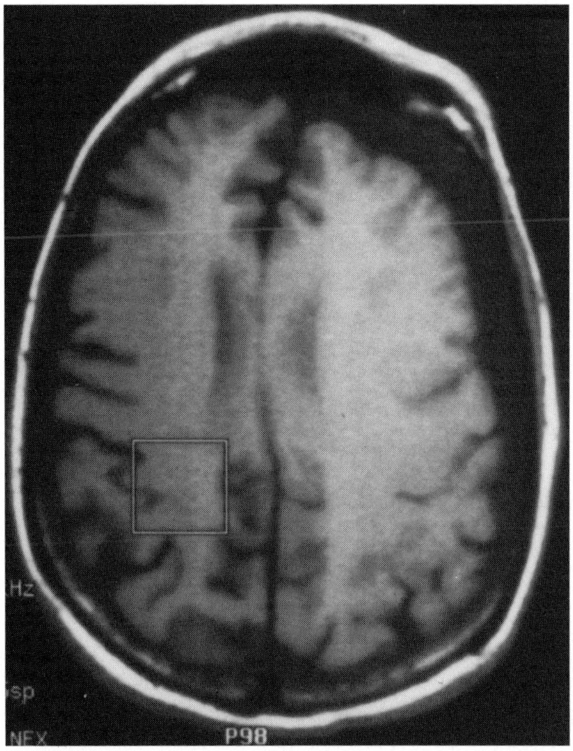

Fig. 5.84 T_1 weighted MRI showing the MRS location in white matter. (Courtesy of Dr. Wang, Rancho Los Amigos Hospital, Downey, California.)

5.4.9 HEAD INJURY—HYPEROSMOLAR STATE VERSUS AIDS DEMENTIA

A. Purpose of MRS

MRS was requested to obtain a prognosis for a patient, who suffered head injury 5 weeks earlier.

B. Summary of Patient's History

Male 41, with a history of AIDS and head trauma 5 weeks earlier. Abnormal CT, MRI showed changes consistent with hemorrhagic contusion and atrophy, no abnormalities that are likely to be related to the HIV infection.

C. MRS Method, Volume of Interest, and Comments about Examination

PROBE-S the automated version of STEAM (TE = 30 ms, TM = 13.7 ms, TR = 1500 ms). HMRI's normal white matter (Fig. 5.84).

D. White Matter Location

1. Peak Identification and Quality Assurance: Reading the Spectrum from Right to Left

This is a well resolved good quality spectrum, good S/N, and no artifacts, but the baseline slopes a little due to the residual water (Fig. 5.85). There is no detectable lactate but significant lipid resonances. The NAA, β,γ-Glx, the second NAA, Cr, Cho, a minor sI peak at 3.36 ppm, mI, α-Glx, and second Cr and an impressive second mI peaks are identified.

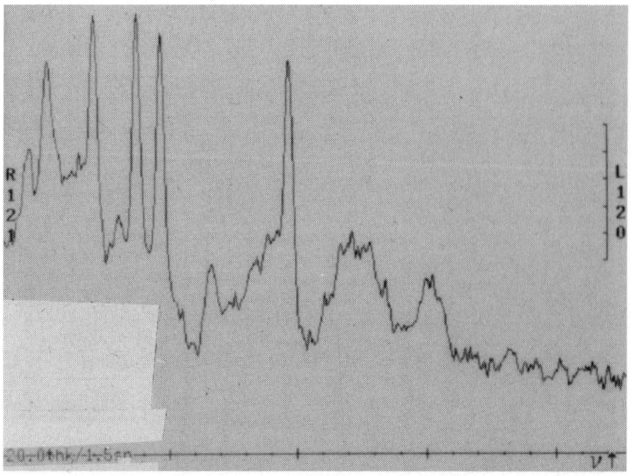

Fig. 5.85 The white matter MRS spectrum from a middle-aged male with head trauma and AIDS. (Courtesy of Dr. Wang, Rancho Los Amigos Hospital, Downey, California.)

2. Interpretation and Analysis

The ratios are estimated, accounting for the baseline, to be NAA/Cr = 1.0 (reduced about 30%), Cho/Cr = 1.0 (increased about 20%), and mI/Cr = 0.9 (increased about 50%). The absolute concentrations are not directly available. But knowing that the patient had a head injury and has AIDS opens two possibilities:

1. One of the patterns of head injury is the hyperosmolar pattern described in Case 5.4.2. It presents elevated mI, Cho, Cr, and even NAA, affecting mI and Cho most and NAA least, so that the mI/Cr and Cho/Cr ratios of about 1 and a decreased NAA/Cr ratio are consistent with the hyperosmolar pattern. If that is the case the prognosis will be good, expecting a recovery over a period of months.
2. On the other hand effects of the HIV infection should be considered. There is no focal lesion in or near the VOI, but if the patient has AIDS dementia, decreased NAA/Cr and increased mI/Cr and Cho/Cr is expected, yielding a less optimistic prognosis (see Case 5.7.5). Absolute quantitation would have shown increased [Cr] for hyperosmolar state, and normal [Cr] for AIDS dementia, and would have been very helpful. The small amount of excess lipids may be consistent both with head injury and AIDS changes in the brain.

E. Conclusion

There is no evidence of severe hypoxic injury, because there is no lactate in the spectra. The spectrum is consistent both with hyperosmolar state of trauma (long term recovery expected) and AIDS dementia (no recovery expected), the first being more likely.

F. Summary of Patient's History Post-MRS

Not available.

G. Other Comments

The spectrum is most likely to be that of hyperosmolar state, a response to trauma, since mI/Cr and Cho/Cr are more elevated than what is usually seen in AIDS dementia. But interpretation must be cautious when more than one disease is present.

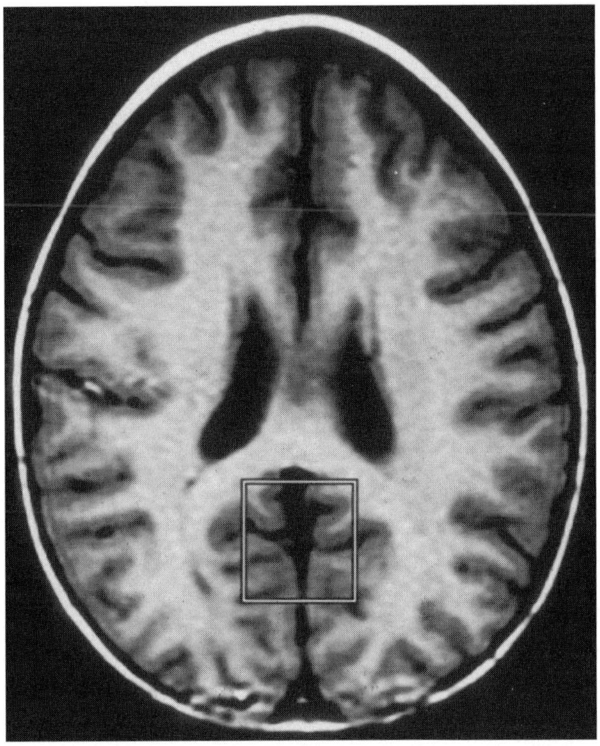

Fig. 5.86 Axial T_1-weighted MRI on day 3 after cardiac arrest during near-drowning shows no abnormal features. Locations of PROBE MRS are identified. Only the grey matter PROBE examination is discussed here.

5.5 Global Hypoxia: Near Drowning, Heart Disease, Asphyxia

5.5.1 NEAR-DROWNING: GOOD OUTCOME

A. Purpose of MRS

To determine prognosis in a comatose child after a near-drowning accident.

B. Summary of Patient's History

Male, age 3 years, day 3 after submersion, with Glasgow Coma Scale score of 6. The child was on ventilator during MRS (PROBE) examination.

C. MRS Method, Volume of Interest, and Comments about Examination

Automated PROBE-STEAM (TE = 30 ms, TM = 13.7 ms, TR = 1500 ms) was performed in HMRI's occipital grey matter (see Fig. 5.86) and parietal white matter (not shown). The examination was performed at a three-day interval after the injury, at which time MRS prognosis with no incidence of false positives is recorded, and no follow-up MRS was necessary.

D. Grey Matter Location

1. Peak Identification and Quality Assurance: Reading the Spectrum from Right to Left

The technical quality of the PROBE spectrum (Fig. 5.87) is excellent, with very satisfactory water suppression, and a small slope from right to left which is easily adjusted upon inspection to obtain metabolite peak ratios.

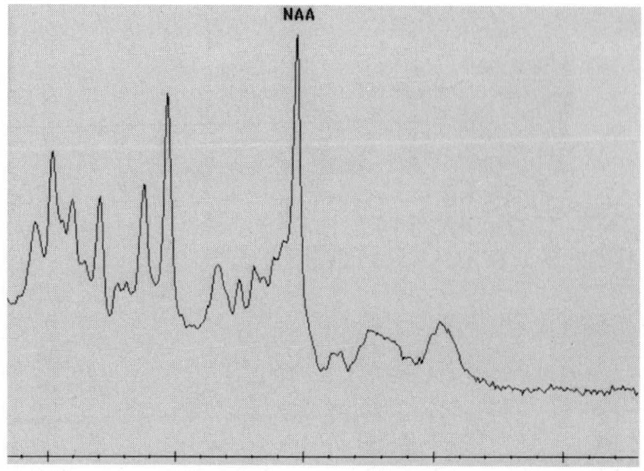

Fig. 5.87 An excellent PROBE examination correctly predictive of good outcome in a comatose patient on life-support, three days after near-drowning. Note that significant abnormalities identified in this PROBE examination are consistent with mild, but recoverable hypoxic injury.

Reading the spectrum systematically from right to left, there is a small excess of lipid at 0.9 ppm and at 1.4 ppm, and possibly 1.8 ppm, with no evidence of a lactate doublet at 1.33 ppm. Further peaks are identified for NAA, β,γ-Glx, second and third NAA peaks, Cr, Cho, sI, glucose, mI, the α-Glx triplet, and second peaks for Cr and mI. The overall appearance of this PROBE spectrum is appropriate for age, and indicates good outcome.

2. Interpretation and Analysis

The metabolite ratios are: NAA/Cr = 1.25 (approximately 10% below three year-old normal); Cho/Cr = 0.52 (normal, implying normal [Cr] in this patient); mI/Cr = 0.42 (normal). β,γ-Glx is slightly elevated, based upon the proportions of $NAA_3/NAA_2 \geqslant 0.5$, and elevated α-Glx appears to confirm this finding. (For a detailed explanation of NAA_3/NAA_2 see p. 49.)

E. Conclusion

Using prognostic features established in more than 20 near-drowning victims, the abnormalities, which are readily detectable on day 3 after hypoxic injury, namely 9% reduction in NAA, a small excess of Glx, but not of lactate, and a small excess of lipid, are all consistent with virtually full neurological recovery. The changes in grey matter occur earlier and are more pronounced than in white matter.

F. Summary of the Patient's History Post-MRS

The patient was successfully weaned from the ventilator and made an uneventful recovery to discharge on day 10. No long-term neurological evaluation is available. However, clinically it is generally accepted that "good" outcome after near-drowning equates with successful neurological development in children.

G. Other Comments

We recommend performing MRS on days 1 and 3 after a near-drowning (or similar acute anoxic episode) accident, but when it is not feasible as in this case, MRS on day 3 or later is still very helpful. Day 2 postinjury is the earliest day an indication about outcome can be obtained with certainty, and the result can be verified on days 3 or 4. Local practice in Intensive Care Units often determines when MRS is most likely to contribute to management in this highly charged clinical emergency.

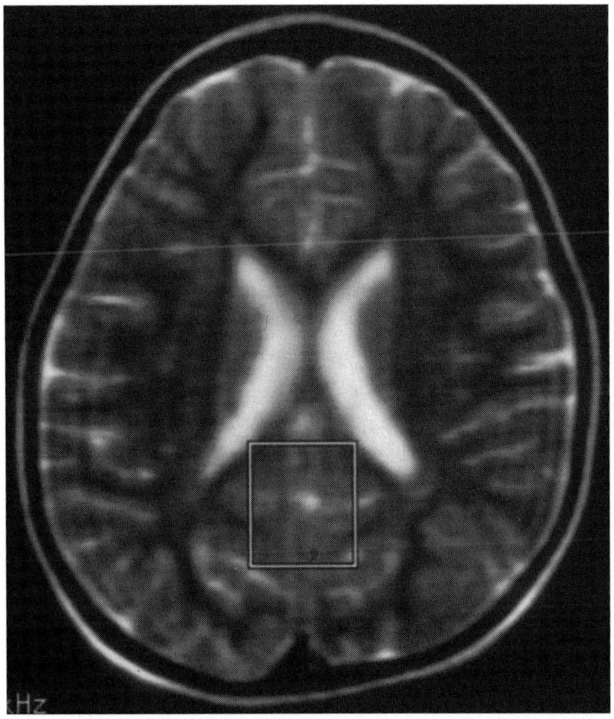

Fig. 5.88 Fast spin echo T_2-weighted axial MRI showing grey matter location of MRS, two days after near-drowning. The MRI is typically uninformative after near-drowning, as in this case.

5.5.2 NEAR-DROWNING

A. Purpose of MRS

Outcome prediction after near-drowning accident.

B. Summary of Patient's History

Female, age 2, submerged unattended for 10 minutes in bathtub. She was admitted to intensive care with Glasgow Coma Scale score 3. She was on dopamine and ventilator support 48 hours later at the time of the first MRS when Glasgow Coma Scale was still equal to 3. MRS was repeated on day 4, and on day 18, after which life support was discontinued.

C. MRS Method, Volume of Interest, and Comments about Examination

Localized STEAM (TE = 30 ms, TM = 13.7 ms, TR = 1500 ms) was performed in HMRI's occipital grey matter (Examinations 1, 2 and 3). VOIs were measured 2 days (Fig. 5.88), 4 days (Fig. 5.89), and 18 days (Fig. 5.90) after the accident. The first examination was performed at 48 hours, because outcome can often be determined with a single examination, and subsequent examinations at day 4 and day 18 to document the poor clinical progress.

D. Examination 1: Two Days After Accident (Grey Matter Location)

1. Peak Identification and Quality Assurance: Reading the Spectrum from Right to Left

Good quality spectrum (Fig. 5.91) apart from possible artifacts in lipid/lactate region and a "dip" in the baseline at 3.5 ppm attributable to eddy-current problem.

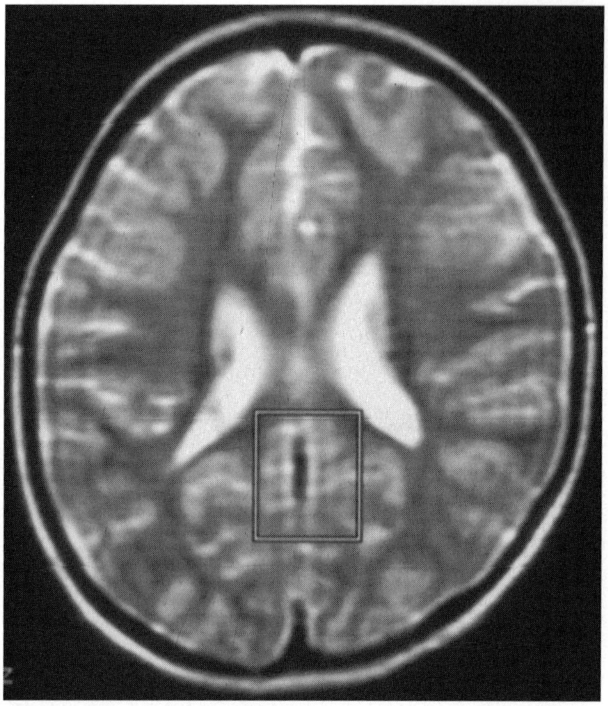

Fig. 5.89 MRI performed at time of second MRS examination on day 4 after near-drowning. The location of grey matter MRS is shown. There is no obvious edema, but the MRS examination suggests brain swelling.

Reading from right to left, there is possible lipid and a trace of excess lactate, although the doublet at 1.33 ppm is *not* convincing. Continuing to read from right to left, the following resonances are identified: NAA, β,γ-Glx, Cr, Cho, mI (3.56 ppm), α-Glx, and the second peaks of Cr and mI. NAA appears reduced, β,γ-Glx is prominent, and Cho is abnormally high for a grey matter location.

2. Interpretation and Analysis

The metabolite ratios are: NAA/Cr = 1.13 (20% reduced, as normal NAA/Cr is close to the adult value at 2 years of age); Cho/Cr = 0.85 (almost 50% increased); mI/Cr = 0.62 (normal). However, [Cr] is 75% of normal, a severe reduction, an indication that NAA is more reduced than indicated by the ratios, while Cho is probably not elevated, as falsely suggested by the ratios. β,γ-Glx/Cr = 0.52 (50% or more elevated), and α-Glx/Cr = 0.49 (apparently normal); the increase in Glx is therefore questionable. Brain water = 81%, which is increased for age.

The prognosis based upon this first examination (day 2) is that even at this early date, on all measures except "lactate" MRS places this patient in poor outcome groups, which include death or persistent vegetative state as the likely outcomes.

E. Examination 2: Four Days After Accident (Grey Matter Location)

1. Peak Identification and Quality Assurance: Reading the Spectrum from Right to Left

The spectral quality is excellent, notwithstanding peaks at 0.9, 1.3 and 1.7 ppm, which are probably not due to artifacts (Fig. 5.92).

Reading from right to left, the lipid peaks at 0.9, 1.3 and 1.7 ppm are of pathological significance. After the lipid peaks follows NAA, very prominent β,γ-Glx, the second NAA, Cr, Cho, possible sI, (a dip at 3.5 ppm is most probably an artifact of eddy-currents and immaterial to the diagnostic MRS), mI, prominent α-Glx (note the truly triplet nature of this metabolite resonance is very obvious here), second Cr, but not a second mI peak.

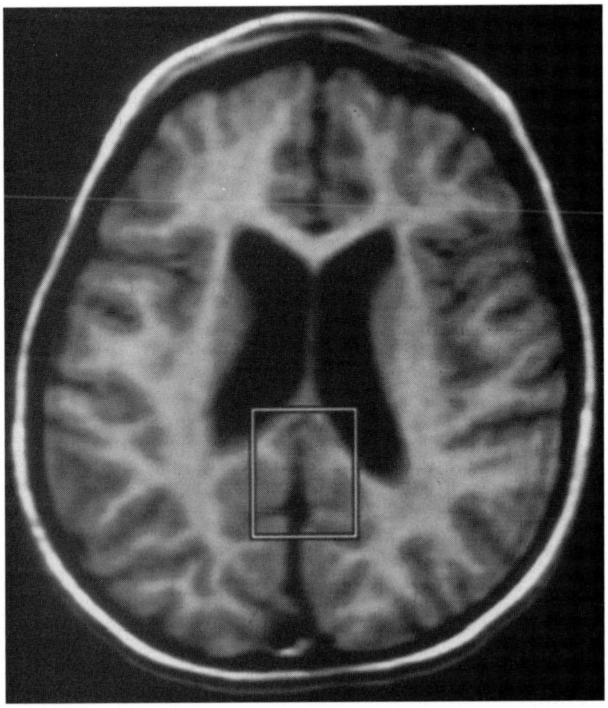

Fig. 5.90 MRI performed after 16 days shows location of MRS examination number 3. Ventricular dilatation and cortical atrophy are noted and are also reflected in MRS results.

2. Interpretation and Analysis

This spectrum (Fig. 5.92) is best considered in terms of its progression from the spectrum acquired 48 hours previously (Fig. 5.91). Excess lipids are in the "juvenile" pattern in the first spectrum (Fig. 5.91).

The metabolite ratios are: NAA/Cr = 0.90; Cho/Cr = 0.73, and mI/Cr = 0.51, all significantly different from Examination 1, indicating clinical deterioration. [Cr] = 72% (not significantly altered from day 2); β,γ-Glx/Cr = 0.57 and α-Glx/Cr = 0.55 probably represent quantitatively increased glutamine compared to day 2. The fraction of CSF in the VOI is 7%, which is reduced from 11%, and is possibly due to brain swelling. For comparison see MRIs from the first and second examinations (Fig. 5.88, 5.89).

The prognosis based upon this second examination (day 4) is that the "index" of MRS findings (which, with no exceptions, the false/positive = 0), is incontrovertibly "poor outcome."

F. Examination 3: Eighteen Days After Accident (Grey Matter Location)

1. Peak Identification and Quality Assurance: Reading the Spectrum from Right to Left

The spectral quality is excellent (Fig. 5.93). There is a small "dip" due to outer-volume lipid which may artificially reduce NAA.

Reading from right to left, lipid and an artifact of outer-volume lipid at 1.7 ppm are present. A significantly reduced NAA peak, unremarkable β,γ-Glx, Cr, Cho, sI, mI, the triplet of α-Glx, and second peaks for Cr and mI are identifiable.

2. Interpretation and Analysis

The metabolite ratios show further changes from the second examination. Reading from right to left, the metabolite ratios are: NAA/Cr = 0.74 (now approximately 50% of normal); Cho/Cr = 0.79 (unchanged, but elevated compared to normal); and mI/Cr = 0.65 (essentially normal). [Cr] = 93% of normal, an apparent recovery between the second and third examinations, which may be accounted for by the consolidation of surviving brain tissue which is identified in the image from this third

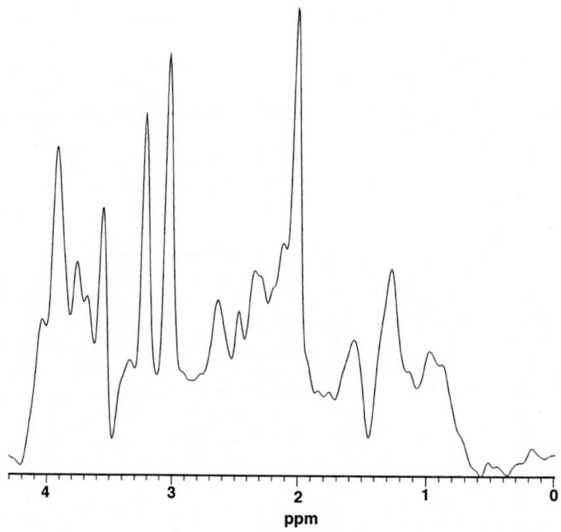

Fig. 5.91 The grey matter spectrum measured two days after the accident is consistent with early hypoxic encephalopathy.

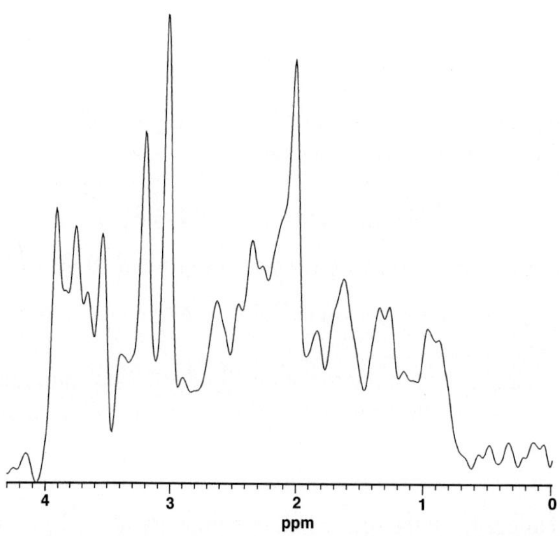

Fig. 5.92 Grey matter measured on day 4 shows progression of hypoxic encephalopathy after near-drowning.

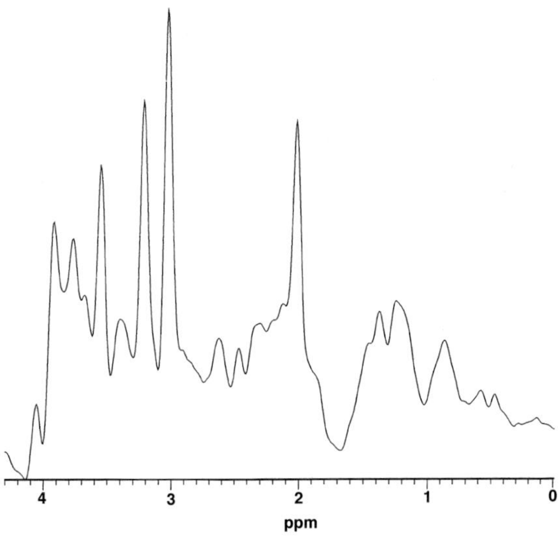

Fig. 5.93 Grey matter on day 16 after near-drowning shows the typical late MRS sequelae of poor neurological outcome after such an hypoxic injury with lipid, reduced NAA, and now normalized Glx.

examination (Fig. 5.90), and from a CSF compartment which is massively increased from 7 to 25%. Another notable change is the β,γ-Glx/Cr = 0.31, and α-Glx/Cr = 0.51, falling back into the normal range within the period of 18 days after the hypoxic insult.

G. Conclusion

Marked progression with cortical atrophy and nearly 50% loss of NAA after near-drowning.

H. Summary of the Patient's History Post-MRS

The outcome was poor. After discontinuing life support, the patient succumbed.

I. Other Comments

Because this is an example of global ischemia with relatively little variation (apart from severity of injury), near-drowning provides a very useful data base upon which to interpret more complex hypoxic insults, such as cardiac arrest, perinatal asphyxia, posttrauma, and after acute stroke. In these more complex situations, prognosis will depend on other variables, but diagnosis is likely to be achieved with added precision using all of the markers of hypoxic injury documented in the survivors of near-drowning accidents.

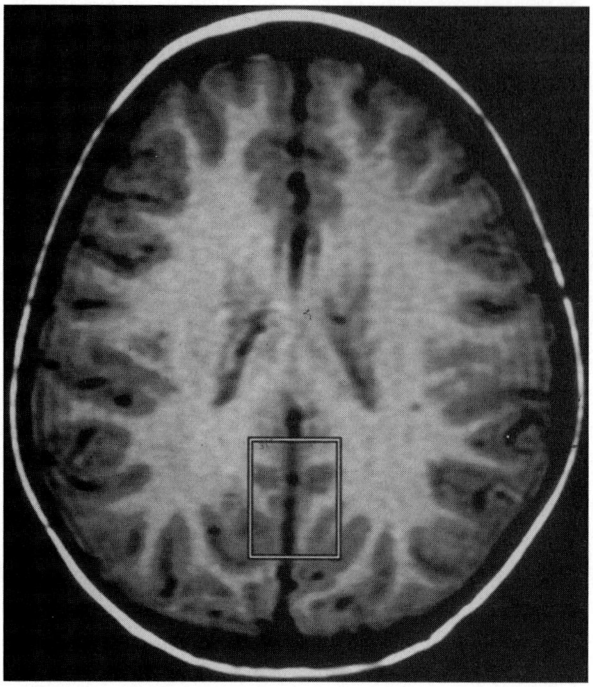

Fig. 5.94 The MRI on day 2 shows normal appearing brain, and the location of the grey matter VOI. MRS was already abnormal on day 2, and deteriorated further by day 4, despite the normal appearing MRI (not shown).

5.5.3 NEAR-DROWNING: FATAL OUTCOME

A. Purpose of MRS

Sequential examination to provide early prognosis after near-drowning.

B. Summary of Patient's History

Male, age 2 years, suffered arrest and near-drowning after falling into a swimming pool; Glasgow Coma Scale score 4–5 on admission. MRS was performed on days 2 and 4, with the infant still on life support.

C. MRS Method, Volume of Interest, and Comments about Examination

Automated PROBE-STEAM (TE = 30 ms, TM = 13.7 ms, TR = 1500 ms) was performed in HMRI's occipital grey matter location (see Fig. 5.94) on day 2, and day 4 after the accident. These times and intervals are established as documenting severity and progression of global hypoxic injury with the greatest precision and at the earliest possible time in the clinical course of near-drowning. The grey matter location is more sensitive to hypoxic injury, and prognosis is readily established as early as day 1, but with certainty by day 4.

In addition to the automated PROBE examination, quantitative, nonautomated MRS was performed in the grey matter location on both days (STEAM: TE = 30 ms, TM = 13.7 ms, TR = 1500 and 3000 ms). Quantitative examinations were also performed in HMRI's white matter location. Only the grey matter spectra acquired at TR 1500 ms are shown for comparison with the result of PROBE.

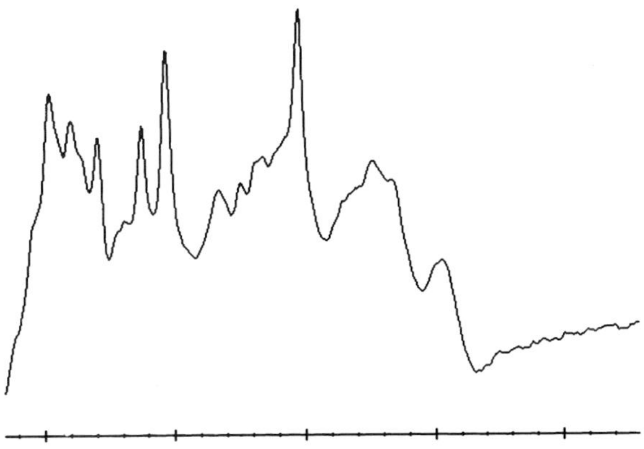

Fig. 5.95 PROBE spectrum of grey matter 2 days after hypoxic brain injury due to near-drowning. There is some water suppression artifact, but excess lipid and Glx, as well as a small but significant reduction in NAA/Cr, are already present.

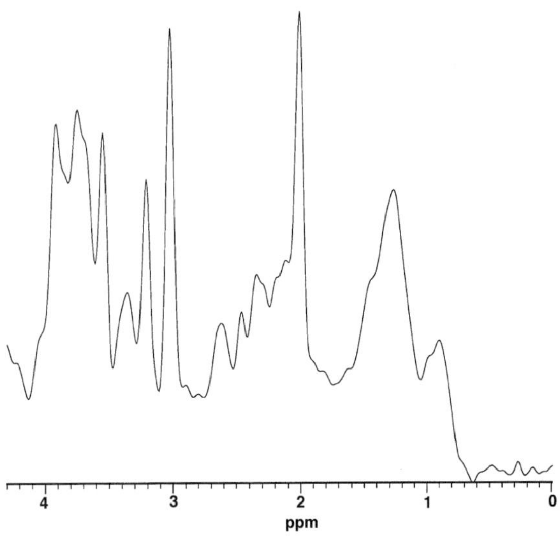

Fig. 5.96 Quantitative STEAM spectrum acquired with manual shimming and water suppression adjustments measured in grey matter two days after near-drowning. The spectrum is very similar to the PROBE spectrum when the over suppression of water in the PROBE is discounted.

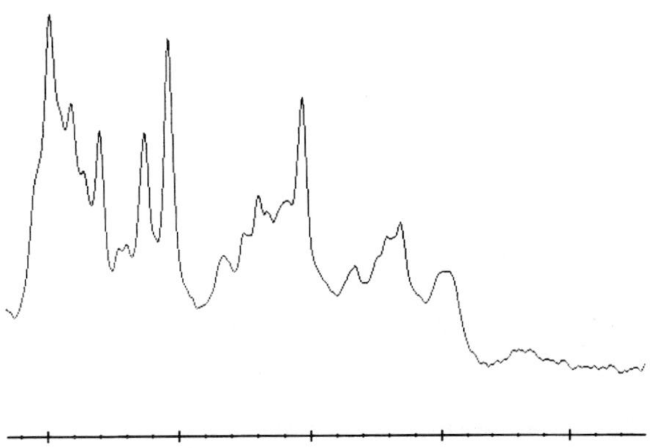

Fig. 5.97 PROBE of grey matter after near-drowning. Further changes in PROBE 4 days after injury. Excess lipid, lactate and Glx, the latter which partly obscures a dramatic loss of NAA, signal fatal outcome after near-drowning.

D. Examination 1 (Grey Matter), Day 2

1. Peak Identification and Quality Assurance: Reading the Spectrum from Right to Left

The technical quality of the PROBE spectrum (Fig. 5.95) is not perfect, but is nevertheless of diagnostic quality. Water is somewhat oversuppressed, giving a baseline which slopes down from right to left, as well as possible artifacts in the lipid region at 0.95 ppm and 1.5 ppm.

Apart from these lipid peaks, which in subsequent PROBE examination proved to be correctly read as excess lipid, reading the spectrum systematically from right to left, the following metabolite peaks are identified: NAA, β,γ-Glx (present in excess), the second NAA peak, Cr, Cho, mI, broadened α-Glx, and a second Cr, but no second mI peak.

Despite the baseline and lipid abnormalities, this PROBE examination is already suggestive of severe hypoxic injury.

2. Interpretation and Analysis

Reading from right to left, the metabolite ratios are: NAA/Cr = 1.27 (slightly reduced); Cho/Cr = 0.57 (normal, implying normal [Cr] at this early stage); mI/Cr = 0.49 (normal). β,γ-Glx is more prominent than the second NAA, and α-Glx is also prominent.

With these abnormalities, no definite prognosis can be offered. A repeat examination at 4 days is essential.

The quantitative, nonautomated STEAM spectrum from the same grey matter VOI (Fig. 5.96) is essentially the same as the PROBE spectrum, except that the slope of the baseline, and the out-of-phase lipid artifacts are avoided. sI/glucose and mI are more pronounced in this spectrum, probably because the water is not oversuppressed, as it is in the PROBE spectrum (Fig. 5.95). Different methods of postprocessing resulted in more line-broadening in the PROBE spectrum, which can explain minor discrepancies between the "visually estimated" PROBE metabolite ratios, and the computer calculated ratios from the quantitative spectrum.

The metabolite ratios for the nonautomated STEAM spectrum (Fig. 5.96) are: NAA/Cr = 1.04; Cho/Cr = 0.66; and mI/Cr = 0.77. Only mI/Cr differs greatly from that obtained with PROBE, as discussed above. The impression of Glx being prominent in the PROBE spectrum is confirmed here by the ratios β,γ-Glx/Cr = 0.54, and α-Glx/Cr = 0.90 (more than 50% increased). From the PROBE spectrum, the [Cr] was implied to be close to normal, based on the fact that Cho/Cr was normal (and Cho is not expected to be abnormal on day 2 after hypoxia). The near normal [Cr] was confirmed; [Cr] is 0.95.

Other results included a CSF fraction of 1.3%, and a water content of 86%, which is high for grey matter, and possibly indicates edema.

E. Examination 2 (Grey Matter), Day 4

1. Peak Identification and Quality Assurance: Reading the Spectrum from Right to Left

The technical spectral quality of the PROBE (Fig. 5.97) is excellent, with good water suppression and no artifacts.

Reading from right to left, lipid (0.9 and 1.33 ppm), a lactate doublet at 1.33 ppm and a further abnormal "lipid" resonance at 1.65 ppm are identified, followed by peaks of NAA (much reduced), β,γ-Glx (much increased), the second NAA (also reduced), Cr, Cho, sI, glucose, mI, two peaks of the α-Glx triplet, and a second peak of Cr, but not of mI.

The spectrum is very obviously abnormal and signifies poor prognosis.

2. Interpretation and Analysis

The metabolite ratios are: NAA/Cr = 0.85 (40% reduced from normal); Cho/Cr = 0.56; mI/Cr = 0.55. As Cho/Cr is normal, [Cr] also is probably normal. Nevertheless, NAA/Cr is overestimated, and allowance must be made for the effect of nearly twice normal β,γ-Glx/Cr which underlies the NAA peak. As indicated in the chart provided (see Fig. 6.3D, p. 308), prognosis is poor for this child.

The quantitative, nonautomated STEAM spectrum from the same grey matter VOI (Fig. 5.98) hardly differs from the PROBE spectrum, except for the line widths caused by different postprocessing, and except for the absence of a 1.65 ppm "lipid" peak. Note especially how similar every detail in the profile of the β,γ-Glx and α-Glx resonances are (accounting for line width), showing how reproducible MRS is when measurement parameters are identical, and artifacts are avoided.

For comparison with the PROBE spectrum, the metabolite ratios of the nonautomated STEAM spectrum are (Fig. 5.98): NAA/Cr = 0.75; Cho/Cr = 0.64; mI/Cr = 0.62; β,γ-Glx/Cr = 0.43; α-Glx/Cr = 0.66; [Cr] = 0.85. These numbers are identical to PROBE, taking into account the above mentioned reservations. Water content was 77% (normal for grey matter), and CSF

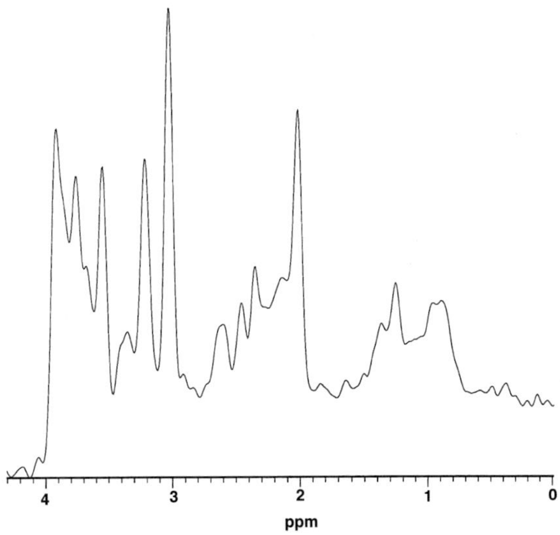

Fig. 5.98 Quantitative STEAM spectrum acquired with manual shimming and water suppression adjustments measured in grey matter four days postinjury. This spectrum and the PROBE spectrum are virtually identical, demonstrating the reproducibility of MRS. The extra quantitative information, [Cr], water content, and CSF fraction is convenient to have measured, but in this case is not essential to the conclusion (very poor, if not fatal, outcome).

was 2.5% (also normal). The quantitative examination accurately measures the exact Cr concentration, and avoids the chance of incorrectly estimating the metabolite ratios. However, in this case, the additional information provided by the quantitative examination was not vital to the conclusion.

F. Conclusion

Progressive neuronal injury after initial hypoxic insult indicative of poor outcome on day 2, and conclusively so on day 4. Lactate together with significantly reduced NAA are the defining changes, as the increase in Glx appears reversible.

G. Summary of the Patient's History Post-MRS

The patient died after ventilator support was discontinued.

H. Other Comments

Prognostic data established with formal MRS is accurately reproduced by the fully automated PROBE examination. The white matter spectra (not shown) showed similar, but less severe, hypoxic changes. The spectra acquired at TR 3000 ms had slightly better S/N, as expected, and minor T_1 effects were observed.

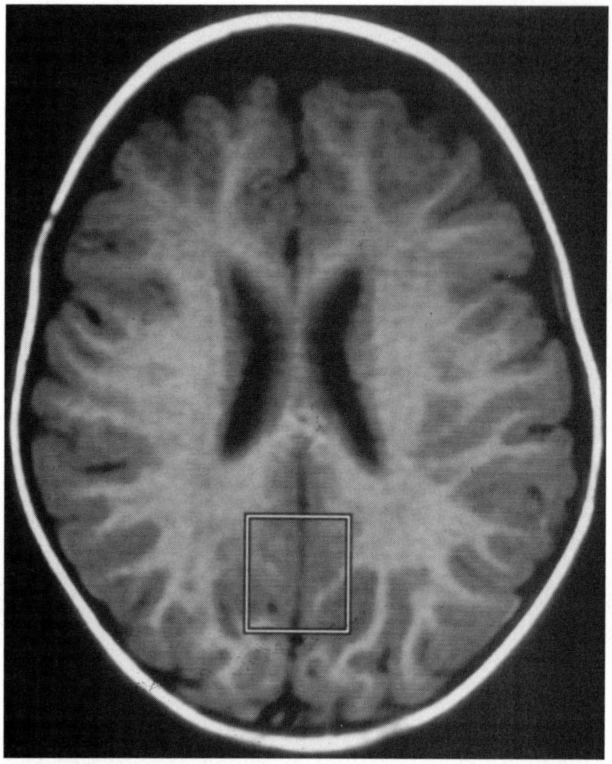

Fig. 5.99 Apparently normal MRI obtained in comatose child 2 days after near-drowning, showing the grey matter PROBE location.

5.5.4 NEAR-DROWNING: POOR OUTCOME

A. Purpose of MRS

Early prognosis after near-drowning.

B. Summary of Patient's History

Male, age 2 years, was submerged in a swimming pool, and admitted to the hospital with GCS score 3–4. MRS was performed on day 2 and day 4.

C. MRS Method, Volume of Interest, and Comments about Examination

Automated PROBE-STEAM (TE = 30 ms, TM = 13.7 ms, TR = 1500 ms) was performed in HMRI's occipital grey matter location on day 2 and day 4 (see Fig. 5.99 and Fig. 5.100).

D. Examination 1, Day 2: Grey Matter Location

1. Peak Identification and Quality Assurance: Reading the Spectrum from Right to Left

This automated PROBE spectrum (Fig. 5.101) is of excellent technical quality apart from slight oversuppression of water. The spectrum, which is from HMRI's grey matter location (not white matter despite its appearance), is strikingly abnormal.

Reading from right to left, there are two lipid peaks at 0.9 and 1.3 ppm, but no lactate doublet at 1.33 ppm. Continuing from right to left, the following resonances are identified: NAA (much reduced), β,γ-Glx (strikingly prominent), the second NAA, Cr, Cho (much increased relative to Cr), mI, a prominent α-Glx triplet, and the second Cr peak.

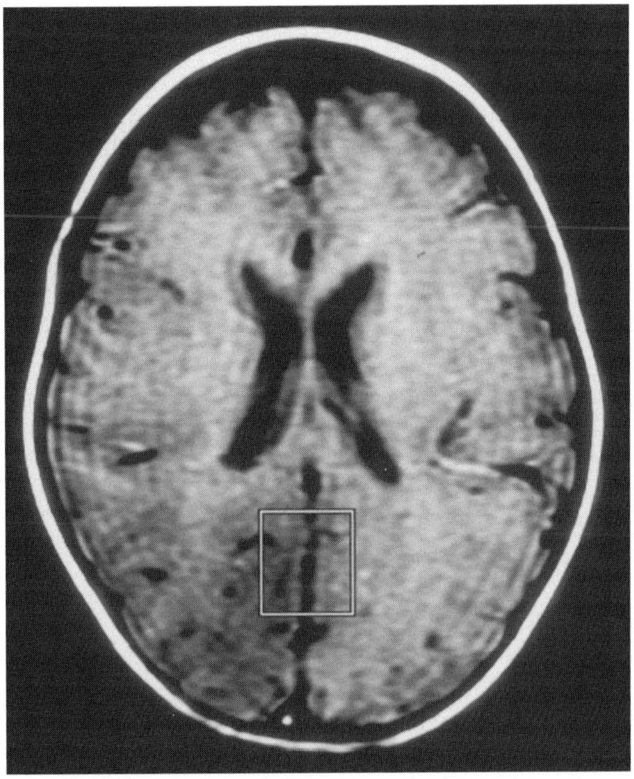

Fig. 5.100 T_1-weighted MRI, 4 days after near-drowning accident, showing possible edema.

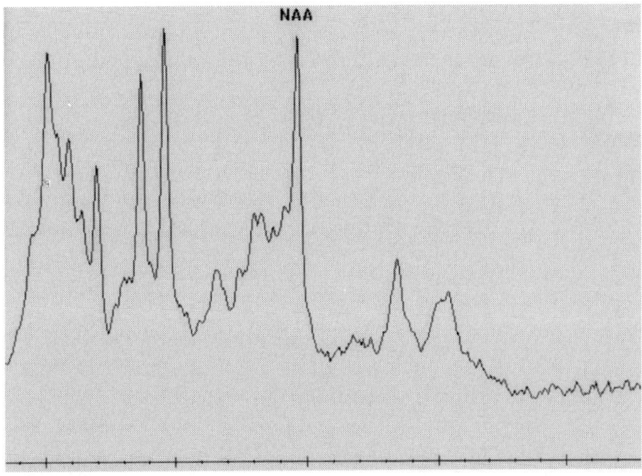

Fig. 5.101 Grey matter automated PROBE spectrum acquired 2 days after near-drowning accident. Prognosis is poor.

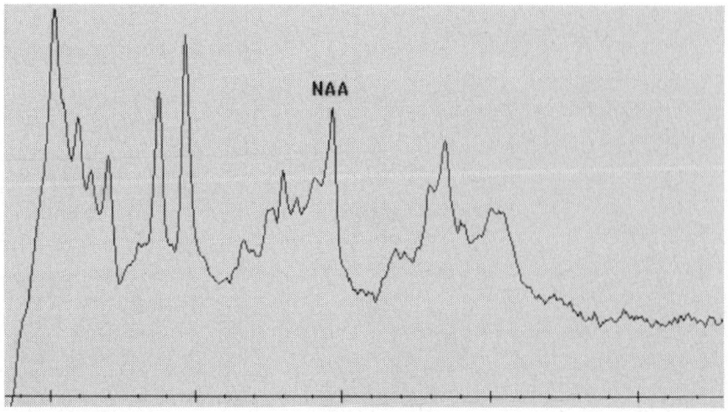

Fig. 5.102 Grey matter PROBE spectrum acquired 4 days after near-drowning accident. Note the profound deterioration of the spectrum which now shows lipid, lactate, Glx, reduced [Cr] and dramatically reduced NAA.

2. Interpretation and Analysis

Reading from right to left, the metabolite ratios are: NAA/Cr = 0.94 (30% reduced); Cho/Cr = 0.86 (suggestive of a reduction in [Cr] of 30%); mI/Cr = 0.56 (normal for age); β,γ-Glx/Cr is markedly elevated, as judged by the relationship to the second NAA peak, and confirmed by prominence of the α-Glx triplet.

The results are highly suggestive of severe hypoxic encephalopathy, and predictive of poor outcome. A follow-up examination at day 4 provided conclusive information.

E. Examination 2, Day 4: Grey Matter Location

1. Peak Identification and Quality Assurance: Reading the Spectrum from Right to Left

Of good technical quality, the automated PROBE spectrum (Fig. 5.102) suffers from oversuppression of water (see far left of PROBE spectrum), but there are no artifacts, and the baseline is virtually horizontal.

Reading from right to left, there are pathological lipid peaks at 0.9, 1.3 and 1.6 ppm, together with a doublet of excess lactate at 1.33 ppm. Continuing, there is a small NAA peak, prominent and detailed Glx pattern with β,γ-Glx region between 2.1 to 2.4 ppm, an appropriate and small second NAA peak, Cr, Cho (elevated for grey matter), mI, prominent α-Glx peaks, a second Cr peak, but no second mI. The latter is lost in the steep, inverted, oversuppressed water peak.

The spectrum is that of severe progressive hypoxic encephalopathy.

2. Interpretation and Analysis

The metabolite ratios are: NAA/Cr = 0.74 (45% reduced from normal, and a further 15% reduction over the two days since the first examination); Cho/Cr = 0.73 (still elevated for grey matter and therefore indicative of low [Cr]); mI/Cr = 0.49 (normal or low); β,γ-Glx/Cr is very dramatically increased as judged by the relative heights of Glx resonances and the adjacent second NAA peak.

F. Conclusion

The spectra show severe and irreversible brain damage secondary to hypoxic encephalopathy. Despite the relatively innocent MRI, prognosis is poor.

G. Summary of the Patient's History Post-MRS

The patient died three days after the second PROBE examination.

H. Other Comments

Sequential PROBE examination shows clinical value in prognosis after hypoxic brain injury. A direct measurement of [Cr] was not available; however this is an example of how one can often give a qualified estimate. Marked changes of Cho are not expected in hypoxic encephalopathy; the increased Cho/Cr, therefore, indicates that [Cr] is low.

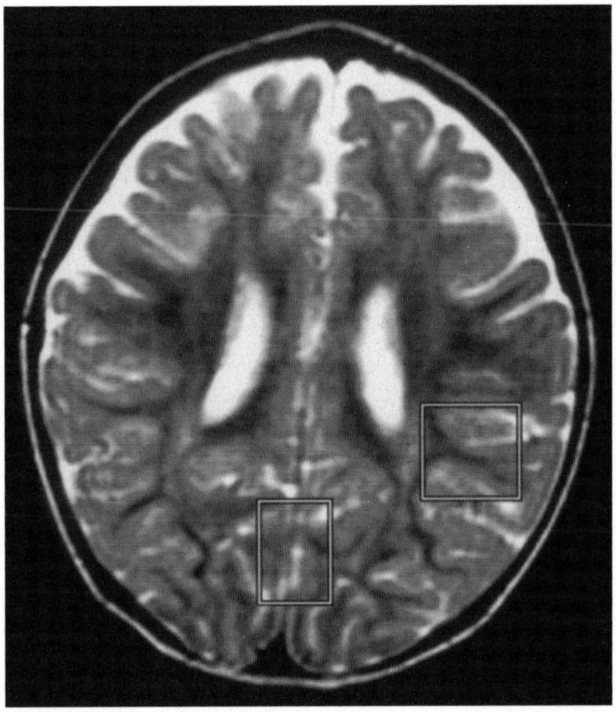

Fig. 5.103 Fast spin echo T_2-weighted MRI showing grey and white matter locations measured six days post accident. There is possible evidence of cortical atrophy. MRS, by determining the fraction of CSF in the grey matter VOI, accurately quantifies the degree of atrophy. No other abnormalities are apparent.

5.5.5 ASPHYXIA: MODERATE HYPOXIC ENCEPHALOPATHY

A. Purpose of MRS

Evaluation of severity, follow-up, and prognosis after hypoxic encephalopathy.

B. Summary of Patient's History

Male, age 10 months, who choked on food and suffered cardiopulmonary arrest. At time of first MRS on day 6, he was hypertonic, hyperreflexic, and possibly decerebrate.

C. MRS Method, Volume of Interest, and Comments about Examination

Localized STEAM (TE = 30 ms, TM = 13.7 ms, TR = 1500 ms) was performed in HMRI's parietal white matter (Locations 1, 3, 5), and occipital grey matter (Locations 2, 4, 6). Measurements were made 6 days after the accident (Fig. 5.103; Locations 1, 2); on day 11 (not shown; Locations 3, 4); and on day 32 (Fig. 5.104; Locations 5, 6). The first examination was performed with the patient on ventilator and i.v. support when showing signs of decerebrate rigidity, and the last examination was performed immediately preceding long-stay care.

D. Examination 1: Location 1 (White Matter), Day 6

1. Peak Identification and Quality Assurance: Reading the Spectrum from Right to Left

The white matter spectrum (Fig. 5.105) is of good technical quality apart from possible outer-volume lipid as indicated by the inverted peaks at 1.6 ppm. The spectrum is strikingly abnormal for age.

Reading from right to left, there is excess lipid (0.9 and 1.3 ppm), but no convincing evidence of a lactate doublet at 1.33

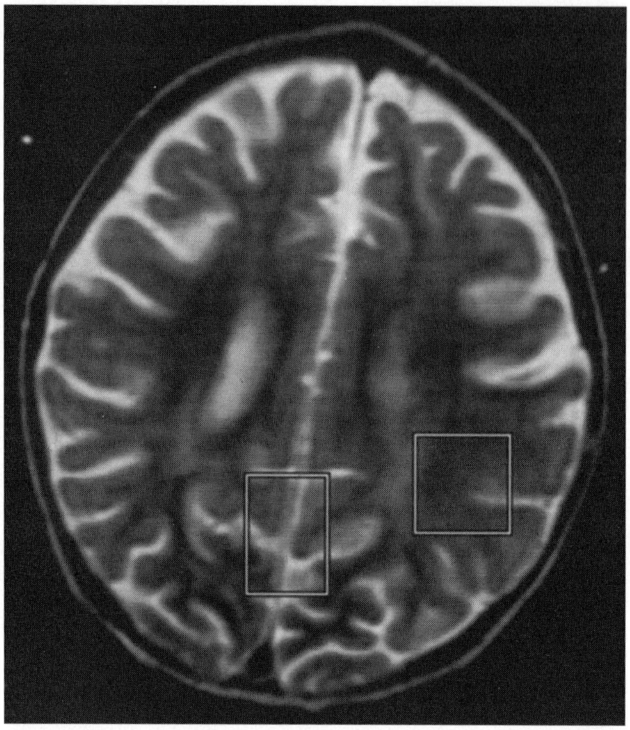

Fig. 5.104 Fast spin echo T_2-weighted MRI of the third examination showing white matter and grey matter VOIs. Apart from further cortical atrophy, no evidence of hypoxic injury is apparent on MRI.

ppm. Continuing right to left, the following metabolites are present: NAA, β,γ-Glx (prominent), Cr, Cho, sI, mI (at 3.56 ppm), α-Glx, and second peaks of Cr and mI. NAA appears reduced, and β,γ-Glx is increased.

2. Interpretation and Analysis

The excess lipid appears to be real. Reading from right to left, the metabolite ratios are: NAA/Cr = 1.00 (20% below normal for a child of gestational age 85 weeks, or 10 months postnatal age); Cho/Cr = 0.92 (normal for age); mI/Cr = 0.60 (normal for age). Absolute [Cr] is 96% of normal, so the metabolite ratios reflect the true concentrations. β,γ-Glx/Cr = 0.42, and α-Glx/Cr = 0.58, both of which are increased significantly. Brain water is 84.3%, and is appropriate for age and location.

E. Examination 1: Location 2 (Grey Matter), Day 6

1. Peak Identification and Quality Assurance: Reading the Spectrum from Right to Left

The grey matter spectrum (Fig. 5.106) is of excellent quality. The abnormalities recognized in white matter are immediately recognizable, together with other significant abnormalities that contribute to the diagnosis.

Reading from right to left, excess lipid (0.9 and 1.3 ppm), together with a highly suggestive doublet for lactate at 1.33 ppm are seen, followed by significantly reduced NAA, prominent β,γ-Glx, the second NAA, Cr, Cho, sI and mI peaks, α-Glx, and the second peak of Cr (but not mI). Despite this, the second Cr peak is in appropriate 2/3 proportion to the principal Cr peak at 3.0 ppm. In addition to lipid, lactate, reduced NAA, and increased β,γ-Glx, Cho may be relatively increased. This is an important finding as it may indicate that [Cr] is reduced, a sinister finding in hypoxic encephalopathy.

2. Interpretation and Analysis

Reading from right to left, the metabolite ratios are: NAA/Cr = 0.95 (25% reduced compared with normal ratio at this age); Cho/Cr = 0.68 (normal for age); mI/Cr = 0.53 (below normal). Absolute [Cr] is 91% of normal; β,γ-Glx/Cr = 0.42 (somewhat elevated), and α-Glx/Cr = 0.50.

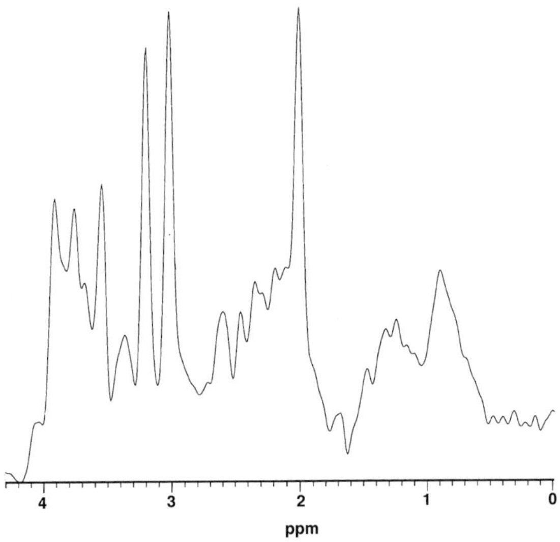

Fig. 5.105 The white matter spectrum measured six days after the accident (examination 1) is consistent with moderate hypoxic injury.

F. Summary of Examination 1

In keeping with the clinical history and decerebrate rigidity at the time of examination, there is significant evidence from MRS of global hypoxic encephalopathy. The changes are more marked in grey matter than white matter, as anticipated from the sensitivity of neurons to hypoxic insult. Prognosis may be estimated from the more systematic studies after near-drowning, and at day 6 appears consistent with a persistent vegetative outcome. Nonetheless, in this infant, two further MRS examinations were performed.

G. Examination 2: Location 3 (White Matter), Day 11

1. Peak Identification and Quality Assurance: Reading the Spectrum from Right to Left

The technical quality of the spectrum is good (Fig. 5.107), apart from a small outer-volume lipid peak, inverted at 1.65 ppm and an unexplained artifact at 2.5 ppm, which distorts the β,γ-Glx and second and third NAA peaks. This is possibly a phasing error, but has little impact upon the diagnostic value of the MRS examination in this case.

Reading from right to left, there is excess lipid at 0.9 and 1.3 ppm, followed by: NAA (obviously reduced); β,γ-Glx (together with artifact), the second NAA, Cr, Cho, sI and mI, α-Glx and the second Cr and mI peaks. The principal feature is the significant further reduction of the NAA peak when compared to that measured five days previously (Fig. 5.105).

2. Interpretation and Analysis

Reading from right to left, the metabolite ratios are: NAA/Cr = 0.91; Cho/Cr = 0.93; mI/Cr = 0.55; β,γ-Glx/Cr (no reliable result); α-Glx/Cr = 0.50. Absolute [Cr] is 102% of normal. There was a further slight, but significant fall in [NAA] compared to the previous examination.

H. Examination 2: Location 4 (Grey Matter), Day 11

1. Peak Identification and Quality Assurance: Reading the Spectrum from Right to Left

The technical quality of the spectrum is good (Fig. 5.108). Once again, outer-volume lipid appears as an inverted peak at 1.7 ppm. It is, however, most probable that this lipid is within the brain, in the immediate vicinity of the VOI, since the latter now contains significant intrinsic lipid, as indicated by peaks at 0.9, 1.3 and 1.5 ppm.

Reading from right to left, there is pronounced lipid, but no obvious lactate, followed by: NAA (very reduced); β,γ-Glx, the second NAA, Cr, Cho, mI, α-Glx, and the second Cr and mI peaks. The latter is indistinct.

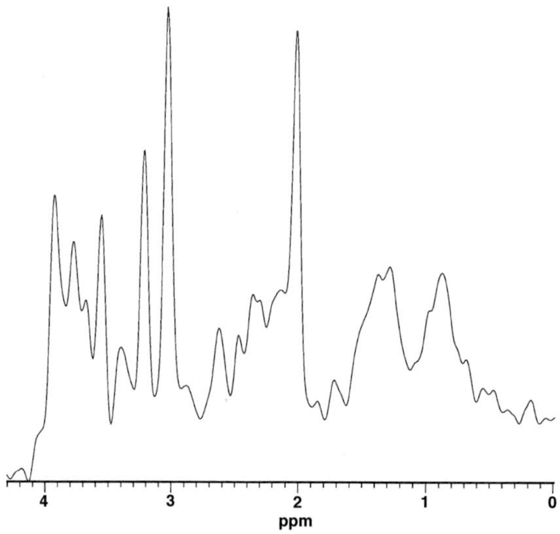

Fig. 5.106 The grey matter spectrum (examination 1) shows more marked changes of hypoxic injury than does the MRS of white matter.

In addition to the increasing evidence of intracerebral lipid, the most notable feature is the further reduction of NAA when compared to the MRS examination performed five days earlier (Fig. 5.106).

2. Interpretation and Analysis

Reading from right to left, the metabolite ratios are: NAA/Cr = 0.87; Cho/Cr = 0.70; mI/Cr = 0.71. Absolute [Cr] is 98% of normal; β,γ-Glx/Cr = 0.33, and α-Glx/Cr = 0.50 (both are within normal limits). While brain water of 81.2% is probably normal and not significantly different from that of 79.0% determined in this location five days prior, the fraction of CSF (atrophy index) is now increased to 15.5%, compared with only 8.8% in the earlier examination. Indeed, cortical atrophy and some ventricular dilatation are apparent in the image (not shown).

I. Summary of Examination 2

There was evidence of worsening of the MRS apparent in the reduction of NAA in both white matter and grey matter. We may speculate that the small increase (from 91% to 98%) in [Cr] is apparent rather than real, and attributable to consolidation of surviving brain within the VOI. This explanation is of importance in considering the results of the following (third) MRS and the eventual clinical outcome.

J. Examination 3: Location 5 (White Matter), Day 32

1. Peak Identification and Quality Assurance: Reading the Spectrum from Right to Left

Spectrum is of good quality apart from a dip at 1.7 ppm (Fig. 5.109). This dip does not compromise the diagnostic information and is indicative of intracerebral lipid, outside the VOI.

Reading the spectrum from right to left, lipid excess with peaks at 0.9, 1.3 and 1.5 ppm is apparent, but there is no lactate doublet at 1.33 ppm. Continuing right to left, the following metabolites are present: NAA (slightly increased compared with Location 3, but nevertheless, well below normal); slightly unusual β,γ-Glx, the second NAA peak, Cr, Cho, sI (possibly overlapping glucose), mI and the appropriate α-Glx peaks. A second Cr and mI peak are present.

2. Interpretation and Analysis

Reading from right to left, the metabolite ratios are: NAA/Cr = 0.97 (identical to examination 1, and 20% below normal for age); Cho/Cr = 0.97; mI/Cr = 0.74; β,γ-Glx/Cr = 0.27; α-Glx/Cr = 0.50; and [Cr] is 99% of normal. Brain water of 77% is

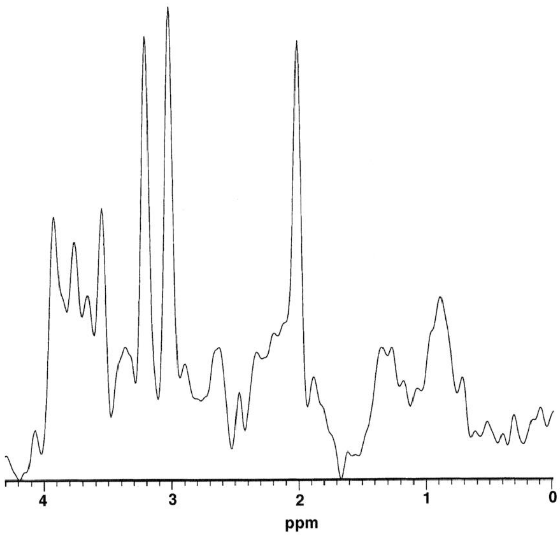

Fig. 5.107 The white matter spectrum acquired at the second examination, 11 days after the accident, shows further modest changes after hypoxic injury. Excess Glx can no longer be reliably identified because of an artifact, but Glx usually recovers to normal after this interval.

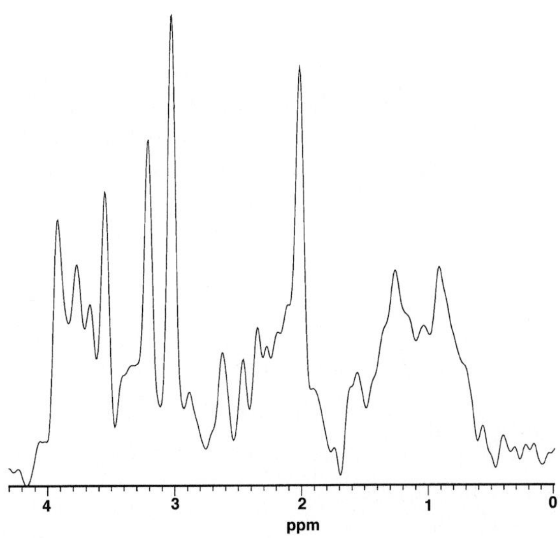

Fig. 5.108 Grey matter spectrum 11 days after the accident. NAA is further reduced and more lipid is identified.

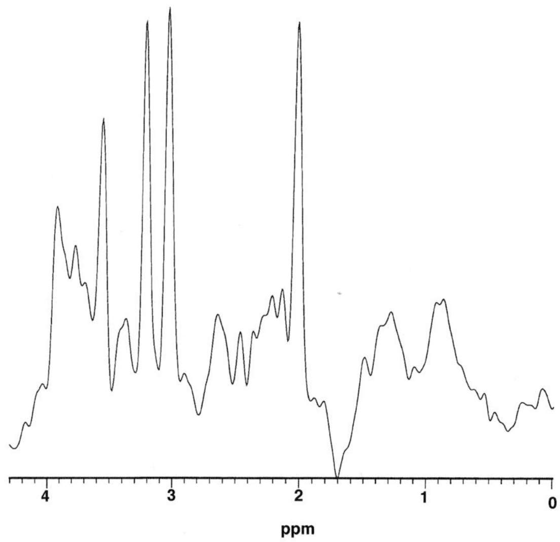

Fig. 5.109 White matter spectrum acquired at third examination, 32 days after the accident. The modest recovery of NAA is spurious, and secondary to atrophy noted in the corresponding MRI.

normal, and the CSF compartment of 11.7% is large, due to the overlap of the VOI with the expanded lateral ventricle (see Fig. 5.104).

K. Examination 3: Location 6 (Grey Matter), Day 32

1. Peak Identification and Quality Assurance: Reading the Spectrum from Right to Left

The spectral quality is excellent (Fig. 5.110).

Reading from right to left, the excess lipid is less prominent than 3 weeks earlier, but still abnormal. Again, no lactate is seen. The following metabolites are identified: NAA ("recovered" towards normal); β,γ-Glx (normal); Cr; Cho (appropriate for grey matter locations, and definitely reduced compared to examination 2); sI and glucose; mI; α-Glx; and the second Cr peak.

2. Interpretation and Analysis

Reading from right to left, in addition to the reduction in the amount of abnormal lipid, the metabolite ratios are: NAA/Cr = 0.99; Cho/Cr = 0.61; mI/Cr = 0.68; β,γ-Glx/Cr = 0.28; α-Glx/Cr = 0.44; and absolute [Cr] is 95% of normal. Brain water is 77%. Atrophy (CSF compartment) is quantifiably increased to 21.9%, from values of 8.8% and 15.5% in earlier MRS examinations. This "consolidation" of brain secondary to hypoxic injury is well recognized, and demonstrated in the sequence of figures (Fig. 5.103–5.104).

L. Conclusion

This is an instructive case by virtue of the subtle and sequential MRS changes. MRS is diagnostic of moderate neuronal loss, secondary to hypoxic injury. Furthermore, MRS provides no encouragement that neuronal "recovery" can occur in the short to medium term, even in this very young child.

M. Summary of the Patient's History Post-MRS

An interview conducted with the child and his parents 12 months after injury revealed a severely handicapped child with cerebral palsy.

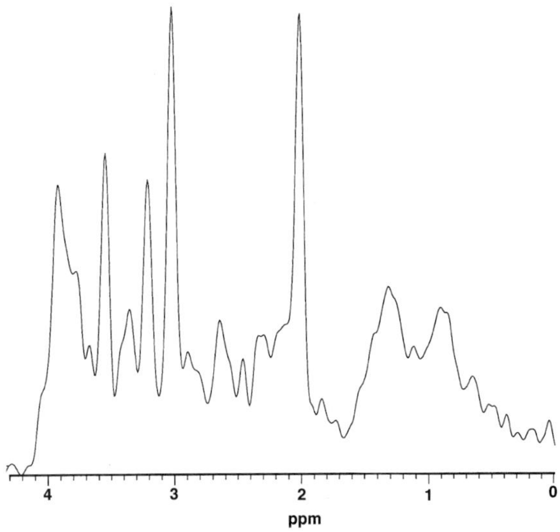

Fig. 5.110 Grey matter spectrum acquired at third examination, 32 days after the accident. Lipid is less prominent, and both NAA and Glx have recovered. While the reduction of Glx is a positive finding, it is suggested that the "recovery" of NAA is only apparent, following atrophy and consolidation of surviving neurons and axons within the VOI.

N. Other Comments

This patient was 10 months old and MRS data should still be corrected with reference to age-related data. Hypoxic injury in very young children, shares with other forms of severe brain injury at this age, the finding of excessive lipid in the brain spectra. This is not a common finding in older children or adults and may indicate a special response to injury of the metabolically immature brain.

Finding lactate is not necessary for the conclusion of poor outcome, and accumulation of Glx is common, but reversible and not itself a poor-prognostic sign. Prognosis with MRS in this setting is reliable, provided there are no special circumstances which are suggestive of a very focal brain injury centered upon the VOI's selected. This caution should be applied in viral, bacterial, or posttraumatic comas.

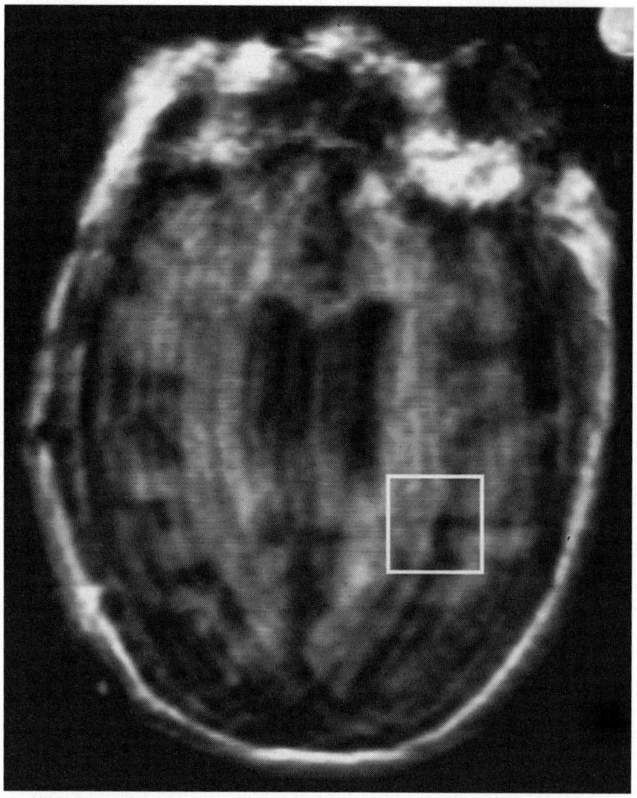

Fig. 5.111 T_2-weighted MRI is severely distorted by motion, but sufficient to show no marked atrophy, and location of single VOI in white matter.

5.5.6 CARDIAC ARREST, RULE OUT HYPOXIC ENCEPHALOPATHY

A. Purpose of MRS

To determine the prognosis after second cardiac arrest.

B. Summary of Patient's History

Male, age 64, admitted to hospital 24 days previously in full cardiac arrest, and unrousable at time of MRS examination. Previous arrest 5 months prior; alcoholic liver disease.

C. MRS Method, Volume of Interest, and Comments about Examination

STEAM (TE = 30 ms, TM = 13.7 ms, TR = 1500 ms) was performed in HMRI's parietal white matter location 24 days post-arrest. As seen by the poor quality of the MR image (Fig. 5.111), the patient moved, and therefore a second VOI was not attempted.

D. White Matter Location

1. Peak Identification and Quality Assurance: Reading the Spectrum from Right to Left

Surprisingly good quality spectrum (Fig. 5.112) despite severe degradation of image by motion; no outer volume lipid, and well-resolved resonances.

Reading the spectrum systematically from right to left, a small, and probably unimportant, excess of two lipid peaks at 0.9

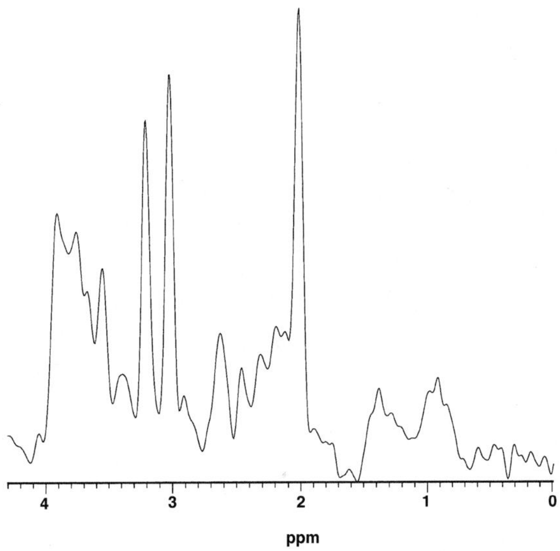

Fig. 5.112 The white matter spectrum 24 days after cardiac arrest shows minimal excess lipid and only 20% reduction in NAA/Cr. Consistent with very mild, recoverable hypoxic encephalopathy. Evidence of subclinical hepatic encephalopathy.

and 1.3 to 1.4 ppm, and no evidence of lactate. Continuing to read right to left, the following resonances are identified: NAA, β,γ-Glx, second NAA, Cr, Cho, glucose or sI, mI, α-Glx, second Cr, and a questionable second mI. NAA appears somewhat reduced, β,γ-Glx and α-Glx appear increased, and mI is somewhat reduced.

2. Interpretation and Analysis

The metabolite ratios are: NAA/Cr = 1.16 (reduced 20%); Cho/Cr = 0.89 (normal); mI/Cr = 0.52 (15% reduced). [Cr] = 99%, so ratios are acceptable. β,γ-Glx/Cr = 0.32 (25% increased) and α-Glx = 0.72 (50% increased). An increased glucose at 3.43 and 3.8 ppm may explain the higher α-Glx.

E. Conclusion

NAA/Cr remains in a satisfactory range to expect good outcome from cerebral ischemia, particularly as there is little excess of lipid and no lactate.

As evidence of liver disease induced by ethanol, mI is low and Glx is elevated, but low Cho, of the classic triad, is not seen (no ethanol triplet is seen in Fig. 5.112).

F. Summary of the Patient's History Post-MRS

The patient made a good (and full) recovery.

G. Other Comments

There appear to be two diagnoses here, with modest hypoxic injury superimposed on subclinical hepatic encephalopathy.

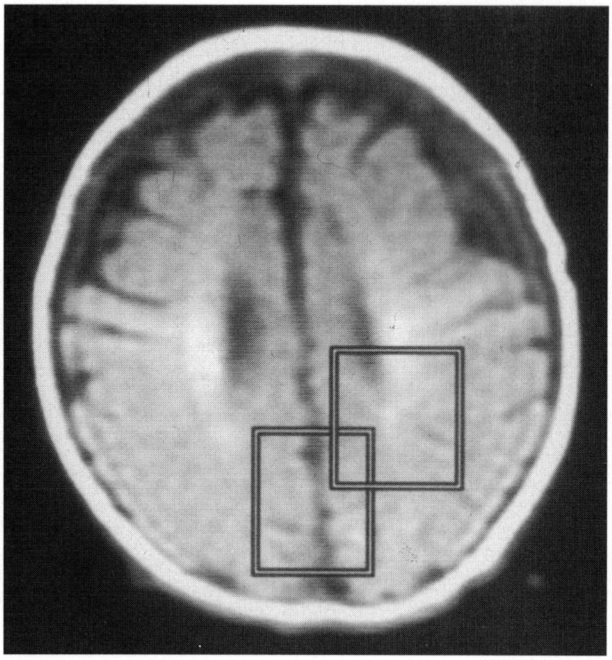

Fig. 5.113 T_1-weighted MRI showing grey and white matter locations measured 2.5 months after graft surgery in infant with complex congenital heart disease. Note possible frontal lobe atrophy.

5.5.7 CONGENITAL HEART DISEASE

A. Purpose of MRS

Rule out hypoxic encephalopathy in candidate for cardiac transplantation.

B. Summary of Patient's History

Male, age 4 months (gestational age, 56 weeks) with complex congenital heart disease (*situs inversus*, AV canal and pulmonary atresia) treated at age 1½ months by graft. Fontana procedure was ruled out, so cardiac transplant was considered a high probability at the time of MRS. MRS was indicated to determine whether the child had already suffered significant hypoxic brain injury.

C. MRS Method, Volume of Interest, and Comments about Examination

Localized STEAM (TE = 30 ms, TM = 13.7 ms, TR = 1500 ms) was performed in HMRI's occipital grey matter (Location 1), and parietal white matter (Location 2) 2.5 months post shunt, and 1 week before transplant (Fig. 5.113). At the time of MRS, the child was cyanosed, but not obtunded.

D. Location 1 (Grey Matter)

1. Peak Identification and Quality Assurance: Reading the Spectrum from Right to Left

The technical quality of the spectrum (Fig. 5.114) is excellent, and appears generally appropriate for a 4 month old.

Reading from right to left, there is little obvious lipid and no lactate at 1.33 ppm. Continuing reading, the following resonances are identified: NAA, β,γ-Glx, Cr, Cho, sI and mI (at 3.56 ppm), α-Glx, and the second peaks of Cr and mI. With reference to the typical spectrum at 4 months, the spectrum appears normal for age.

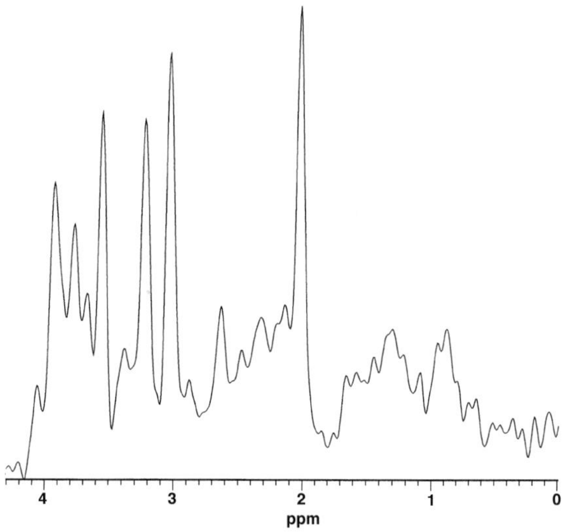

Fig. 5.114 This grey matter spectrum is entirely appropriate for a 4 month old and shows no evidence of acute or chronic hypoxic injury.

2. Interpretation and Analysis

The metabolite ratios are: NAA/Cr = 1.11 (normal for age and location); Cho/Cr = 0.83 (normal); mI/Cr = 0.85 (normal). [Cr] is 90% of adult, and normal for a child of this age. With the possible exception of slight elevation of Glx/Cr, the grey matter spectrum is normal. Brain water = 82%, normal for an infant at this age.

E. Location 2 (White Matter)

1. Peak Identification and Quality Assurance: Reading the Spectrum from Right to Left

The spectral quality of the white matter location (Fig. 5.115) is also of excellent quality. The lipid peak at 0.9 ppm appears real, but the region between 1.0 and 1.7 ppm may be artifactual. There is no evidence of a lactate doublet at 1.3 ppm.

Reading from right to left, there is a possible lipid excess, but no lactate, and the following resonances are identified: NAA, β,γ-Glx, Cr, Cho, possible sI, mI (at 3.56 ppm), α-Glx, and the second peaks of Cr and mI. The relative peak intensities appear appropriate for age and location.

2. Interpretation and Analysis

The metabolite ratios are: NAA/Cr = 1.09 (within normal range for age); Cho/Cr = 1.07 (normal); mI/Cr = 0.86 (normal). Absolute [Cr] = 97% of adult, and within normal range for an infant. β,γ-Glx/Cr = 0.40, and α-Glx/Cr = 0.67 (both are slightly increased for this white matter location). Brain water = 80%, which is normal for infant white matter.

F. Conclusion

Despite severe hypoxemia since birth, MRS was normal and showed none of the MRS hallmarks associated with acute or chronic hypoxic ischemic encephalopathy. MRI on the other hand did show some frontal cortex atrophy, possibly due to persistent hypoperfusion.

G. Summary of the Patient's History Post-MRS

Outcome unavailable.

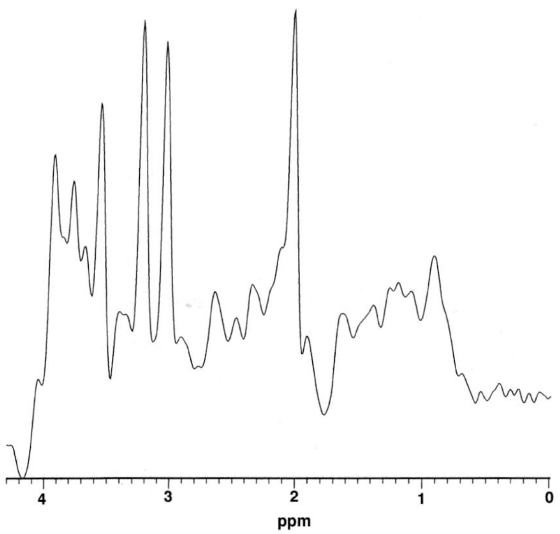

Fig. 5.115 Normal white matter spectrum with no evidence of hypoxic brain injury. Together, the two spectra provide strong incentive for the needed cardiac transplantation to proceed.

H. Other Comments

The patient was 4 months old (term infant) at the time of MRS. An accurate gestational age was essential to draw the optimistic conclusion that the status of the neurochemistry as shown by MRS was not a contraindication to urgent cardiac transplantation.

The modest elevation of cerebral Glx/Cr is consistent with persistent cerebral hypoxia, even in the absence of lactate. But, unlike elevated lactate, this appears to be fully reversible.

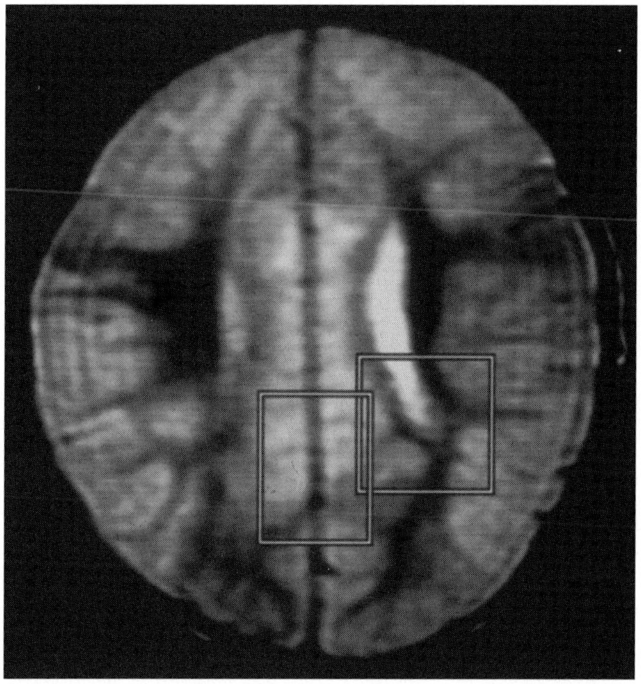

Fig. 5.116 Fast spin echo T_2-weighted axial image showing locations of MRS. The brain is strikingly abnormal with edema and loss of sulci as well as loss of grey-white contrast.

5.5.8 AGONAL PROGNOSIS AFTER HYPOXIC INJURY

A. Purpose of MRS

Differential diagnosis and prognosis in complex disorder in neonate.

B. Summary of Patient's History

Male, age 2 months (Moribund), with hypernatremia and status epilepticus following dehydration and sepsis.

C. MRS Method, Volume of Interest, and Comments about Examination

STEAM (TE = 30 ms, TM = 13.7 ms, TR = 1500 ms) was performed in HMRI's occipital grey matter (Location 1) and parietal white matter (Location 2). VOIs, shown on the image in Fig. 5.116, were measured 18 hours after admission to pediatric intensive care.

D. Location 1 (Grey Matter)

1. Peak Identification and Quality Assurance: Reading the Spectrum from Right to Left

The technical quality of the spectrum (Fig. 5.117) is very good, but S/N is low for all metabolites, despite 256 acquisitions from VOI > 11 cm^3. This suggests that intrinsically low metabolite concentrations are present.

Reading the spectrum systematically from right to left, there is little obvious lipid, but a dominant doublet from lactate at 1.33 ppm. Continuing right to left, reduced peak intensities scaled down by the high lactate peak, are noted for NAA, β,γ-Glx, no obvious second NAA peak, Cr, a broader than expected Cho, glucose, mI, α-Glx, and second peaks for Cr and mI. The two middle peaks of the lactate quartet centered at 4.11 are also seen, one of them overlapping the second mI at 4.06 ppm.

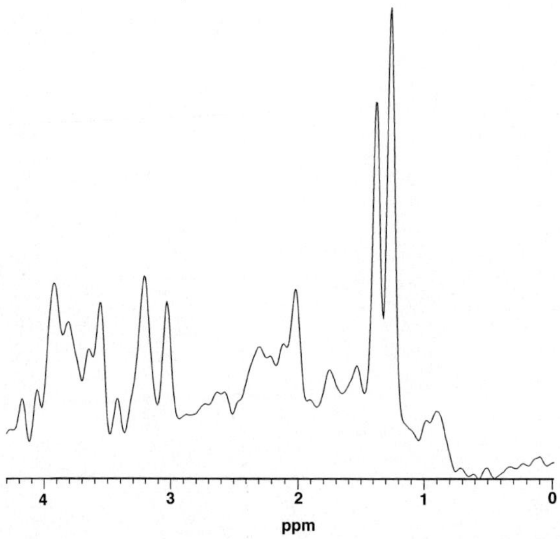

Fig. 5.117 MRS from grey matter location shows a relatively normal metabolite profile for age, but is dominated by a vast excess of lactate. MRS is consistent with terminal disease or acute oxygen deprivation.

Lactate is exceptionally high, while low NAA and other peaks may be accounted for by one of two general factors: the spectrum is automatically scaled to the tallest peak, "suppressing" all others, or [Cr] is much reduced within the VOI. This distinction, which is crucial, can be easily made.

2. Interpretation and Analysis

The metabolite ratios are probably invalidated by the finding that [Cr] is 46% of adult normal. Reading from right to left: NAA/Cr = 1.10 (normal); Cho/Cr = 1.16 (normal); mI/Cr = 1.03; β,γ-Glx/Cr = 0.73 (elevated); α-Glx/Cr = 0.83 (elevated). H_2O is 99% of pure water concentration (normal 90%).

E. Location 2 (White Matter)

1. Peak Identification and Quality Assurance: Reading the Spectrum from Right to Left

This spectrum from white matter (Fig. 5.118) is of excellent quality, and essentially similar in all respects to that from grey matter; lactate dominates the spectrum.

2. Interpretation and Analysis

At 76% of adult normal, [Cr] is better conserved than in grey matter of this patient. Reading from right to left, the metabolite ratios are: NAA/Cr = 1.05; Cho/Cr = 1.26; mI/Cr = 0.90; β,γ-Glx/Cr = 0.57; α-Glx/Cr = 0.53. An estimated lactate concentration in this and the grey matter locations is in excess of 20 mM (normal $\leq$ 1.0 mM). H_2O is also 99% in white matter.

F. Conclusion

This MRS examination, performed very late in the course of an unidentified cerebral catastrophe, reflects agonal events with massive lactate accumulation. There is no obvious distortion of the primary brain metabolites, the concentrations of which have fallen in a coordinated way.

MRI shows cerebral edema, and brain water is correspondingly increased. Both may be secondary to cerebral energy failure.

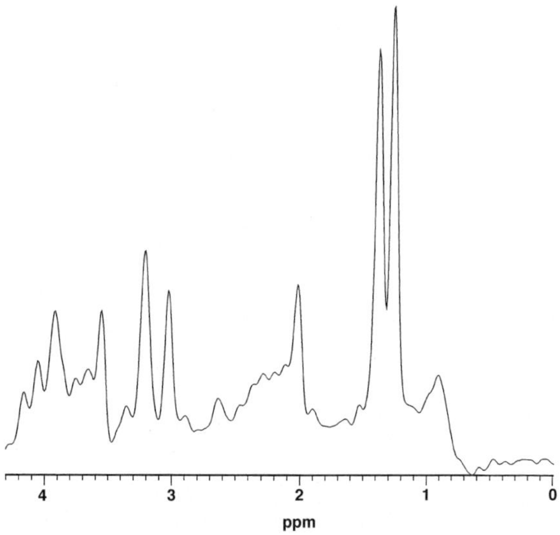

Fig. 5.118 MRS of white matter is comparable to the grey matter spectrum with agonal changes of excess lactate and relatively normal metabolite ratios for the infant brain. Metabolite concentrations (apart from ~20 mM lactate) are low in this and the preceding spectrum.

G. Summary of the Patient's History Post-MRS

The patient died a few hours after MRS.

H. Other Comments

The broader than normal Cho may indicate the presence of other metabolites near Cho which are not seen in normal brain.

At first, the water content of 99% raises the question of an technical error, but the fit of the water signal amplitudes were perfect and there was no indication of calibration problems. In this patient the 99% water content is consistent with cerebral edema preceding death. A similarly high water content has been seen in life, in white matter, in a patient with Canavan's disease (see Case 5.9.5).

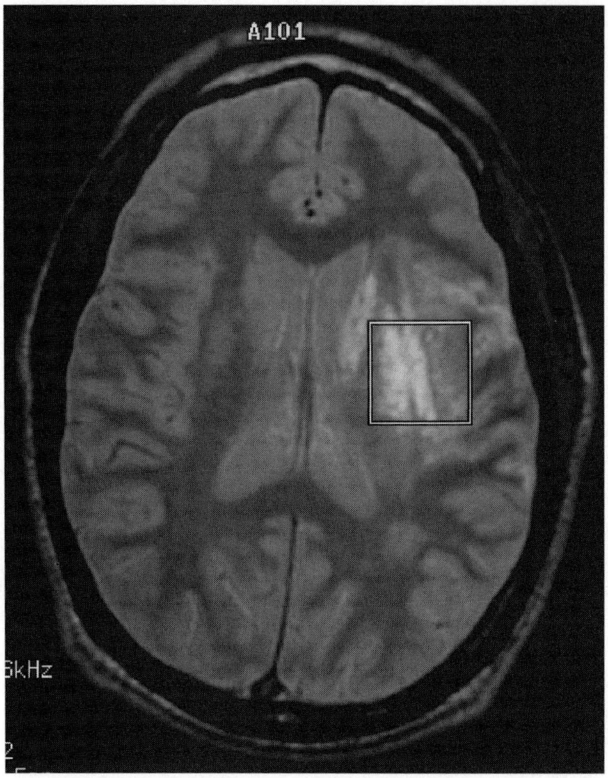

Fig. 5.119 The fairly large VOI was prescribed so that it best overlapped with the lesion, and an excellent spectrum was obtained. It can be tempting to choose smaller VOIs, but the price paid in terms of S/N is high. (Courtesy of Dr. Wang, Rancho Los Amigos Hospital, Downey, California.)

5.6 Focal Hypoxia: Vascular Diseases, Stroke, Infarcts, and Nonvascular Focal Hypoxia

5.6.1 HEMORRHAGIC INFARCT

A. Purpose of MRS

The purpose of the MRS (and the MRI) was to further investigate a patient who had been diagnosed with cerebral vascular accident.

B. Summary of Patient's History

Male, age 44, who was diagnosed with cerebral vascular accident. A more detailed history is not available.

C. MRS Method, Volume of Interest, and Comments about Examination

PROBE-STEAM (TE = 30 ms, TM = 13.7 ms, TR = 1500 ms) was performed in the lesion as shown on the image (Fig. 5.119). The second location 2 was contralateral (Fig. 5.120). Both VOIs were 25 × 25 × 20 mm^3, and all other measurement parameters, such as number of averages and signal amplification, were identical.

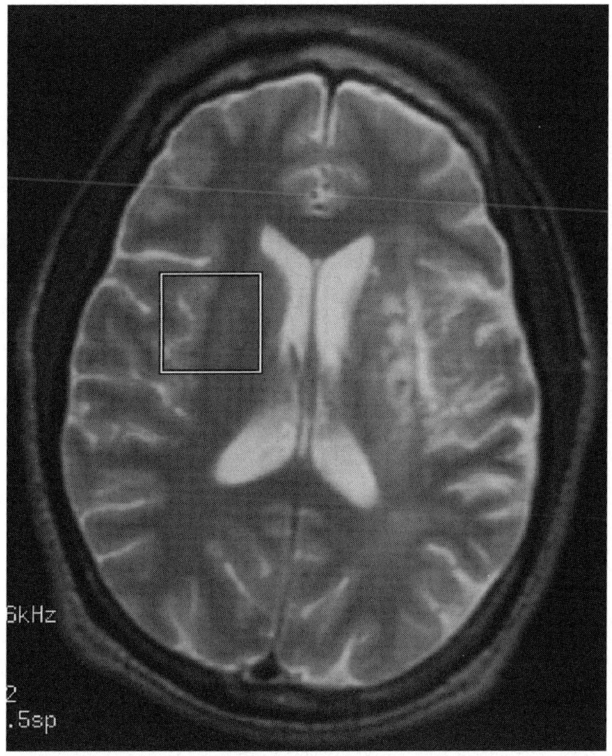

Fig. 5.120 The same size VOI was prescribed contralaterally for the purpose of getting a control spectrum. In some cases the contralateral location can be assumed normal, in other not. So this approach has to be used with some caution. (Courtesy of Dr. Wang, Rancho Los Amigos Hospital, Downey, California.)

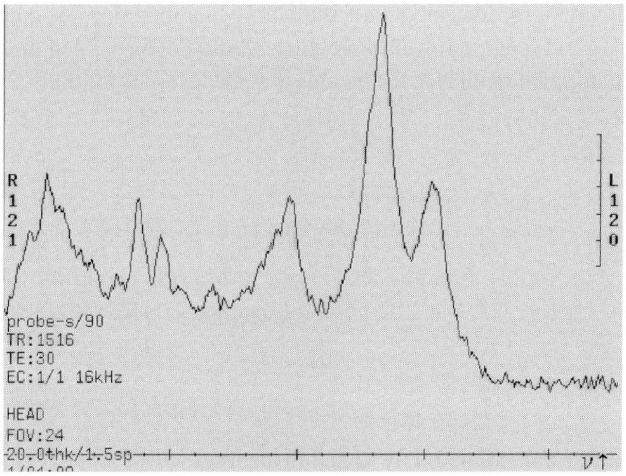

Fig. 5.121 The spectrum from the lesion is consistent with a nonacute hemorrhagic infarct. (Courtesy of Dr. Wang, Rancho Los Amigos Hospital, Downey, California.)

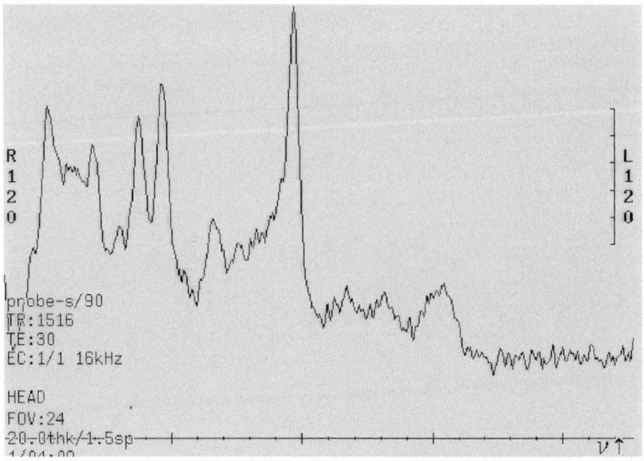

Fig. 5.122 The normal white matter spectrum from the opposite side. (Courtesy of Dr. Wang, Rancho Los Amigos Hospital, Downey, California).

D. Location 1 (Within Lesion)

1. Peak Identification and Quality Assurance: Reading the Spectrum from Right to Left

The good S/N, resolution, baseline, and the absence of artifacts guarantees the high quality of the spectrum (Fig. 5.121).

Reading from right to left, first the typical patterns of the CH_3 and CH_2 groups of lipids (fatty acid) are seen at 0.9 and 1.3 ppm. There is no lactate; the resolution is so good that a doublet would have been clearly visible. A small NAA peak is seen, followed by hardly any β,γ-Glx, a small second NAA, Cr, Cho, and possibly sI, but mI is absent. There is a small amount of α-Glx, and the second Cr; both relate well to their first resonances.

2. Interpretation and Analysis

[Cr] is not available from this automated MRS examination, but a qualified estimate is that [Cr] is reduced to less than half of normal, and that Cho is normal despite the apparently inverse Cho/Cr ratio, (compare the signal-to-noise ratio of Cr in this and the contralateral VOI, which were acquired under identical conditions). The major findings are: a) the dominant lipids consistent with membrane breakdown; b) the absent lactate, indicating that the infarct is not acute; c) elevated Cho/Cr, which may indicate guarded prognosis; d) good spectral resolution, which would not have been the case if there had been fresh blood in the VOI; and e) reduced, but not absent Cr and NAA indicates that the lesion is unlikely to be a tumor (despite some partial volume effect).

E. Location 2 (Contra Lateral)

1. Peak Identification and Quality Assurance: Reading the Spectrum from Right to Left

The spectral quality is excellent (Fig. 5.122). Reading from right to left, the spectrum is basically a normal white matter spectrum, with no significant signals before NAA at 2.02 ppm, followed by β,γ-Glx, the second NAA, Cr, Cho, sI, mI, α-Glx, and the second Cr.

2. Interpretation and Analysis

[Cr] is not available. The peak heights are estimated by visual inspection with a ruler. The residual water causes a baseline, which is adjusted for in the inspection process. The ratios obtained are: NAA/Cr = 1.5, Cho/Cr = 0.8, and mI/Cr = 0.6, which are all within the normal range.

F. Conclusion

MRS in the lesion is consistent with a nonacute hemorrhagic infarct. The findings were normal in the contralateral location.

G. Summary of Patient's History Post-MRS

Not available.

H. Other Comments

Note the NAA, Cr, Cho and mI peaks in the unaffected spectrum follow a nice slope which is typical for white matter. Although the spectrum is not acquired from HMRI's standard occipitoparietal VOI, the appearance of the white matter spectrum (Fig. 5.122) is very similar.

5.6.2 HEMORRHAGIC INFARCT, RULE OUT SECONDARY CARCINOMA: NEGATIVE STUDY

A. Purpose of MRS

To differentiate secondary carcinoma from hemorrhage. The case demonstrates a successful and a "failed" PROBE Q.

B. Summary of Patient's History

Female, age 68 years, with history of breast cancer, developed a small, nonspecific lesion in right basal ganglia.

C. MRS Method, Volume of Interest, and Comments about Examination

PROBE-Q STEAM (TE = 30 ms, TM = 13.7 ms, TR = 1800 ms—*not* 1500 ms) with numerical result was performed. VOI was placed in basal ganglia, a notoriously difficult target for PROBE, PROBE-Q or MRS. On the right, the partial volume occupied by the lesion is probably less than 50%.

D. Basal Ganglia and Contralateral Control Locations

1. Peak Identification and Quality Assurance: Reading the Spectrum from Right to Left

The PROBE-Q spectrum (Fig. 5.123, upper) is successful, with broad resonances, and undersuppression of water clearly visible. PROBE-Q automated peak integration, nevertheless, failed. The immediate repetition of PROBE-Q (Fig. 5.123, lower) was successful, and provided excellent spectral quality and automated peak integrals (numbers shown). PROBE-Q in the contralateral control VOI (Fig. 5.123, middle) was equally successful, and of excellent quality, despite the difficult location.

2. Interpretation and Analysis

Despite the poor spectral appearance, no excess of lipid or lactate doublet at 1.33 ppm would be inferred from the initial PROBE-Q examination. Reading from right to left, peak ratios by inspection are: NAA/Cr = 1.5; Cho/Cr = 0.73; and mI/Cr = 0.6. These ratios are not pathological.

In the subsequent, successfully integrated PROBE-Q MRS, NAA/Cr = 1.41, Cho/Cr = 0.81, and mI/Cr = 0.49; all are completely within normal limits. The lipid, protein, lactate region is also entirely unremarkable.

Comparison with the contralateral VOI, both visual and by the calculated peak ratios, is impressively similar.

No pathology was detected with MRS in the obviously pathological basal ganglia.

E. Conclusion

The striking normality of MRS strongly militates against the visible lesion being a secondary neoplasm.

F. Summary of Patient's History Post-MRS

Imaging studies performed two weeks later confirmed there was hemorrhage in this lesion, but could not totally exclude an underlying metastasis.

G. Other Comments

Failures of automated shimming occur in the presence of blood, in difficult locations of magnetic field inhomogeneity, and (but not in this case) when VOIs of very small volumes are attempted. As in this case, repeated examinations probably will succeed because shim files established in the prior study are much closer to the desired values.

Partial volume should not be overlooked as a cause of negative MRS findings when lesions are obvious on the images. Another possibility, as in this case, is that the metabolite profile of the MRI lesion is indeed essentially normal. Hemorrhage alone does not signify hypoxia, necrosis, or other pathology. In short, hemorrhage on MRI is not a major indication of MRS.

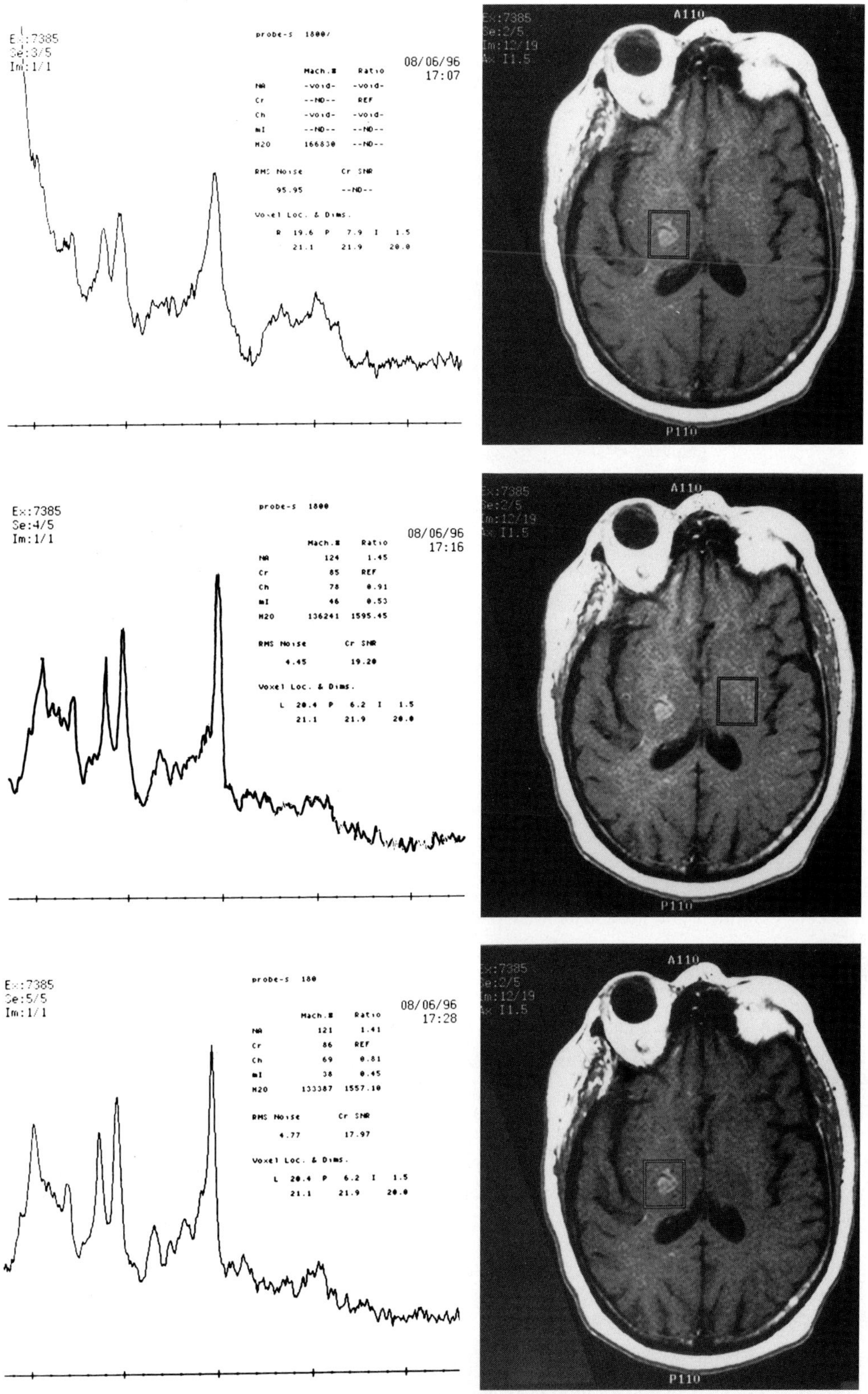

Fig. 5.123 A series of PROBE-Q examinations performed to rule out secondary carcinoma in the basal ganglia. The small lesion identified in the axial MRI was subjected to two PROBE-Q examinations: the first (upper) "failed" (the resolution is poor and PROBE-Q did not produce metabolite ratios); the second (lower) is of excellent quality, and not different from the normal basal ganglia (middle). (Courtesy of Dr. D. Hungerford, Mobile Infirmary, Alabama.)

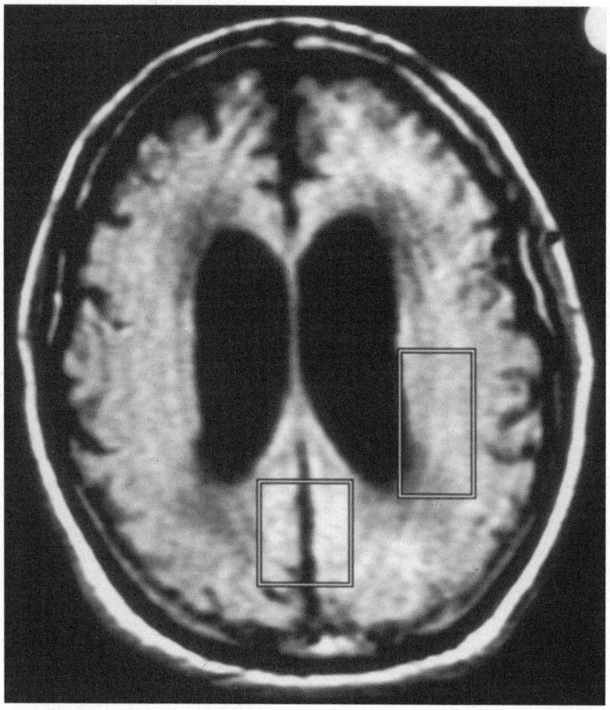

Fig. 5.124 Abnormal MRI, 14 days after cardiac arrest, shows marked ventricular dilatation and cortical atrophy, but no focal abnormalities.

5.6.3 ASTHMATIC CARDIAC ARREST

A. Purpose of MRS

MRS was requested to determine the prognosis after possible cardiac arrest in an asthmatic.

B. Summary of Patient's History

An asthmatic smoker, with a history of inhalant abuse, this 17-year-old male was found in ventricular fibrillation and probable cardiac arrest. At the time of the MRS examination (14 days after accident) he was unresponsive, and on ventilator support.

C. MRS Method, Volume of Interest, and Comments about Examination

STEAM (TE = 30 ms, TM = 13.7 ms, TR = 1500 ms) was performed in HMRI's parietal white matter (Location 1), and in the occipital grey matter (Location 2). The MRI showed marked ventricular dilatation and cortical atrophy, consistent with long-term effect of global hypoxic insult, (Fig. 5.124).

D. Location 1 (White Matter)

1. Peak Identification and Quality Assurance: Reading the Spectrum from Right to Left

The MR spectrum (Fig. 5.125) is of excellent technical quality.

Reading from right to left, there is a small lipid peak at 0.9 ppm, and a very well-resolved excess of lactate doublet at 1.33 ppm, with 7 Hz (0.11 ppm) splitting. Inversion of the baseline at 1.75 ppm is due to possible outer-volume (but still conceivably, intracerebral) lipid, but does not limit diagnostic quality of the MRS. Continuing, there is essentially no NAA, with perhaps β,γ-Glx, an unidentified peak (possibly aspartate) at 2.6 ppm, Cr, Cho, sI, mI (at 3.56 ppm), indistinctly resolved α-Glx, a second Cr and mI (at 4.06 ppm).

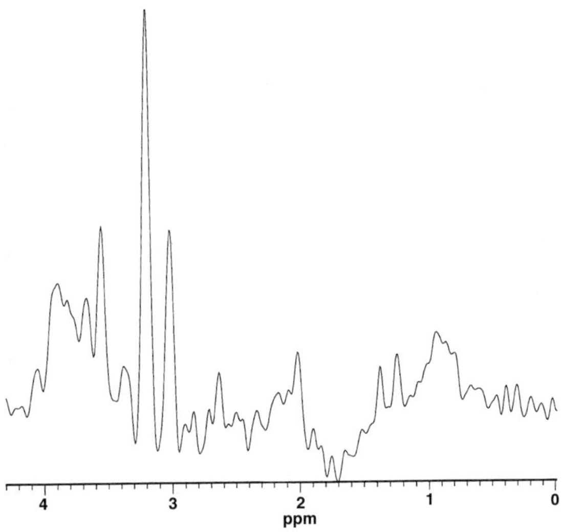

Fig. 5.125 The spectrum obtained from white matter 14 days after cardiac arrest in a 17 year-old asthmatic. There is notable lactate, and almost complete loss of NAA consistent with hypoxic encephalopathy. The elevated Cho/Cr and mI/Cr probably represent gliosis, which develops only after a week or more.

2. Interpretation and Analysis

A strikingly abnormal spectrum, with excess lactate clearly separated from lipid, which is probably also pathological, and essentially no NAA. The calculated peak ratios are: NAA/Cr = 0.37 (25% of normal, but almost certainly an overestimate compared with NAA/Cr < 0.1, by visual inspection); Cho/Cr = 2.05 (150% above normal); and mI/Cr = 0.97 (59% above normal). [Cr] is no more than 70% of normal, so that [NAA] is 15% of normal (or less), [Cho] is 177% of normal, and [mI] is 27% above normal. β,γ-Glx/Cr is not increased, and while α-Glx/Cr is increased significantly, this should be read as an excess of glucose, supported by the second glucose resonance which is prominent at 3.43 ppm. Brain water is 64.4% (normal).

The result is characteristic of late hypoxic encephalopathy with very poor prognosis.

E. Location 2 (Grey Matter)

1. Peak Identification and Quality Assurance: Reading the Spectrum from Right to Left

The MR spectrum (Fig. 5.126) is of good quality, with the exception of an inversion of the baseline at 1.55 ppm (due, probably to intracerebral, but outer volume, lipid) which distorts the expected lactate doublet at 1.33 ppm.

Reading from right to left, the lipid excess is prominent at 0.9 ppm, and the lactate doublet at 1.33 ppm is distorted. Continuing right to left, the following metabolites are identified: NAA, β,γ-Glx, Cr, Cho, glucose, mI, and α-Glx. The second peaks of Cr and mI are present.

2. Interpretation and Analysis

This strikingly abnormal spectrum, exhibiting possible lactate and lipid, is compatible with the clinical vegetative status at the time of the examination. The calculated NAA/Cr is 0.54, but by inspection is effectively < 0.3; Cho/Cr = 1.75 (twice normal); mI/Cr = 0.85 (significantly increased); β,γ-Glx/Cr = 0.17 (reduced); α-Glx/Cr = 0.68 (probably increased cerebral glucose). [Cr] is determined as only 55% of normal. Thus, all metabolite concentrations are only half those estimated above from the ratios of peaks. Brain water of 67.4% is normal, while CSF is 12.3% of the VOI, an increase secondary to cortical atrophy (MRI is normal).

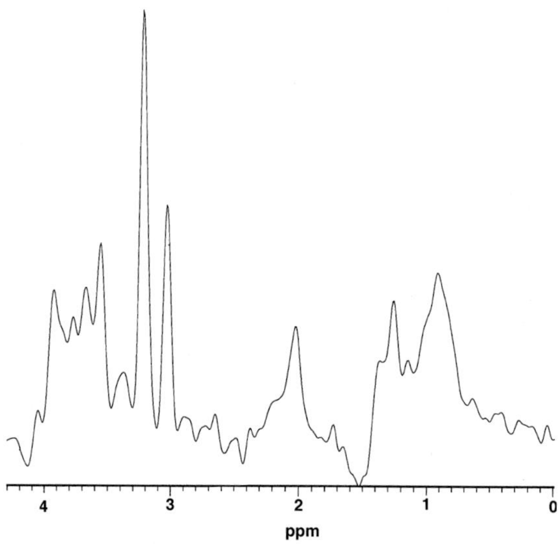

Fig. 5.126 Grey matter spectrum. Even with the absence of lactate, the MRS results are conclusive.

F. Conclusion

The almost total loss of NAA and reduced [Cr], even without lactate, is indicative of severe hypoxic brain injury. The elevation of Cho/Cr and mI/Cr appear as late events, consistent with the two-week interval which elapsed between injury and the MRS examination.

Prognosis is grim, with persistent vegetative state inevitable. An earlier examination would have yielded the same information, and no follow-up MRS is required.

G. Summary of Patient's History Post-MRS

The patient was discharged to long-term care. An interview confirmed the hopeless neurological status 12 months later.

H. Other Comments

MRS performed 24 or 48 hours after admission, would have reduced the need for many other tests. In younger patients with, global hypoxia the MRS criteria for determining clinical outcome have been established. Loss of more than 30% of NAA has uniformly poor prognosis. In this case, the loss was virtually 100%. This case should be compared with the following case, where a similar clinical history had a contrasting, favorable outcome, readily predicted by MRS.

5.6.4 APNEA VERSUS CEREBRAL VASCULAR ACCIDENT (PART 1) AND FOCAL HYPOXIC BRAIN INJURY (PART 2)

PART I

A. Purpose of MRS

MRS was requested to determine prognosis (survival) after the sudden, unexplained onset of a coma.

B. Summary of Patient's History

Male, age 24 years, admitted in deep coma. Patient was undergoing out-patient analgesic therapy after an accidental, self-inflicted gun shot wound to the pelvic girdle. The patient, having been previously normal, was found unrousable in bed one morning. Initial differential diagnosis was between cardiac arrest/hypoxic encephalopathy (question as to cause), overdosage of analgesics, including fentanyl, a cerebrovascular accident, or fat embolism.

Twelve days after admission to hospital, the *first* MRS was performed while patient was still comatose (Glasgow Coma Scale 6–7). Involuntary movements by the patient made MRS technically difficult. Clinical prognosis was deemed very poor (with one dissenting opinion). The complexity of the clinical history is described in the Comments section of this case study.

C. MRS Method, Volume of Interest, and Comments about Examination

STEAM (TE = 30 ms, TM = 13.7 ms, TR = 3000 *and* 1500 ms) was performed in HMRI's occipital grey matter (Location 1), and parietal white matter (Location 2) VOIs, Fig. 5.127. The patient was not perfectly motionless during the examination. Therefore, only 64 scans were acquired in the grey matter (at TR = 3000), and 128 in the white matter (at TR = 1500).

D. Location (Grey Matter)

1. Peak Identification and Quality Assurance: Reading the Spectrum from Right to Left

This is a well-resolved spectrum of excellent technical quality, Fig. 5.128. Reading from right to left, small but probably abnormal lipids or macromolecules are seen at 0.9 ppm and 1.3–1.4 ppm. There is definitely no excess of lactate.

Continuing reading from right to left, the following resonances are identified: NAA, β,γ-Glx, the second NAA, Cr, Cho, sI or glucose, mI, α-Glx, and the second Cr and mI.

2. Interpretation and Analysis

The metabolite ratios are: NAA/Cr = 1.08 (27% below normal); Cho/Cr = 0.78 (25% above normal); mI/Cr = 0.51 (normal). Brain water was determined to be 87.1% (this is about 10% above normal; possible artifact from patient motion), with a CSF compartment of 4% (normal). β,γ-Glx = 0.33, and α-Glx = 0.51; both are in the normal range. [Cr] is 89% of normal. [NAA] is 0.73, which is appropriately and significantly reduced, but the ratio of NAA to Cr (NAA/Cr) is less severely affected than it *should* be. Cho/Cr is best explained by a small (10%) reduction in the denominator, together with a small but significant elevation of [Cho] in this location.

E. Location (White Matter)

1. Peak Identification and Quality Assurance: Reading the Spectrum from Right to Left

Considering patient motion (reflected in the poor quality of MRI), and the small number of acquisitions, MRS, Fig. 5.129, is of acceptable diagnostic quality. A "shoulder" on the Cr peak before line broadening is the only sign of motion artifact. In the lipid region, there are two prominent triglyceride resonances (0.85 and 1.25 ppm), which are unusual for brain in that they are present as a ratio ~4:1, more like free triglyceride. This is a clue to question the result, and possibly to repeat the examination. There is no clear evidence of a lactate doublet at 1.33 ppm.

Reading from right to left, the unusual lipid pattern and probable absence of lactate are noted. Furthermore, the following resonances are identified: NAA, a singlet at 2.12 ppm (unidentified), β,γ-Glx, second NAA, Cr, Cho, possible sI, mI, α-Glx, second Cr, and mI peaks.

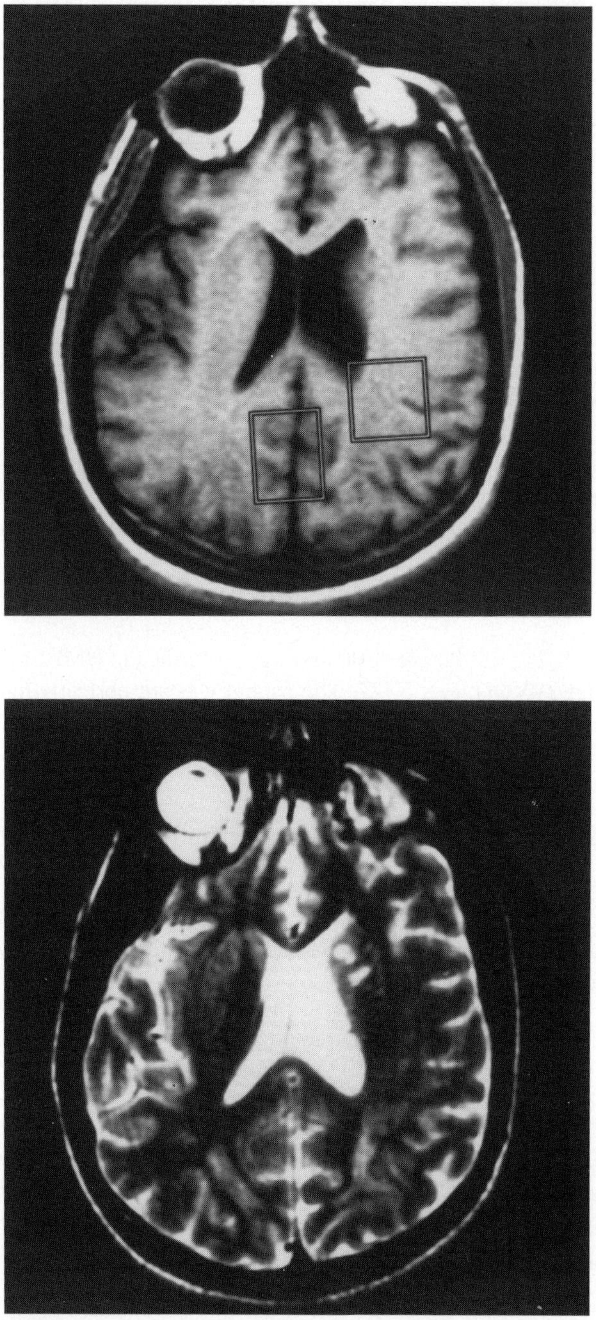

Fig. 5.127 Axial MRI localizers, although asymmetric as a result of inadequate patient positioning show no pathology, even in the posterior parietal cortex at the time of this examination (see the next figure of MRS in grey matter on Day 12). Locations of MRS were selected to optimize detection of global hypoxic encephalopathy. The white matter location is far from scalp. The lower image is included to look for evidence of a lipid droplet in the white matter VOI that could explain the dominant lipid peaks detected in this VOI.

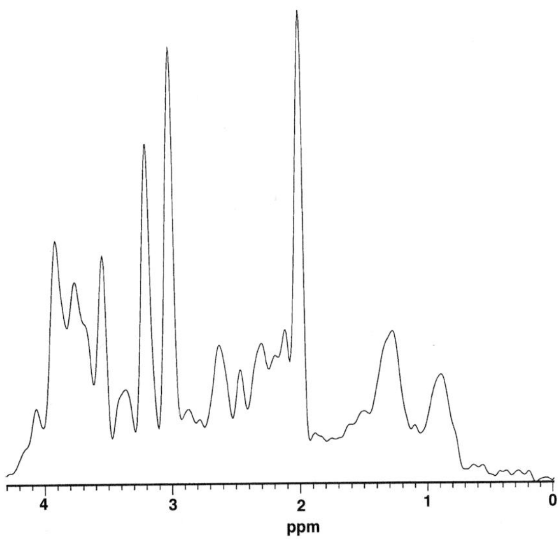

Fig. 5.128 MRS of grey matter Day 12. The first MRS examination, 12 days after presentation as "comatose post cardiac arrest," shows no evidence of severe hypoxic injury. NAA/Cr is reduced; no lactate is present. Cho/Cr increase probably reflects below-normal Cr (see Text).

2. Interpretation and Analysis

Lipid is apparently intracerebral and very prominent (half the height of NAA), and consistent with "necrosis" after hypoxic injury. The metabolite ratios are: NAA/Cr = 1.12 (reduced 20%); Cho/Cr = 0.97 (increased 17%); mI/Cr = 0.51 (normal). β,γ-Glx = 0.33 (normal), and α-Glx = 0.64 (slightly elevated). This result may arise from excess glucose. [Cr] is 81% of normal, making the ratios an overestimate.

F. Conclusion

A relatively good MRS result in both locations, making it unlikely that the patient suffered prolonged anoxia. Nevertheless, reduction in [NAA] and [Cr] together imply significant neuronal injury. In the absence of lactate or of elevated Glx, a guarded, but generally good prognosis may be given. The result is nevertheless puzzling; lipid accumulation is generally a "bad" prognostic finding, but the absence of lactate or glutamine, and NAA/Cr that remains > 1.00 is optimistic. The increase in Cho/Cr is largely the result of reduced Cr, which must contribute to the relatively poor prognosis.

A caveat is the unknown clinical diagnosis. These interpretations are based upon widespread or global changes. If evidence of "localized" hypoxia becomes available, this points to better recovery with only focal neurological deficits. A speculative cause of lipid in the brain is fat-embolism. The patient's history of fracture lends some support to this speculation.

G. Summary of Patient's History Post-MRS

The patient awoke, and over the course of several months became alert, able to walk with weakness only in the left leg, but suffered significant visual field deficits. A series of MRI and subsequent MRS examinations explain the initial findings (see Part II below).

There were several possible pitfalls elucidated by this case. First, in patients with stroke, heterogenous MRS results are obtained depending upon where the VOI is placed. Generally, a grey matter and white matter VOI will allow the correct prognosis. In this patient, hypoperfusion determined in part by gravity, affected the posterior ⅓ of the cerebrum, disproportionately.

Second, unknown drug effects may contribute to MRS findings. Fentanyl may be such a drug. Finally, the rare finding of lipid-triglyceride may point to fat-embolism from an earlier pelvic fracture.

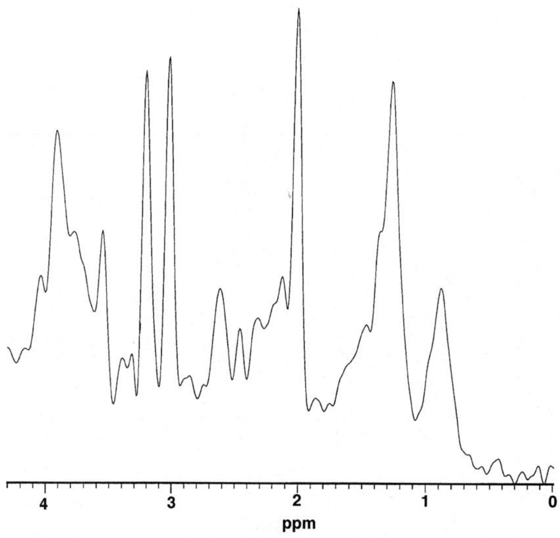

Fig. 5.129 MRS of white matter Day 12. Narrow line-width and modest reduction of NAA/Cr suggests little hypoxic injury. Line-shapes of the dominant lipid peaks point to artifactual origin or unusual brain lipid.

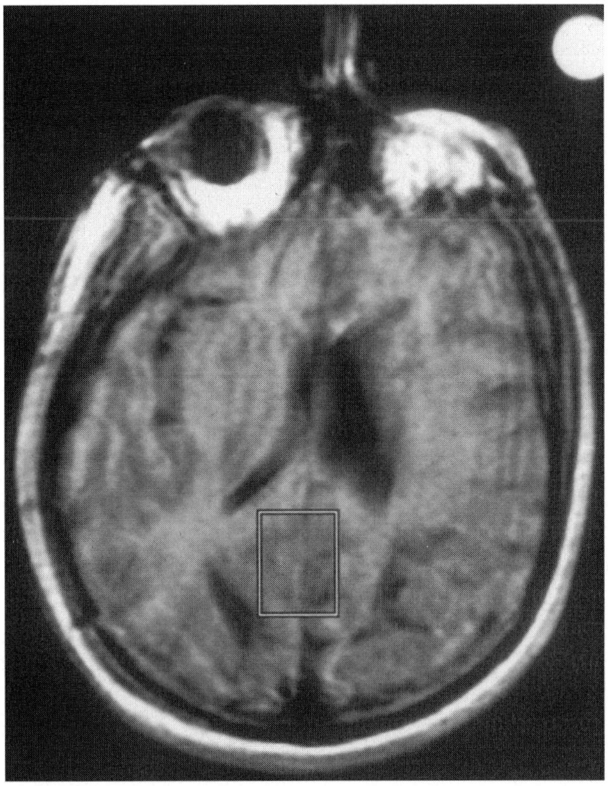

Fig. 5.130 Axial MRI at Day 26. Despite severe motion-artifact, possible widening of posterior-partial sulci is evidence of early cortical atrophy.

PART II

A. Purpose of MRS

MRS was requested by the family for prognosis after incomplete recovery from hypoxic encephalopathy.

B. Summary of Patient's History

The patient was 24 at the time of the *second* MRS examination 26 days after admission to hospital. He was confused, but recovering. Clinical prognosis was guarded, but optimistic. Further follow-up MRS examinations were performed seven and ten months after initial onset of coma.

C. MRS Method, Volume of Interest, and Comments about Examination

STEAM (TE = 30 ms, TM = 13.7 ms, TR = 3000 ms) was performed in HMRI's occipital grey matter Fig. 5.130; examination was terminated prematurely.

D. Location (Grey Matter)

1. Peak Identification and Quality Assurance: Reading the Spectrum from Right to Left

This is a well-resolved spectrum of good quality, Fig. 5.131. Reading from right to left, unremarkable "lipids," and no excess lactate. Continuing, the following resonances are identified: NAA, β,γ-Glx, second NAA, Cr, Cho, glucose, mI, α-Glx, and second Cr, and mI.

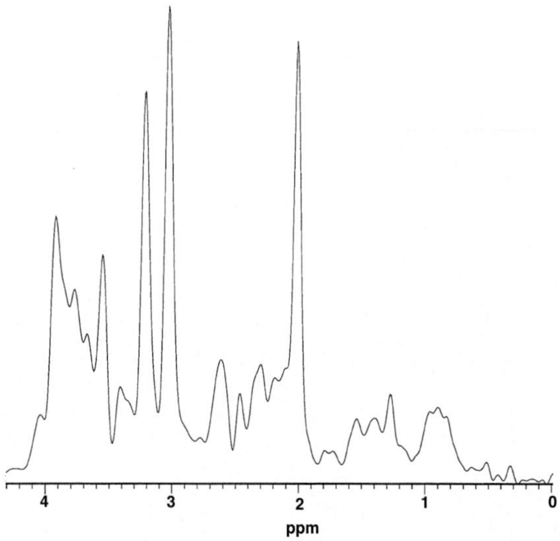

Fig. 5.131 MRS of grey matter at Day 26. [Cr] has recovered towards normal but NAA/Cr is significantly below normal; evidence of significant neuronal loss in this region.

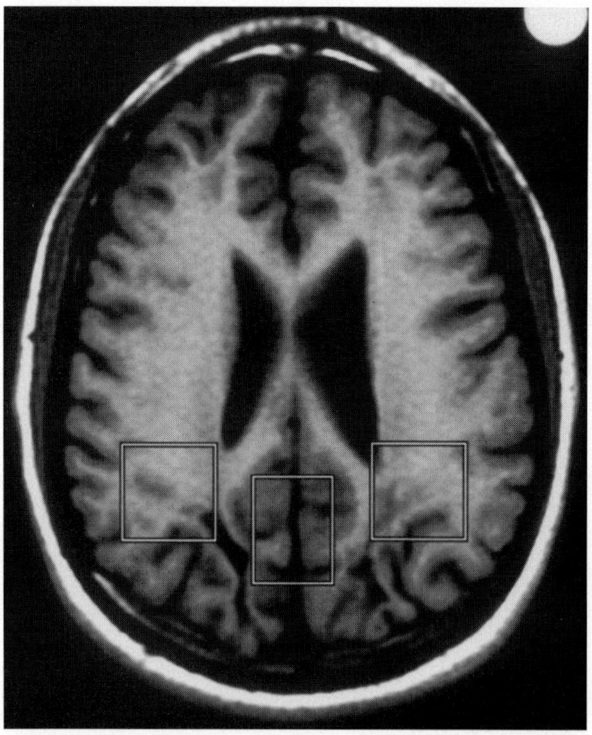

Fig. 5.132 Axial MRI after 7 months. Cortical atrophy is increased in the posterior parietal cortex; also shown are three VOI's examined by MRS, see the next figure.

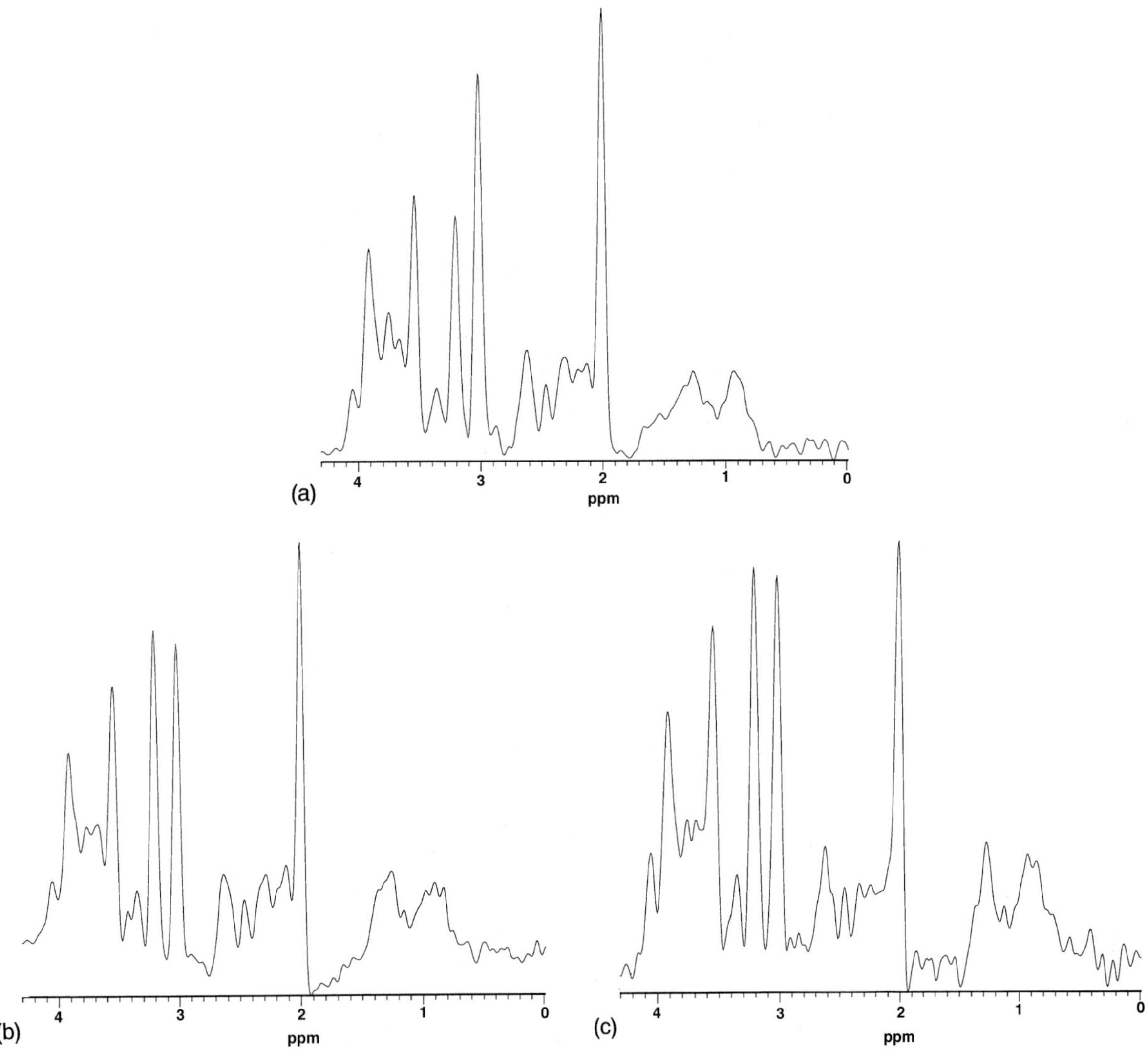

Fig. 5.133 MRS after 7 months of grey matter (a) and right and left parietal white matter (b) and (c). Three MRS examinations performed at follow-up 7 months after the original admission show persistent reduction of NAA and increase in mI/Cr (particularly in left parietal white matter (c)). MRS is thus consistent with neuronal loss and gliosis indicated by cortical atrophy in the corresponding MRI.

2. Interpretation and Analysis

The metabolite ratios are: NAA/Cr = 0.92 (32% reduced); Cho/Cr = 0.82 ("increased" 39%); and mI/Cr = 0.49 (reduced by 10–15%). [Cr] is 100%, and brain water approaching normal at 77%.

E. Conclusion

Compared to an examination in the same location 14 days previously, significant changes have occurred. First, the normalized [Cr], more normal brain water content, and increased Cho/Cr ratio are all indicative of considerable *improvement*. Nevertheless, in this location, NAA is 30–35% below normal, providing evidence of severe injury in *this region of the brain*.

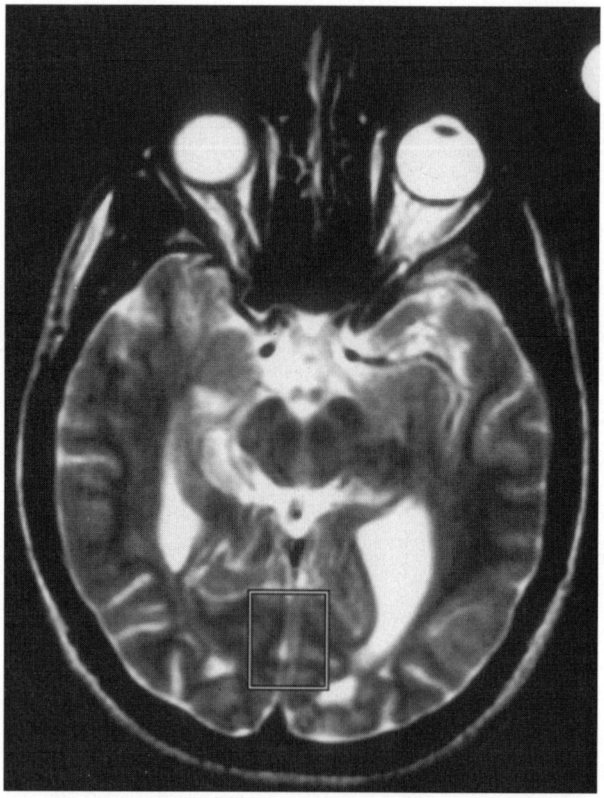

Fig. 5.134 Axial MRI after 10 months. There is again evidence of cortical thinning. Location of the VOI in visual cortex is indicated, inferior to the standard grey matter location.

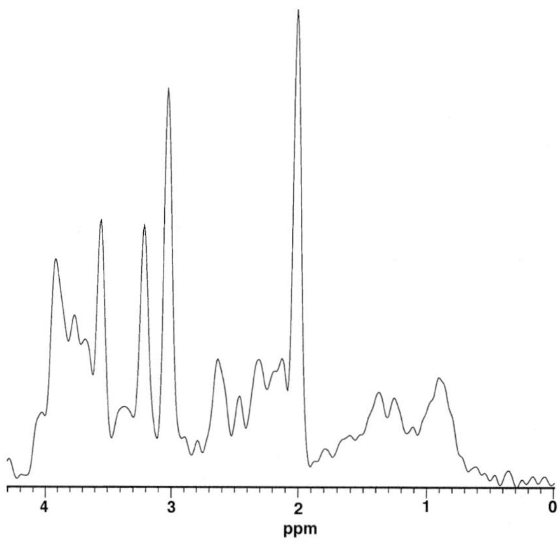

Fig. 5.135 MRS of grey matter at 10 months. The result is essentially indistinguishable from that of the grey matter spectrum acquired 3 months earlier.

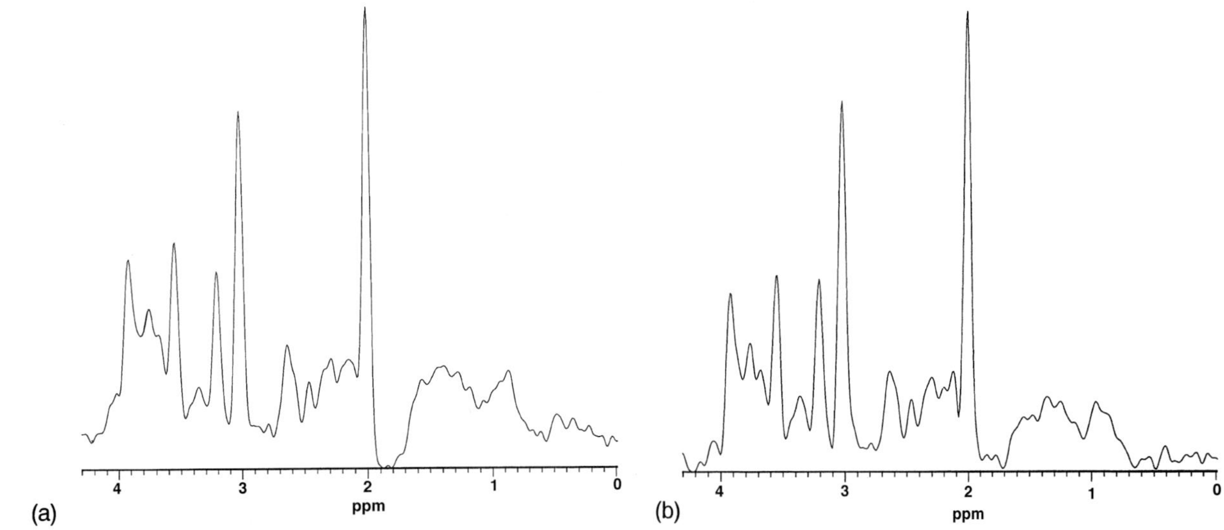

Fig. 5.136 MRS of visual cortex. (a) The examination was performed in an attempt to explain homonymous hemianopia. Cho/Cr < 0.5 and NAA/Cr do not indicate local neuronal loss. (b) Visual cortex in age-matched normal subject.

Lipid is not present in the grey matter location; the critical reexamination of left parietal white matter was not possible due to patient movement.

F. Summary of Patient's History Post-MRS

The patient was transferred to rehabilitation where a neurological examination showed the following, three months after MRS: Alert and no focal neurological signs apart from partial homonymous hemianopia and bilateral upgoing toes.

Subsequent examinations at seven (Fig. 5.132 and Fig. 5.133) and ten months (Fig. 5.134, Fig. 5.135, and Fig. 5.136) were performed to compare left and right parieto-occipital white matter. Comparable mild abnormalities with reduced NAA/Cr and increased Cho/Cr, mI/Cr and absolute concentrations of both [Cho] and [mI] appear to indicate that recovery from hypoxic injury extended into both hemispheres. Fig. 5.134 acquired at 10 months shows the location of a VOI within the grey matter of visual cortex may also now show some ventricular dilatation of the posterior horns. Occipital grey matter (at 10 month, Fig. 5.135), and an occipital location selected to include visual cortex, showed almost normal metabolite ratios and concentrations, and no specific neuronal loss from the primary visual cortex, Fig. 5.136.

G. Conclusions (Parts I and II)

The history and MRS are quite atypical of "arrest." We conclude that some "recovery" is to be explained by cortical atrophy and consolidation. Another element, severe overdose of long acting analgesics, may confuse the picture. Can these drugs reversibly depress NAA? Finally, it is tempting to call this fat-embolism for the following reasons:

(a) The patient has a complex fracture.
(b) The onset of "coma" was atypical.
(c) A strong lipid peak with intensities more appropriate for long-chain triglyceride was identified early, in normal-appearing white matter. At the same time, there is a small but significant excess of lipid in grey matter.
(d) The clinical course was surprisingly benign (despite residual neurologic deficits).

5.7 Focal Lesions: Ectopic Grey Matter, HIV Related Lesions, Tumors, and Tuberculoma

5.7.1 ECTOPIC GREY MATTER

A. Purpose of MRS

Quality assurance only.

B. Summary of Patient's History

Female, age 29 years, with a focus of ectopic grey matter bulging into the left trigone, but unchanged on numerous imaging studies over many years.

C. MRS Method, Volume of Interest, and Comments about Examination

Automated PROBE Q-STEAM (TE = 30 ms, TM = 13.7 ms, TR = 1800 ms—*not* 1500 ms) was performed in a region to include ectopic grey matter, and an equivalent contralateral VOI (Locations 1 and 2) (see image in Fig. 5.137).

D. Locations 1 and 2 (Ectopic Grey Matter and Contralateral)

1. Peak Identification and Quality Assurance: Reading the Spectrum from Right to Left

These spectra are of excellent quality Fig. 5.137, and can almost be superimposed. This is an excellent demonstration of reproducibility in a carefully conducted PROBE-Q examination.

2. Interpretation and Analysis

PROBE-Q ratios of metabolites are representative of normal white matter, with a small difference only in NAA/Cr; visible differences in the printed values. However, even small details, such as the similar lipid profiles, and the differences in the β,γ-Glx profile between 2.2 to 2.6 ppm, are immaterial.

E. Conclusion

This negative study serves to illustrate the value of PROBE-Q MRS when combined with a contralateral control.

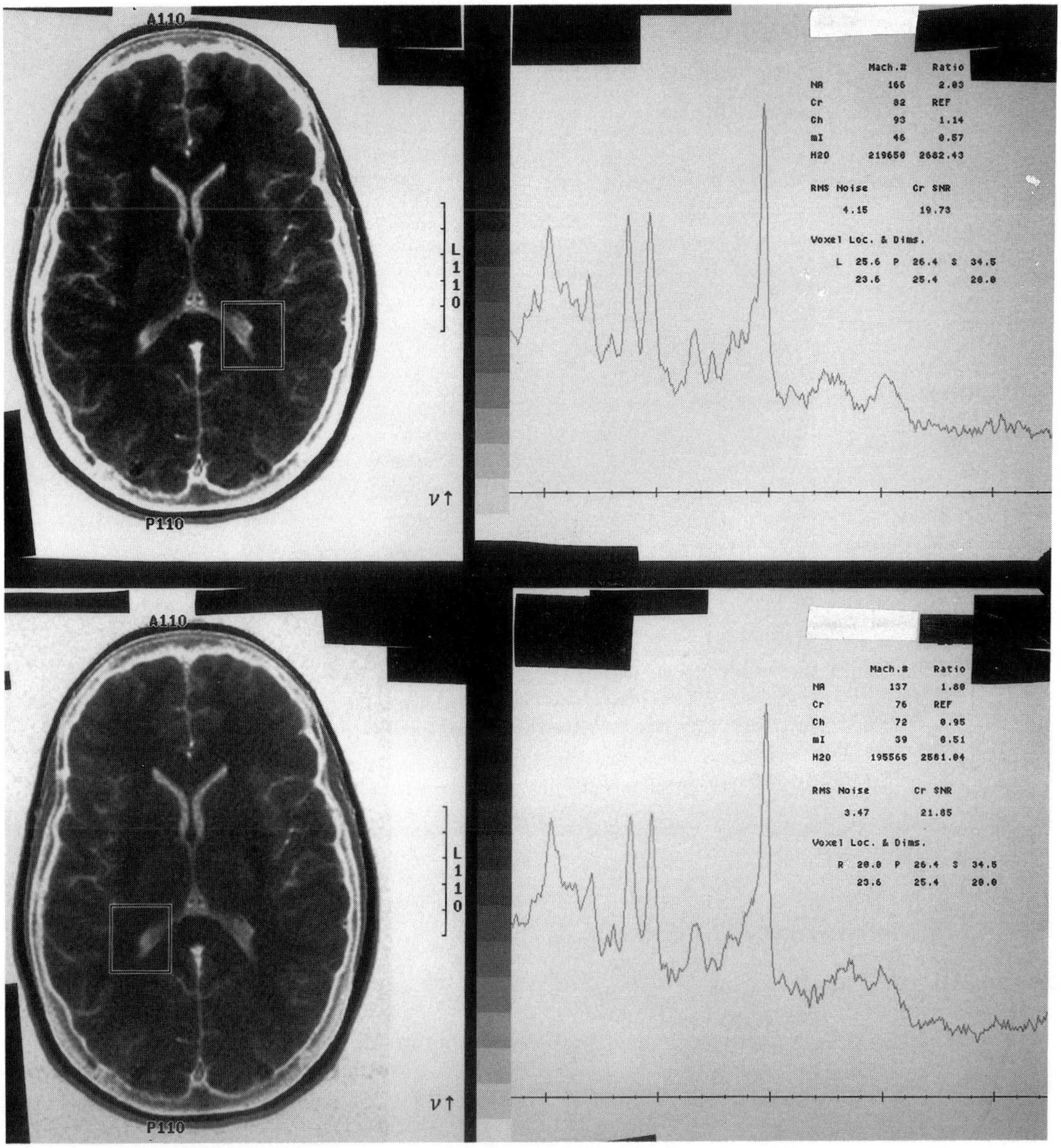

Fig. 5.137 Axial MRI and bilateral PROBE-Q illustrates normal profiles in a patient with ectopic grey matter (poorly visualized here) at the left trigone. (Courtesy of Dr. D. Hungerford, Mobile Infirmary, Alabama.)

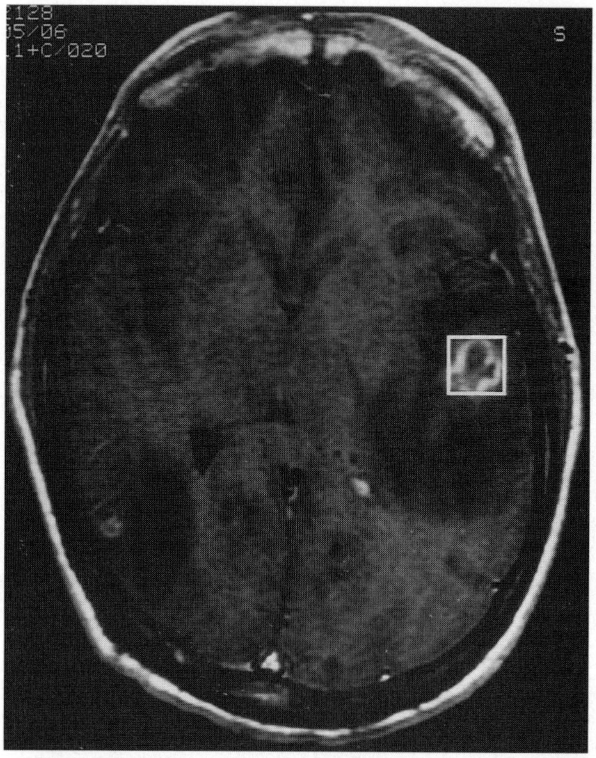

Fig. 5.138 T_1-weighted MRI showing the MRS location in the lesion. (Courtesy of Dr. J. Cousins, University Hospital, Albany, New York.)

5.7.2 AIDS BRAIN MASSES, TOXOPLASMOSIS

A. Purpose of MRS

MRI and MRS was requested to evaluate for intracerebral abnormality in this patient with neurological complications of AIDS.

B. Summary of Patient's History

Male, age 31 years, HIV positive, presenting with headaches, nausea, and vomiting.

C. MRS Method, Volume of Interest, and Comments about Examination

Automated PROBE-STEAM (TE = 30 ms, TM = 13.7 ms, TR = 3000 ms) was performed in a VOI ($1.2 \times 1.1 \times 1.5$ cm^3) within the lesion as shown on the image (Fig. 5.138). A contralateral VOI and VOIs from other lesions were also sampled (not shown). The MRS was performed following a routine MRI using gadolinium contrast (13 cc Magnevist®).

D. Location: Within Lesion

1. Peak Identification and Quality Assurance: Reading the Spectrum from Right to Left

The first impression is a markedly abnormal spectrum with no significant signal from the metabolites seen in normal spectra (Fig. 5.139). The S/N is good, and two broad resonances seen at 0.9 and 1.3 ppm are assigned to lipid. Only very small residual signals close to the noise level are seen at the chemical shift positions where the normal metabolites resonate.

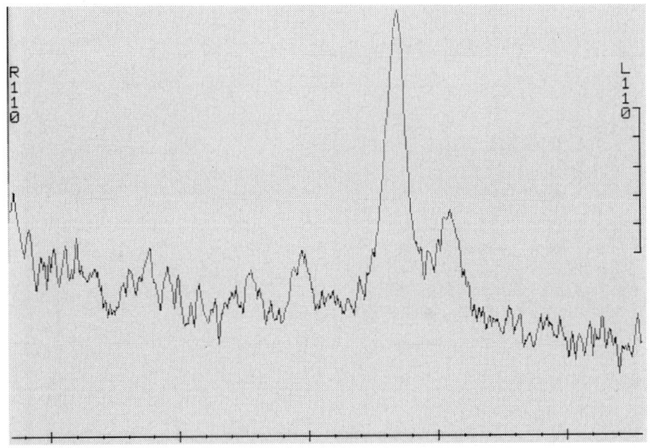

Fig. 5.139 The MR spectrum is consistent with toxoplasmosis. (Courtesy of Dr. J. Cousins, University Hospital, Albany, New York.)

2. Interpretation and Analysis

Metabolite ratios are not available. The strong lipid signal is the "fingerprint" of this spectrum. A comparison to the MRS from the contralateral VOI (not shown) indicated that the lipid signal is very intense; the peak at 1.3 ppm seems to be similar in size to the assumed normal NAA peak. In this case it cannot be ruled out that part of the 1.3 ppm resonance is due to lactate, despite the lack of the characteristic doublet. The general rule is that if the Cho and Cr are convincingly separated, then the resolution should also allow lactate to be seen as a doublet. But in this case, Cho and Cr are absent, and the intrinsic resolution may not be sufficient to show lactate as a doublet superimposed on the lipid peak.

E. Conclusion

The MRS result showing the presence of lipid and possibly lactate, with no other metabolites seen, is consistent with toxoplasmosis. Lymphoma can be ruled out because Cho is increased in lymphoma.

F. Summary of Patient's History Post-MRS

The patient received drug therapy to treat the toxoplasmosis. No follow-up MRS is available.

G. Other Comments

This case is an example of how easily the automated MRS examination can be combined with routine MRI in the population of HIV positive patients. In the cases where an AIDS brain mass is found, MRS can immediately differentiate between lymphoma, toxoplasmosis, PML, and cryptococcoma, hereby avoiding brain biopsies. The identification of toxoplasmosis is of special interest because a successful treatment for this lesion is available, albeit potentially toxic. Therefore, treatment should be reversed for those patients in whom it is not appropriate.

The differentiation between lymphoma and toxoplasmosis can be particularly difficult based on the MRI alone, but the MRS results are very distinct.

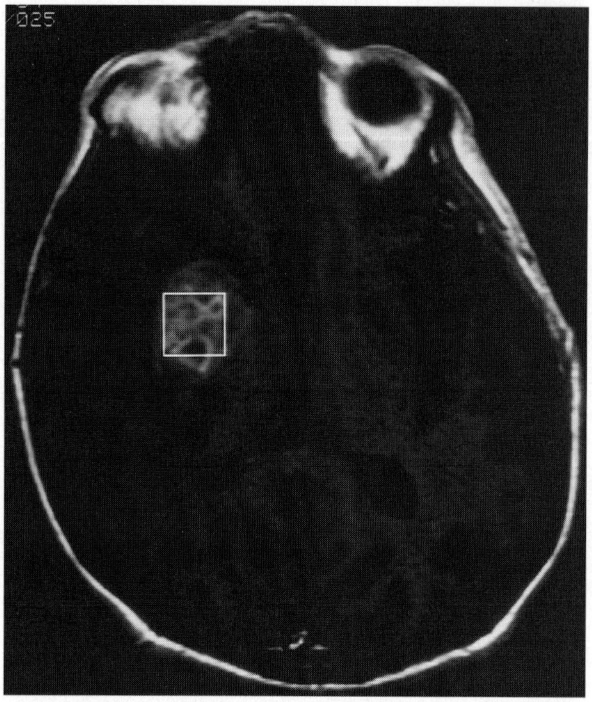

Fig. 5.140 T_1-weighted MRI showing the MRS location in the lesion. (Courtesy of Dr. J. Cousins, University Hospital, Albany, New York.)

5.7.3 AIDS BRAIN MASSES, CNS LYMPHOMA

A. Purpose of MRS

MRS was requested to evaluate for CNS lymphoma versus toxoplasmosis lesions, indistinguishable with MRI.

B. Summary of Patient's History

Male, age 49 years, HIV positive, CNS lesions.

C. MRS Method, Volume of Interest, and Comments about Examination

Automated PROBE-STEAM (TE = 30 ms, TM = 13.7 ms, TR = 2000 ms) was performed to obtain a spectrum from a $17 \times 17 \times 15$ mm^3 VOI in the gadolinium-enhanced lesion as shown on the image (Fig. 5.140). A second, contralateral VOI was measured (not shown), but the remaining smaller lesions were not measured due to increased patient discomfort and movement during the measurement of the second VOI. The MRS was performed following a routine MRI using gadolinium contrast (14 cc Magnevist®).

D. Lesion Location

1. Peak Identification and Quality Assurance: Reading the Spectrum from Right to Left

The first impression is a markedly abnormal spectrum (Fig. 5.141). The overall quality of the spectrum is good, judged by S/N, resolution, and the absence of artifacts. Two broad resonances, seen at 0.9 and 1.3 ppm, are assigned to lipid. Part of the 1.3 ppm is assigned to lactate, due to the 7 Hz splitting (0.11 ppm at 1.5 T). Neither of the NAA peaks are seen; the signal

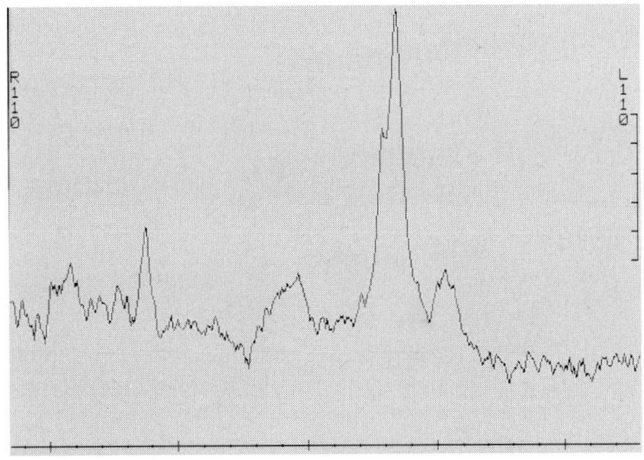

Fig. 5.141 The MRS spectrum is consistent with lymphoma. (Courtesy of Dr. J. Cousins, University Hospital, Albany, New York.)

between approximately 2.0 and 2.4 is due to β,γ-Glx. Apart from the lipid/lactate peaks, the major feature of this spectrum is a large narrow Cho peak. Cr and mI are not observed.

2. Interpretation and Analysis

Metabolite ratios are not available because Cr is absent. The intense lipid/lactate signal together with the presence of Cho is the "fingerprint" of this spectrum. The Cho intensity in the lesion is about twice the intensity of Cho in the "normal contralateral" spectrum (not shown), and the lipid/lactate signal is correspondingly large.

The narrow linewidth of the Cho peak indicates good spectral resolution, so in this case part of the 1.3 ppm resonance can be assigned to lactate, because the doublet with the correct splitting is seen at the correct chemical shift position, 1.33 ppm, and because the resolution is sufficient to show the splitting.

E. Conclusion

The MRS result, large amounts of lipid, lactate, and Cho present in the lesion, is consistent with lymphoma. Toxoplasmosis can be ruled out, because all the usual metabolites including Cho would have been suppressed to nil in toxoplasmosis.

F. Summary of Patient's History Post-MRS

Not available.

G. Other Comments

This case is an example of how easily the automated MRS examination can be combined with routine MRI in the population of HIV positive patients. In the cases when an AIDS brain mass is found, MRS can immediately differentiate between lymphoma, toxoplasmosis, PML, and cryptococcoma, hereby avoiding brain biopsies.

The differentiation between lymphoma and toxoplasmosis can be particular difficult based on the MRI alone, but the appearances of the MRS results are very distinct, and the differentiation is of special interest, because the alternative to MRS is biopsy, or trying to treat for toxoplasmosis, wasting valuable time if the lesion turns out to be lymphoma.

In differentiating AIDS masses, the nomogram published by Miller, and reproduced here (see p. 314), avoids errors in borderline cases.

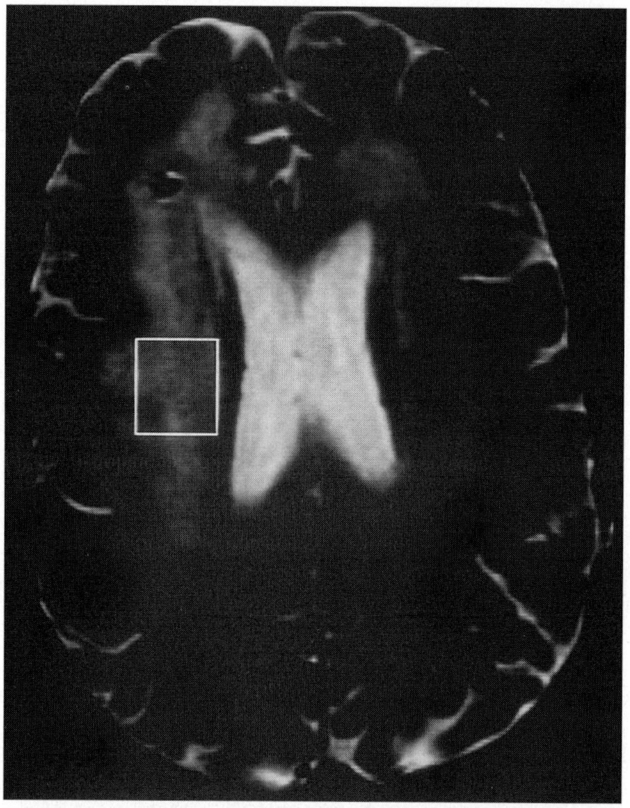

Fig. 5.142 T_2-weighted MRI showing the MRS location in a hyperintense region of the right centrum semiovale. (Courtesy of Dr. J. Cousins, University Hospital, Albany, New York.)

5.7.4 AIDS MASSES: MULTIFOCAL LEUKOENCEPHALOPATHY (PML)

A. Purpose of MRS

MRI/MRS was requested to evaluate for AIDS dementia or CNS lesions.

B. Summary of Patient's History

Male, age 38 years, HIV positive, neuropsychological changes. Reduced Mini Mental State score (details not available).

C. MRS Method, Volume of Interest, and Comments about Examination

The automated MRS method PROBE-S, (TE = 20 ms, TM = 13.7 ms, TR = 3000 ms). The $18 \times 20 \times 15$ mm^3 VOI was placed in a hyperintense region of the right centrum semiovale (Location 1) as shown on Fig. 5.142. A contralateral VOI of the same size in a region with only minor evidence of hyperintensity (Location 2) was measured to get the best possible internal control, Fig. 5.143.

The MRS was performed following a routine MRI using gadolinium contrast (17 cc Magnevist™). There was no evidence of patient motion.

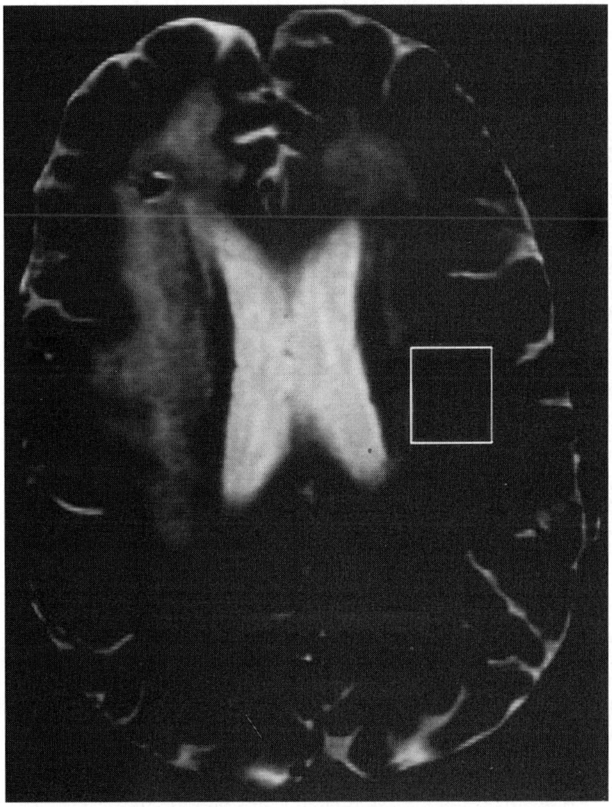

Fig. 5.143 T_2-weighted MRI showing the contralateral MRS location. (Courtesy of Dr. J. Cousins, University Hospital, Albany, New York.)

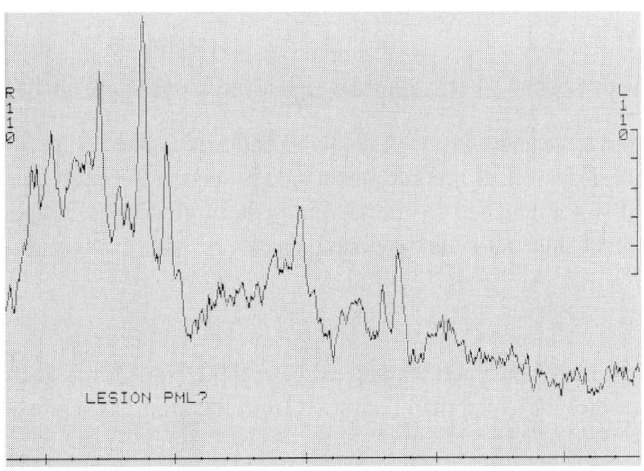

Fig. 5.144 The MRS spectrum (Location 1) consistent with multifocal leukoencephalopathy (PML). (Courtesy of Dr. J. Cousins, University Hospital, Albany, New York.)

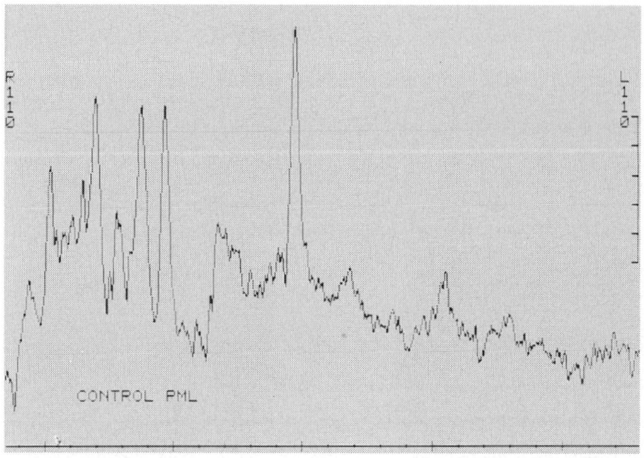

Fig. 5.145 The MRS spectrum from the contralateral VOI, Location 2. (Courtesy of Dr. J. Cousins, University Hospital, Albany, New York.)

D. Location 1 (right centrum semiovale)

1. Peak Identification and Quality Assurance: Reading the Spectrum from Right to Left

The first impression is that the spectrum is of high quality, and is markedly abnormal. The resonances are well resolved and S/N is good, Fig. 5.144. Lactate and NAA are observed, as well as a somewhat unusual β,γ-Glx region; may be "other amino acids (questionable)," or the result of using TE = 20 ms as opposed to TE = 30 ms. Continuing reading the Cr, Cho, mI, α-Glx and the second Cr peaks are observed. No other unusual peaks are assigned.

2. Interpretation and Analysis

The ratios are estimated to be NAA/Cr = 0.60 (less than half of normal), Cho/Cr = 1.80 (approximately double the normal value), and mI/Cr = 1.28 (approximately double the normal value). The [Cr] and compartmental data are not available, but [Cr] seems to be lower than normal based a comparison to the contralateral MRS, thus possibly contributing to the altered metabolite ratios.

E. Location 2 (contralateral VOI)

1. Peak Identification and Quality Assurance: Reading the Spectrum from Right to Left

This spectrum is of high quality, the resonances are well resolved and S/N is good, Fig. 5.145. NAA, β,γ-Glx, Cr, Cho, mI, α-Glx and the second Cr peaks are observed. A marked resonance is seen at 3.4 ppm half way between Cho and mI. This cannot be glucose (3.43 ppm) as it is not matched by the second peak of glucose at 3.8 ppm, and it is not convincingly the resonance of sI (3.36 ppm). No other unusual peaks are seen.

2. Interpretation and Analysis

The estimated ratios are: NAA/Cr = 1.1, Cho/Cr = 0.95, and mI/Cr = 0.90. NAA/Cr is decreased, and Cho/Cr and mI/Cr are increased, but the ratios are less severely affected than the ratios from location 1. The absolute signal of Cr compared to the noise is larger than in location 1.

F. Conclusion

The pattern of decreased NAA/Cr, increased Cho/Cr and mI/Cr, together with a moderate presence of lactate is consistent with multifocal leukoencephalopathy (PML), location 1. The metabolite ratios from location 2 are consistent with early stages of AIDS dementia.

G. Summary of Patient's History Post-MRS

Not available.

H. Other Comments

This case is an example of how easily the automated MRS examination can be combined with routine MRI in the population of HIV positive patients. In the cases when an AIDS brain mass is found, MRS can immediately differentiate between lymphoma, toxoplasmosis, PML, and cryptococcoma, hereby avoiding brain biopsies.

AIDS dementia, like Alzheimer's disease and other dementias, is diagnosed more readily from MRS in grey matter location. Inclusion of the focal abnormality in the VOI reduces NAA/Cr, and may increase mI/Cr. This contributes to some confusion since in the absence of focal lesions, reduced NAA/Cr and increased mI/Cr is otherwise pathognomonic of AIDS dementia.

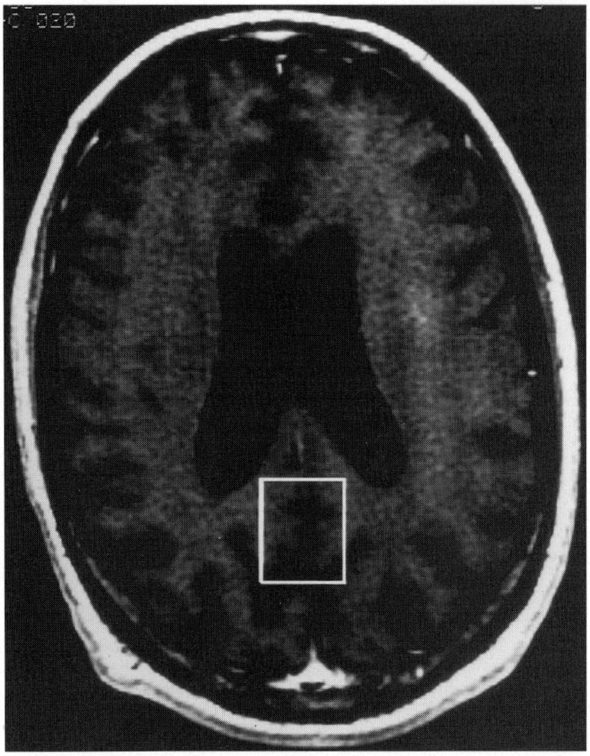

Fig. 5.146 T_1-weighted MRI showing the MRS location. (Courtesy of Dr. J. Cousins, University Hospital, Albany, New York.)

5.7.5 AIDS DEMENTIA

A. Purpose of MRS

MRI/MRS was requested to evaluate for AIDS dementia or CNS lesions.

B. Summary of Patient's History

Male, age 55, HIV positive, and neuropsychological changes.

C. MRS Method, Volume of Interest, and Comments about Examination

Automated PROBE-STEAM (TE = 30 ms, TM = 13.7 ms, TR = 1500 ms) was performed in HMRI's occipital grey matter VOI ($23 \times 27 \times 20$ mm^3) as shown on the image (Fig. 5.146). The MRS was performed following a routine MRI using gadolinium contrast (17 cc Magnevist®). There was no evidence of patient motion.

D. Grey Matter Location

1. Peak Identification and Quality Assurance: Reading the Spectrum from Right to Left

This is a well-resolved spectrum with good S/N (Fig. 5.147). NAA, β,γ-Glx, Cr, Cho, and mI are observed. No unusual resonances are observed.

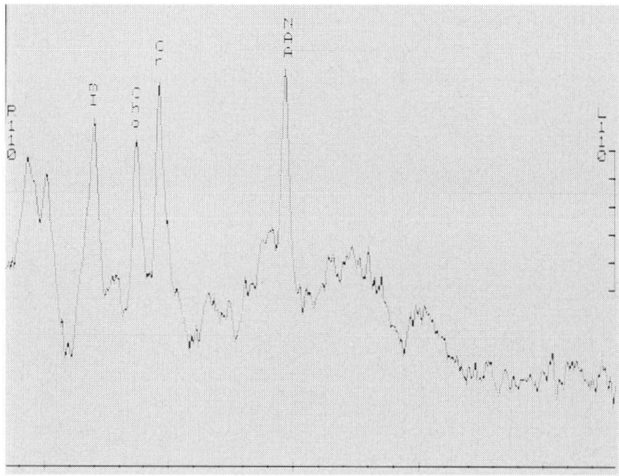

Fig. 5.147 The MR spectrum is consistent with AIDS dementia. (Courtesy of Dr. J. Cousins, University Hospital, Albany, New York).

2. Interpretation and Analysis

The metabolite ratios are estimated to be: NAA/Cr = 1.03 (ca 25% reduced), Cho/Cr = 0.75 (ca 25% increased), and mI/Cr = 0.83 (ca 40% increased). The [Cr] and compartmental data are not available.

E. Conclusion

NAA/Cr is decreased, which is consistent with diffuse AIDS encephalitis/AIDS dementia.

F. Summary of Patient's History Post-MRS

Not available.

G. Other Comments

The early detection of a decreased NAA/Cr ratio indicative of AIDS dementia is a very important application of MRS. Significant changes of the MRS can be observed before any abnormalities are seen on MRI, and in time for the patient to have a chance to benefit from treatments. Considering the various side effects of the treatments currently available, it is valuable to have a tool that can detect abnormalities early.

The increased mI/Cr ratio together with low NAA/Cr is usually associated with dementia of the Alzheimer's type, and an increased Cho/Cr ratio is usually associated with multi-infarct dementia. The mI/Cr and Cho/Cr ratios have, however, also been noted to be increased in patients with AIDS dementia. The precise significance thereof is not known, but it raises a question of whether there are any similarities of the biochemical background for the two disease processes.

A PROBE spectrum (not shown) from a primarily white matter VOI in the semiovale also showed decreased NAA/Cr.

The coexistence of PML with AIDS dementia may contribute to abnormalities of both NAA, and mI in this patient population. It will become of importance to differentiate between the two conditions with MRS if therapeutic trials now being instituted are to be optimally controlled.

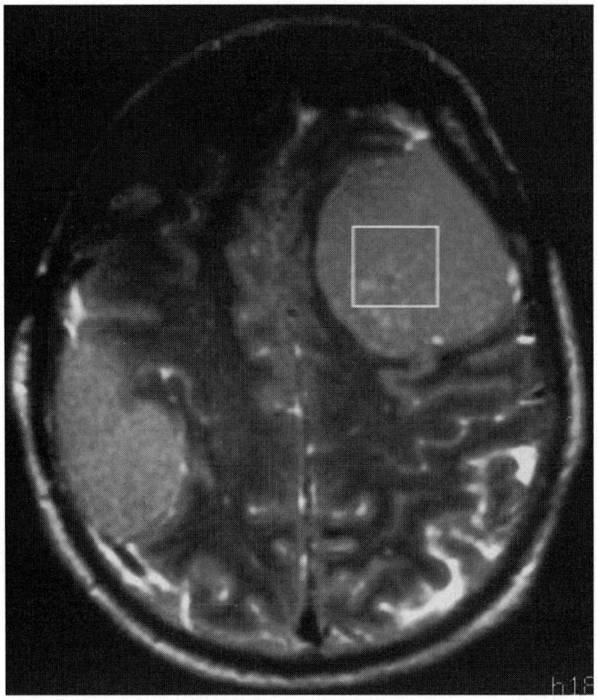

Fig. 5.148 MRI indicates multiple space occupying lesions in the brain, including the one used for PROBE examination. (Courtesy of Dr. Joseph Norfray, Chicago Northside MRI Center, Chicago, IL.)

5.7.6 RECURRENT MENINGIOMA

A. Purpose of MRS

To establish extent of meningioma recurrence after surgery, and confirm histological type prior to biopsy or repeat surgery.

B. Summary of Patient's History

Male, age 26, with meningioma proven by biopsy 10 months previously, and recurrent at time of MRS. MRI confirms suggestion of meningioma.

C. MRS Method, Volume of Interest, and Comments about Examination

Automated PROBE-STEAM (TE = 30 ms, TM = 13.7 ms, TR = 1500 ms) was performed within large frontal mass as shown on the image (Fig. 5.148).

D. Location in Frontal Mass

1. Peak Identification and Quality Assurance: Reading the Spectrum from Right to Left

Very well-resolved spectrum by automated sequence (Fig. 5.149). Lipid and lactate and other metabolites are well-resolved.

Reading from right to left, there is a lipid peak at 0.9 ppm. Next there is a well-resolved doublet centered at 1.33 ppm (lactate), overlapping with a second doublet, centered at 1.48 ppm (alanine), such that the center peak is appropriately greater than its partners. Continuing to read from right to left, the following resonances are identified: NAA (broad or absent); β,γ-Glx; a very small Cr; Cho; mI (at 3.56 ppm); elevated α-Glx triplet (at 3.8 ppm); no second Cr, or mI peaks.

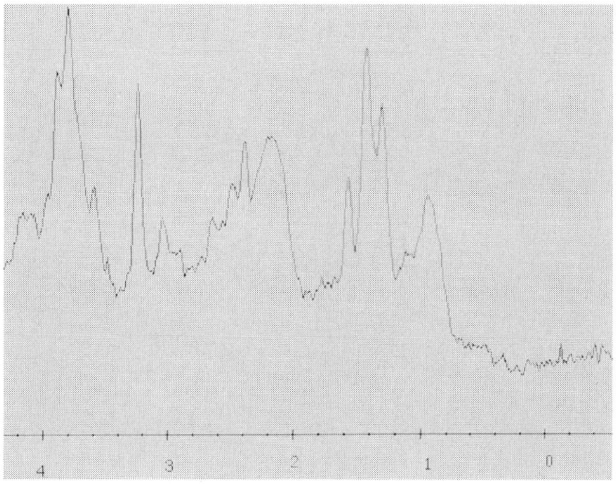

Fig. 5.149 Automated PROBE spectrum of recurrent meningioma. The characteristic features of a tumor include excess lipid, lactate, and Cho, with much reduced NAA and Cr. The presence of alanine, and glutamine, while not specific, are deemed highly suggestive of meningioma. (Courtesy of Dr. Joseph Norfray, Chicago Northside MRI Center, Chicago, IL.)

2. Interpretation and Analysis

Reading from right to left, the NAA/Cr ratio is not relevant. NAA is not definable, and Cr is very much reduced. Other metabolite ratios: Cho/Cr is greater than 5.0, as Cho is most likely elevated (compared to noise level), and Cr is reduced, in keeping with a tumor; mI is identified at 3.56 ppm, but not increased; [Cr] is not available from this nonquantitative PROBE examination; β,γ-Glx/Cr is probably > 1; α-Glx/Cr is clearly very much increased, indicating glutamine *or* glutamate is elevated. It is interesting to compare the phantom spectra of glutamine and glutamate (p. 27) with the pattern of peaks in the region 2.0 to 2.6 ppm. This resembles glutamine rather than glutamate. Since glutamine synthetase is present only in astrocytes (not neurons), this finding is appropriate.

E. Conclusion

This is an example of brain tumor, with absent NAA, high Cho, lipid, and lactate. The large excess of alanine is highly suggestive of meningioma. The great excess of glutamine, although not previously described, would be consistent with glutamine and alanine production in this tumor.

F. Summary of Patient's History Post-MRS

Recurrent meningioma was confirmed by histology.

G. Other Comments

Differential diagnosis of tumors by histological type is not straightforward. However, meningioma appears safely distinguishable upon the basis of frequent finding of alanine, together with the usual features of brain tumor MRS.

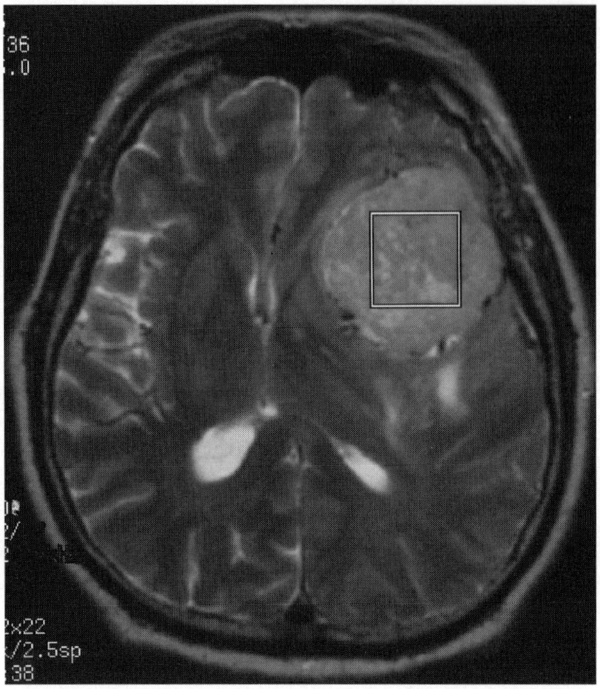

Fig. 5.150 A T_2-weighted MRI showing the location of the VOI totally within the 5 cm, roughly spherical lesion, found in a patient who was referred to rule out dementia, in particular Pick's disease. (Courtesy of Dr. Wang, Rancho Los Amigos Hospital, Downey, California.)

5.7.7 ACCIDENTALLY DISCOVERED TUMOR

A. Purpose of MRS

To evaluate for dementia, in particular to rule out Pick's disease, in a patient with a three-year history of personality changes and memory loss. The plan was to measure first in HMRI's occipital grey matter location, the standard approach for dementias, and then attempt the more difficult frontal lobe location, which is more sensitive to frontal lobe dementias such as Pick's disease. These plans were altered when the MRIs demonstrated a 5 cm roughly spherical lesion in the left frontal lobe, consistent with a large left sphenoid wing meningioma.

B. Summary of Patient's History

Male, age 65, with a three-year history of personality changes and memory loss. There is no history of other diseases, and the patient takes no medications.

C. MRS Method, Volume of Interest, and Comments about Examination

PROBE-STEAM (TE = 30 ms, TM = 13.7 ms, TR = 1500 ms) was performed in the lesion (Location 1), as shown on the image (Fig. 5.150), and in a contralateral VOI (Location 2) using identical measurement parameters (Fig. 5.151). The patient did not need sedation and was cooperative.

D. Location 1 (Within Frontal Tumor)

1. Peak Identification and Quality Assurance: Reading the Spectrum from Right to Left

The S/N of the spectrum (Fig. 5.152) is not very good, and there are no distinct narrow peaks that can give a clear idea of the resolution. Very small Cr and Cho peaks may, however, indicate an acceptable resolution, but the evidence is weak. Artifacts are not seen, except a steep slope on the left-hand side of the spectrum due to inadequate water suppression.

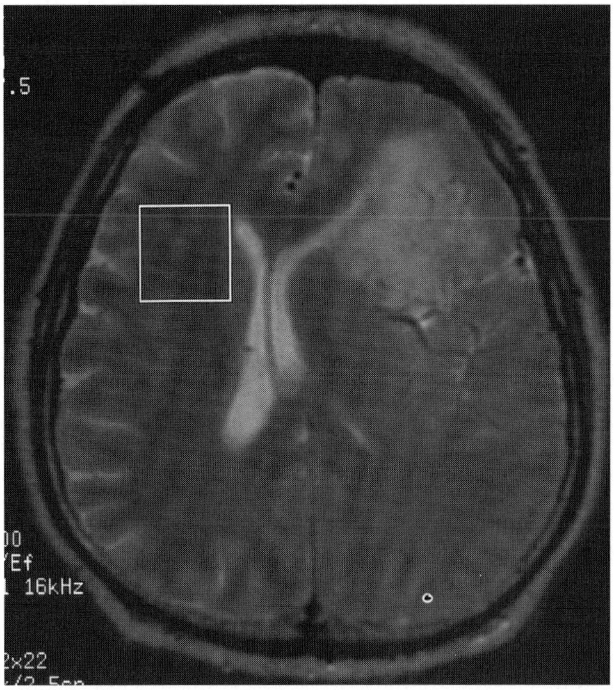

Fig. 5.151 The MRI showing the location of the contralateral VOI. (Courtesy of Dr. Wang, Rancho Los Amigos Hospital, Downey, California.)

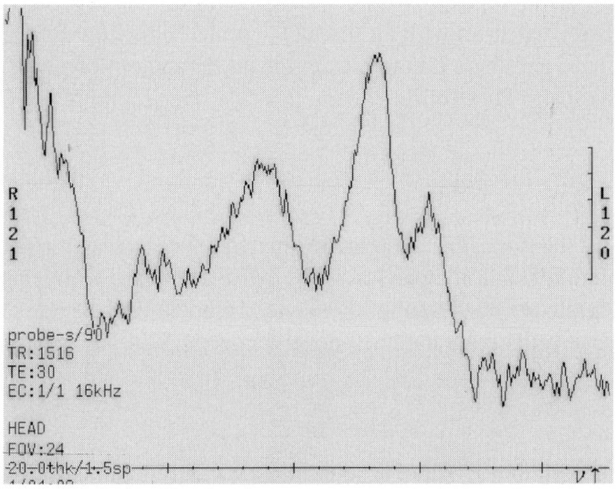

Fig. 5.152 The spectrum from the tumor, showing reduced S/N, mainly necrotic lipid, but also some remaining Cho and Cr. (Courtesy of Dr. Wang, Rancho Los Amigos Hospital, Downey, California.)

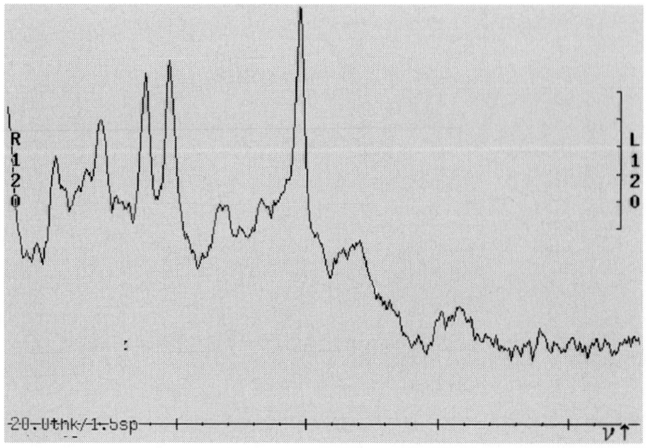

Fig. 5.153 The almost normal spectrum from the contralateral VOI. (Courtesy of Dr. Wang, Rancho Los Amigos Hospital, Downey, California.)

Reading the spectrum from right to left typical lipids are seen at 0.9 ppm and 1.3 ppm. Although the resolution may be sufficient to show lactate as a doublet, the evidence of good resolution is weak, so lactate or other doublets such as alanine (seen in some meningiomas) cannot be totally ruled out. There is no NAA, but some β,γ-Glx, no second NAA, minute Cr and Cho peaks, and no further assignable peaks.

2. Interpretation and Analysis

The metabolite ratios are meaningless, and [Cr] is not available, but compared to the contralateral VOI, it is clear that the signals in this VOI are severely reduced. The main finding is necrotic lipid. There is no clear evidence of lactate, but a broadened lactate overlying the lipid cannot be ruled out.

E. Location 2 (Contralateral)

1. Peak Identification and Quality Assurance: Reading the Spectrum from Right to Left

The spectrum has a small, nonproblematic slope up towards water, and the quality is good (Fig. 5.153). The first impression is that of an almost normal white matter spectrum with all the major peaks following the typical slope down toward the left.

Reading the spectrum from right to left, there is some insignificant in- and out-of-phase lipids from 0.9 to 1.9 ppm, most likely outer volume signal from the scalp. Then follows NAA, β,γ-Glx, the second NAA, Cr, Cho, mI, a somewhat unusual α-Glx, and the second Cr.

2. Interpretation and Analysis

[Cr] is not available. Accounting for the slope, the ratios are estimated to be: NAA/Cr = 1.4 (normal), Cho/Cr = 0.9 (slightly above normal), and mI/Cr = 0.6 (normal). The comparison is to HMRI's standard white matter location, which is not too far from the normal values in frontal white matter. The slightly elevated Cho/Cr may be due to a different grey to white matter ratio in the VOI, or it may be because the contralateral "normal control VOI" is not truly normal.

F. Conclusion

The MRS in the lesion is consistent with tumor. There is some evidence of remaining Cr and Cho, so some remaining tumor activity cannot be ruled out. Lactate and alanine cannot be ruled out, or identified, and as they are not seen in all meningiomas the MRS is consistent with, but not specific for, meningioma.

G. Summary of Patient's History Post-MRS

Not available.

H. Other Comments

The resolution cannot always be judged from the spectrum itself, if none of the normally narrow peaks are seen. But in such a case, the linewidth of the unsuppressed water signal can be used if it has been noted. When MRS is done manually, one usually writes this value down, but with the automated sequence the information may be overlooked. In this case a handwritten note indicates that the water linewidth was approximately twice the normal in the lesion. This would correspond to partially overlapping feet of the Cr and Cho peaks, but with still enough resolution that lactate would have shown some doublet nature.

5.7.8 GLIOBLASTOMA: THERAPEUTIC MONITORING AFTER BRACHYTHERAPY

A. Purpose of MRS

To determine necrosis versus recurrence after brachytherapy of malignant glioblastoma.

B. Summary of Patient's History

Female, age 49, with glioblastoma in right parietal hemisphere, which was treated with brachytherapy. MRI was interpreted as necrosis—a good therapeutic response. MRS was necessary to distinguish this from recurrence.

C. MRS Method, Volume of Interest, and Comments about Examination

Four examinations were performed over nine months. The first three examinations were automated PROBE (Locations 1–7) using a localized STEAM sequence of TE = 30 ms, TM = 13.7 ms, and TR = 1500 ms. The first examination was performed two weeks after brachytherapy (Examination 1), in three VOIs: mid-tumor (Location 1), frontal tumor (Location 2), and a contralateral "control" area (Location 3) located on the corresponding images (Fig. 5.154).

PROBE was also performed 4.5 months after treatment (Examination 2) in the frontal tumor (Location 4), the midtumor (Location 5), and a "control" area (Location 6) located on the corresponding images (Fig. 5.155).

Examination 3 was performed six months after treatment in the frontal tumor (Location 7) (Fig. 5.156).

PROBE-Q (automated plus peak ratio calculation) was used in the last of the four examinations (Examination 4), eight months after treatment. Two spectra were acquired using a localized STEAM sequence of TE = 30 ms, TM = 13.7 ms, and TR = 1500 ms, in the midtumor (Location 8, an approximate repeat of Locations 1 and 5), and occipital tumor (Location 9) shown on the corresponding images (Fig. 5.157). Acquisition and display of PROBE-Q is identical to that of PROBE. On-line data processing, which subtracts water and derives integrals of metabolite peaks, is not displayed. However, the data processing method for obtaining numerical results is essentially identical to that employed for HMRI's MRS examinations. These metabolite ratios are presented in Fig. 5.158 and Fig. 5.159, and discussed more fully elsewhere (see Location 8).

D. Location 1 (Mid-Tumor): Examination 1, Post Brachytherapy (2 weeks)

1. Peak Identification and Quality Assurance: Reading the Spectrum from Right to Left

This automated PROBE spectrum (Fig. 5.160) has poor S/N, and several artifacts in the spectral region to the left of Cho. Water-suppression is incomplete, so that the baseline slopes steeply up from right to left. All of these "problems" arise from the combination of an exceedingly small VOI ($< 1\ cm^3$) and intrinsically low metabolite concentrations in tumors and many other focal lesions.

Despite these limitations, it is immediately clear that the spectrum is very abnormal. Reading systematically from right to left, there is a markedly elevated lipid peak at 0.9 ppm and a well-resolved lactate doublet at 1.33 ppm, followed by reduced NAA, which is "elevated" by the adjacent, poorly resolved β,γ-Glx. The second NAA peak is visible above a noise-background. There follow some noise peaks, from which Cr and Cho are easily resolved. The remainder of the spectrum, however, should not be interpreted as being metabolically or pathologically significant.

In summary, the abnormalities are consistent with a region of necrosis and residual and/or recurrent tumor within an already treated tumor.

Fig. 5.154 Axial MRIs of a recently treated glioblastoma. This first examination was performed two weeks after completion of a course of treatment. Axial T_2-weighted image (A) shows the location of the corresponding small VOI of the PROBE examination, at the margin of the visible tumor and edema. Axial T_1-weighted image (B) shows location of larger VOI to contain a greater proportion of the tumor bed. There is some mass effect. The final image (C) shows the location of the contralateral VOI selected for control. Note that despite normal MRI appearances, MRS often shows, as in this case, subtle abnormalities in regions remote from the tumor. (Courtesy of Dr. D. Hungerford, Mobile Infirmary, Alabama.)

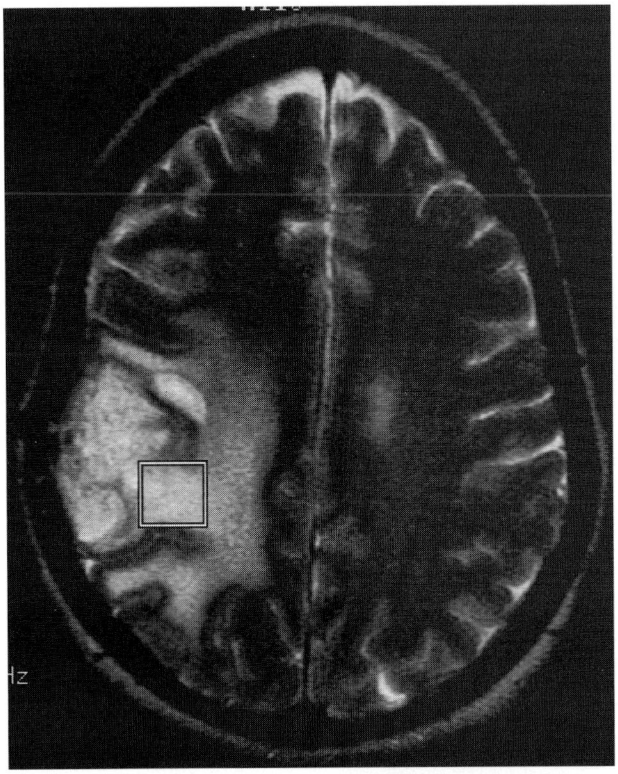

(A)

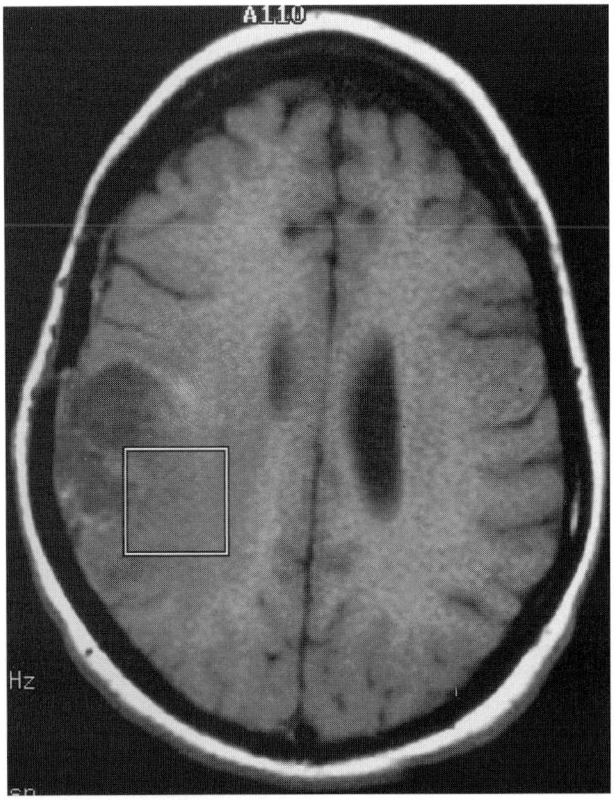

(B)

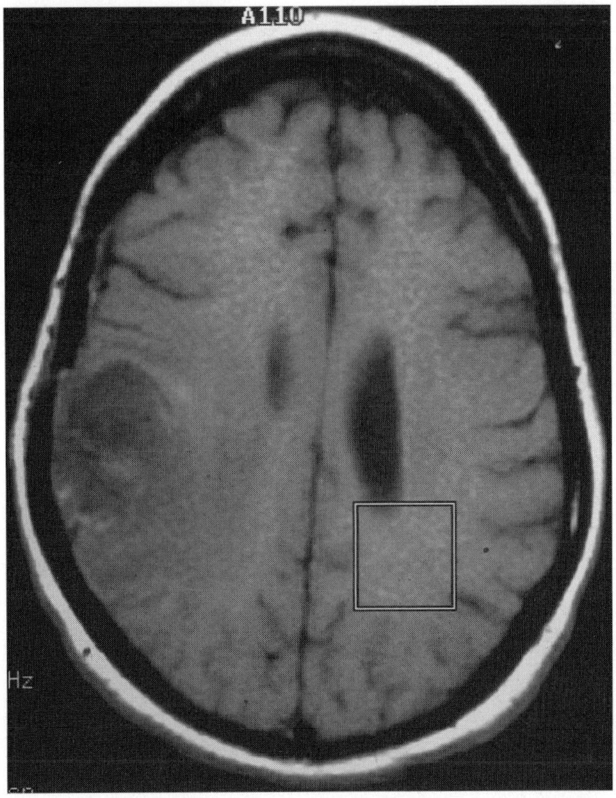

(C)

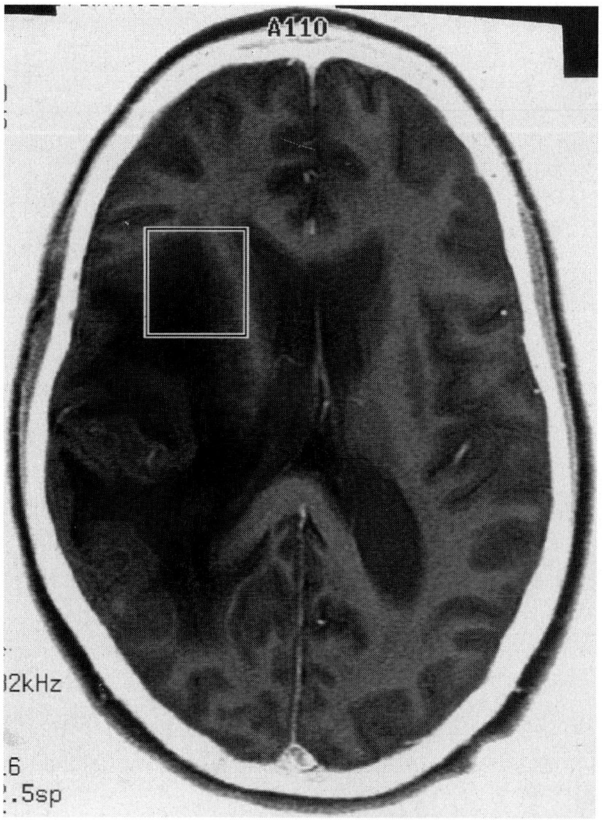
A110

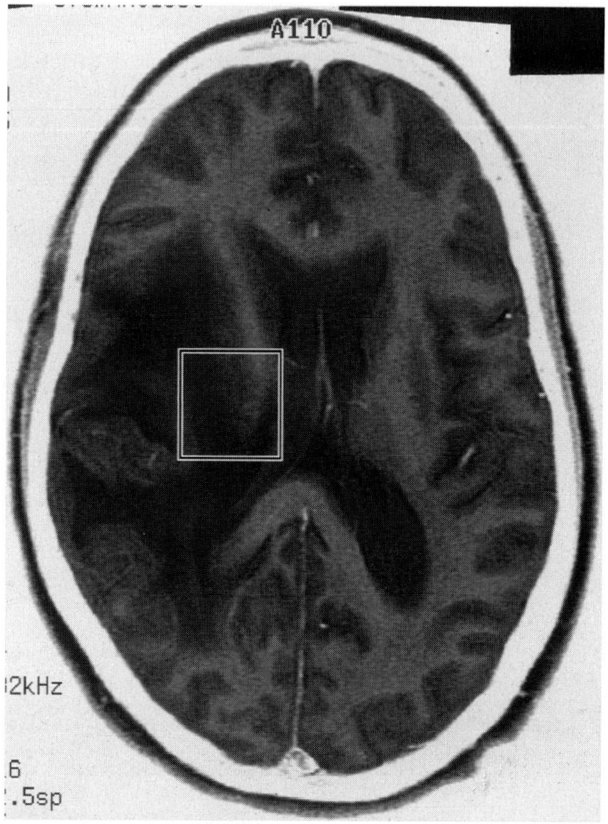
A110

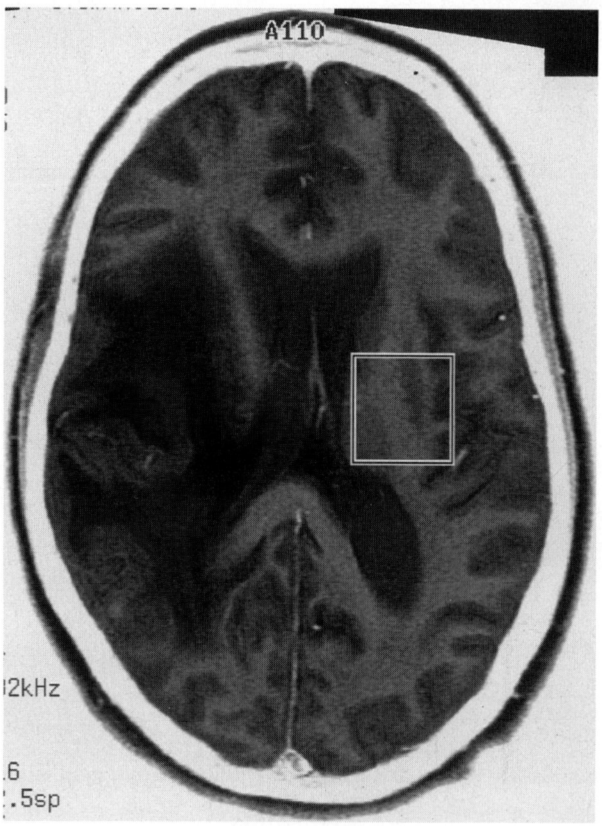
A110

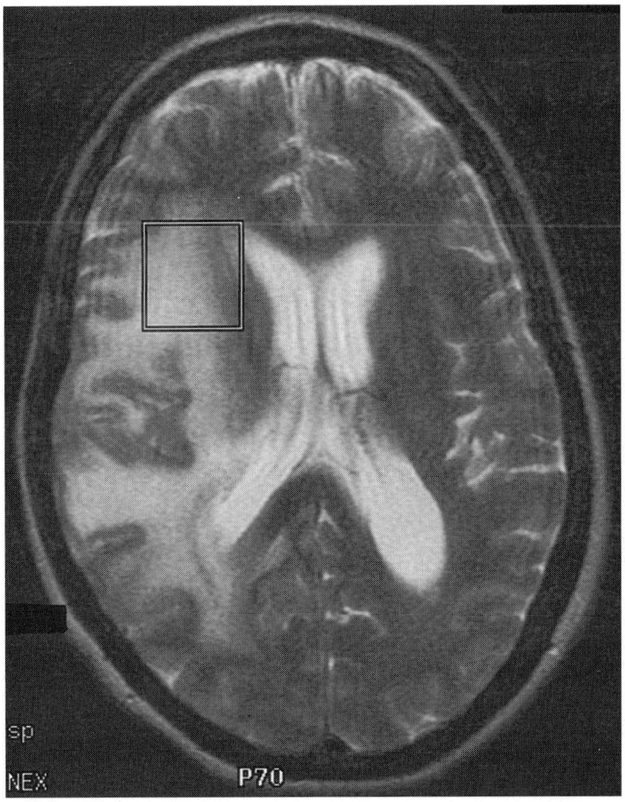

Fig. 5.156 Axial fast spine echo MRI after 6 months. Edema is pronounced but the tumor bed appears smaller. Despite this, mass effect is now greater. The single PROBE examination was performed in the same frontal tumor margin as previously. (Courtesy of Dr. D. Hungerford, Mobile Infirmary, Alabama.)

2. Interpretation and Analysis

Reading from right to left, lipid and lactate are highly significantly elevated. Continuing right to left, the approximate metabolite ratios are: NAA/Cr = 1.1 (significantly reduced); Cho/Cr = 1.5 (much increased, and possibly indicating relative reduction in [Cr]).

This information can be interpreted with increased confidence after examining the following PROBE spectrum acquired from a larger VOI of which the MRI-visible lesion (Fig. 5.154B) forms about 15% by volume.

E. Location 2 (Frontal Tumor): Examination 1, Post Brachytherapy (2 weeks)

1. Peak Identification and Quality Assurance: Reading the Spectrum from Right to Left

This PROBE spectrum (Fig. 5.161) acquired from significantly greater tissue volume (tumor and surrounding edema; ~8 cm^3 VOI) has appropriately superior S/N and is well resolved from artifacts of phasing (which result in broadening of lactate and NAA peaks), with somewhat oversuppressed water.

Reading from right to left, this tumor spectrum exhibits lipid and lactate doublet at 1.33 ppm the latter riding on additional lipid. These peaks are followed by NAA, significantly reduced but artificially elevated by the "phasing" artifact. β,γ-Glx is not pronounced, the second NAA peak, Cr, Cho, mI and α-Glx are all resolved, but the second Cr and mI peaks are in a region altered by oversuppression of water.

Fig. 5.155 Three versions of the same axial fast spin echo MRI are shown to demonstrate location of the 3 PROBE examinations performed 4.5 months after brachytherapy (Examination 2). MRI shows a well demarcated and probably expanded tumor bed, together with some increase in the modest mass effect. (Courtesy of Dr. D. Hungerford, Mobile Infirmary, Alabama.)

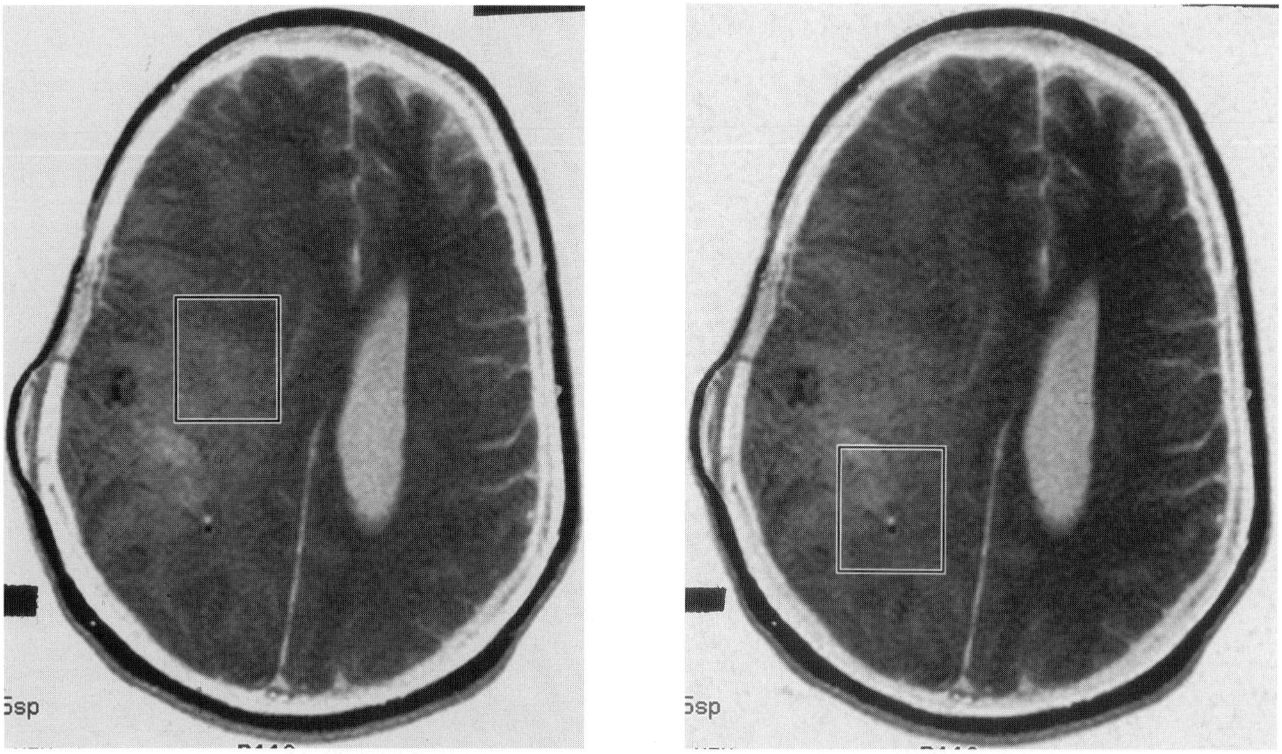

(A) (B)

Fig. 5.157 The locations of two PROBE-Q examinations within the previous areas of edema (A) and tumor bed (B) are shown. (Courtesy of Dr. D. Hungerford, Mobile Infirmary, Alabama.)

2. Interpretation and Analysis

Reading from right to left, the metabolite ratios can be estimated as: NAA/Cr = 0.98 (significantly reduced), Cho/Cr = 1.0 (elevated above normal, but less so than in the tumor bed, Location 1), and mI/Cr = 0.32 (reduced).

Together with lipid and lactate, these changes are consistent with necrosis within a tumor, and a contribution from more normal brain within the extended VOI.

F. Location 3 (Contralateral Control): Examination 1, Post Brachytherapy (2 weeks)

1. Peak Identification and Quality Assurance: Reading the Spectrum from Right to Left

This automated PROBE spectrum (Fig. 5.162) is of excellent quality apart from a water suppression artifact which results in a prominent sloping baseline and loss of resonances beyond mI. Nevertheless, the result is of diagnostic quality. Acquired with identical parameters of Location 2 (Fig. 5.161), the much superior S/N confirms the impression of overall loss of metabolites from within the tumor bed.

Reading the spectrum systematically from right to left, there is normal appearance of brain protein–lipid at 0.9 and 1.4 ppm, and no evidence of a lactate doublet at 1.33 ppm. Peaks of NAA (possibly reduced), reduced β,γ-Glx, a second NAA, Cr, Cho, sI and mI are all readily identified. The remaining "peaks" are artifactual.

The impression is of a moderately abnormal PROBE spectrum from a normal appearing brain region after treatment with brachytherapy.

2. Interpretation and Analysis

Peak ratios are estimated to be NAA/Cr = 1.44 (normal for location); Cho/Cr = 1.0 (slightly elevated); mI/Cr = 0.58 (normal). β,γ-Glx/Cr is possibly reduced, but this is unsupported in the absence of identifiable α-Glx resonances.

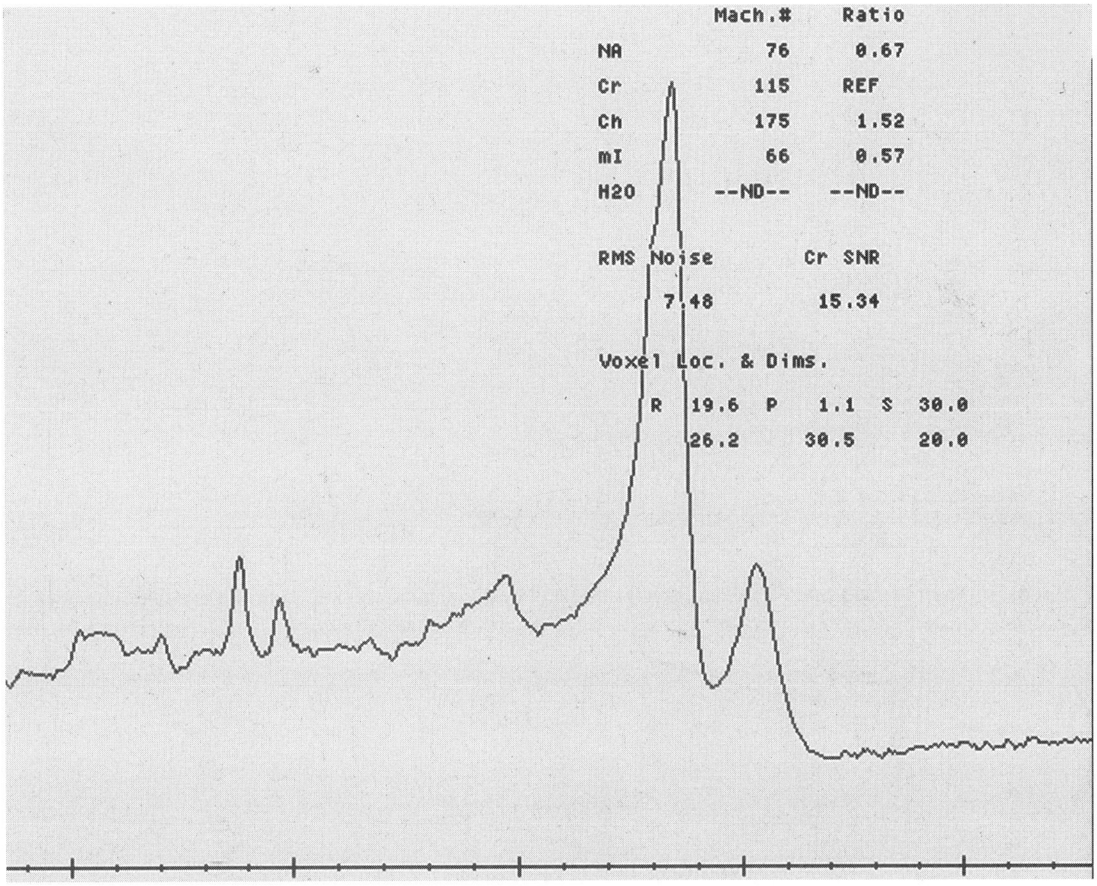

Fig. 5.158 PROBE-Q spectrum and automated numerical print-out of mid-tumor region. Major new features of this automated PROBE spectrum first seen after 8 months are the massive elevation of lipid. Necrosis is also accompanied by further increase in Cho, an indicator in this case of tumor recurrence or progression. (Courtesy of Dr. D. Hungerford, Mobile Infirmary, Alabama.)

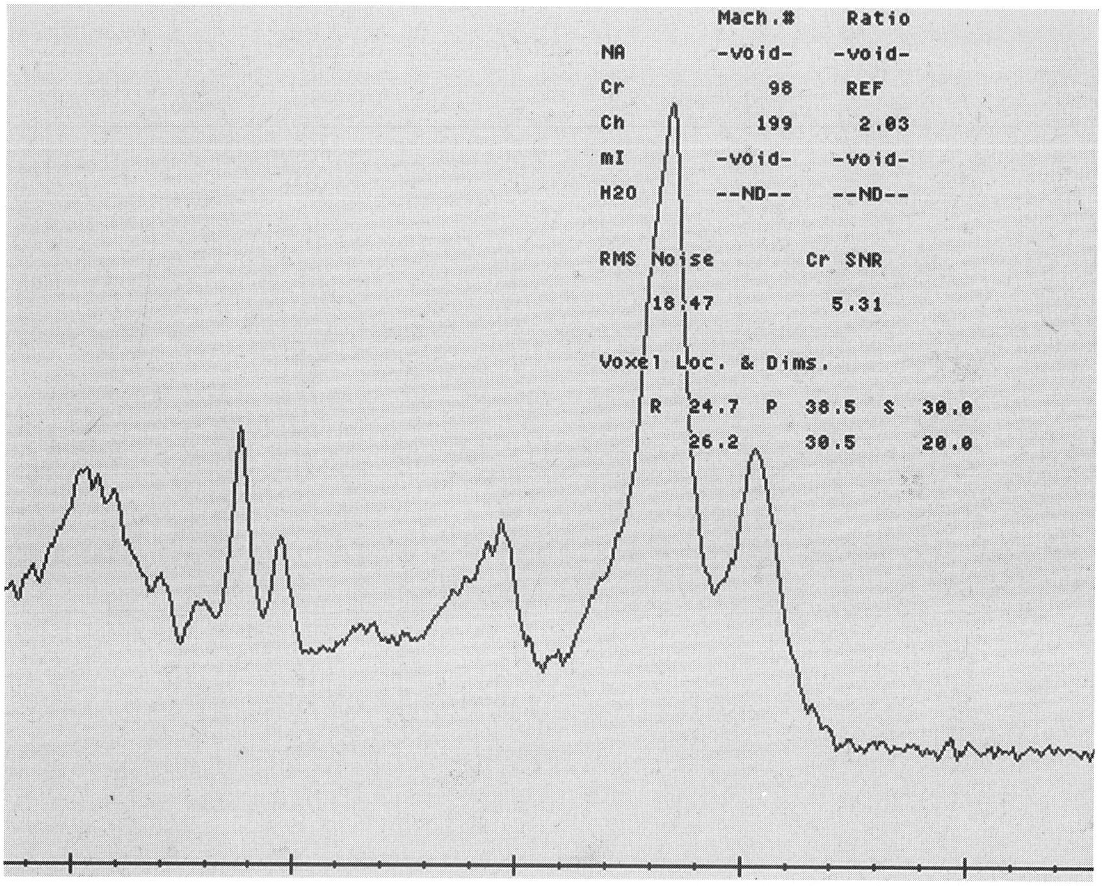

Fig. 5.159 PROBE-Q spectrum and automated numerical print-out of occipital tumor region. This PROBE-Q examination confirms the more widely distributed nature of the malignant and necrotic activity described above. (Courtesy of Dr. D. Hungerford, Mobile Infirmary, Alabama.)

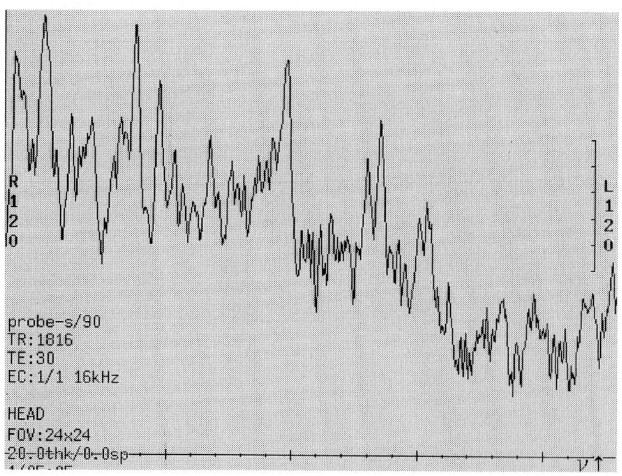

Fig. 5.160 Automated PROBE of 1 cm^3 volume. The spectrum is recognizably that of tumor, despite expected poor S/N. (Courtesy of Dr. D. Hungerford, Mobile Infirmary, Alabama.)

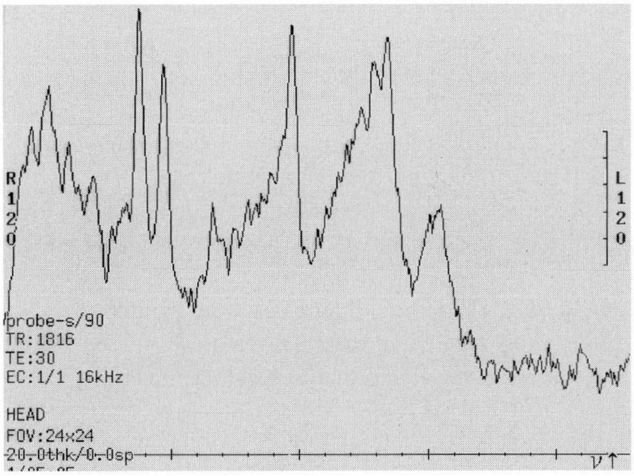

Fig. 5.161 The improved S/N PROBE spectrum shows essentially all of the features of the previous more localized acquisition, and is consistent with active tumor, or recurrence, in the presence of necrosis. (Courtesy of Dr. D. Hungerford, Mobile Infirmary, Alabama.)

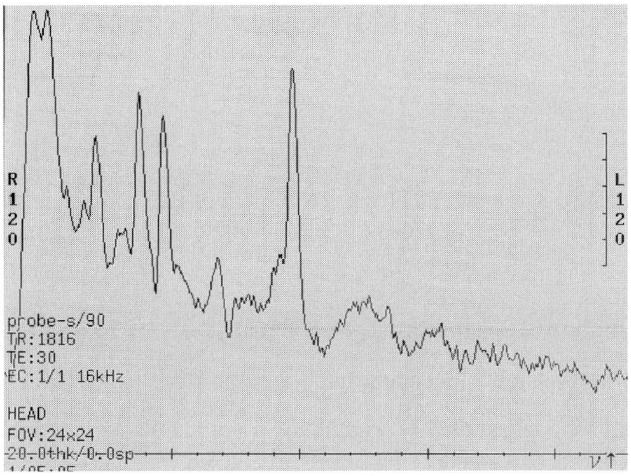

Fig. 5.162 Automated PROBE examination of uninvolved hemisphere, two weeks after brachytherapy. This excellent PROBE spectrum shows no features of tumor, but is subtly abnormal. (Courtesy of Dr. D. Hungerford, Mobile Infirmary, Alabama.)

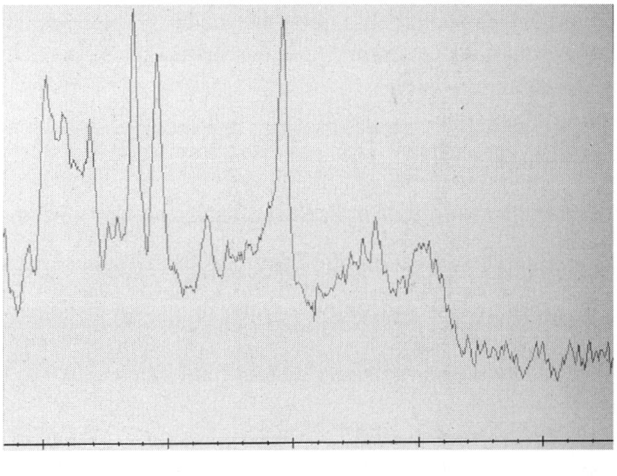

Fig. 5.163 Location 4 shows a well-resolved PROBE spectrum from the frontal margin, not previously involved. The spectrum is subtly abnormal and consistent with necrosis in an active area of tumor growth. (Courtesy of Dr. D. Hungerford, Mobile Infirmary, Alabama.)

G. Summary of Examination 1, Post Brachytherapy (2 weeks)

The clinical question posed is: Is there residual malignant glioblastoma after brachytherapy in this patient, or are we observing the evolution of tissue necrosis as was implied by the therapeutic team?

Three features of the spectrum support the first, more pessimistic conclusion of this PROBE examination: the presence of lipid, lactate and a relative excess of Cho. Loss of NAA and possibly of Cr, although characteristic of tumor, are not inconsistent with necrosis.

The noisy, but highly localized spectrum is supported by the spectrum from location 1, which although contaminated by signal from surrounding normal brain, shows the same essential features.

The contralateral hemisphere shows modest abnormalities and should not be viewed as an appropriate control to the pathological study.

Follow-up studies are strongly indicated.

H. Location 4 (Frontal Tumor): Examination 2, Post Brachytherapy (4.5 Months)

1. Peak Identification and Quality Assurance: Reading the Spectrum from Right to Left

This automated PROBE spectrum (Fig. 5.163) is of excellent quality and without artifact.

Reading the spectrum systematically from right to left, there is lipid at 0.9 ppm, an identifiable lactate doublet at 1.33 ppm, NAA, β,γ,Glx, a second NAA, Cr, Cho, sI, glucose, mI, α-Glx triplet, and second peaks of Cho and ml. The spectrum is not normal, although the abnormalities are subtle.

2. Interpretation and Analysis

In addition to a small, pathological excess of lipid and lactate, by inspection the peak ratios are estimated to be: NAA/Cr = 1.24 (10% below normal); Cho/Cr = 1.04 (35% above normal); mI/Cr = 0.64 (normal); β,γ-Glx is not elevated. These are the features of residual tumor.

I. Location 5 (Midparietal Tumor): Examination 2, Post Brachytherapy (4.5 months)

1. Peak Identification and Quality Assurance: Reading the Spectrum from Right to Left

Another PROBE (Fig. 5.164) is of the highest quality. Pathological lipid and a lactate doublet (0.9 ppm and 1.33 ppm) are followed by peaks of NAA, β,γ-Glx, second and third NAA peaks, Cr, Cho, sI, glucose, mI (at 3.56 ppm), α-Glx as a distinct, but somewhat broadened triplet, and the second Cr and mI peaks are identified.

The spectrum is noticeably abnormal, and may represent persistent tumor at this location.

2. Interpretation and Analysis

This VOI includes more lipid and lactate. The metabolite ratios are modestly, but significantly, abnormal: NAA/Cr = 1.11 (20% reduced); β,γ-Glx appears significantly increased in relation to the second NAA peak; Cho/Cr = 1.16 (55% increased); mI/Cr = 0.49 (normal or slightly reduced).

The spectrum is consistent with necrosis (lipid and lactate) and activity in this treated tumor. The relative reduction of NAA/Cr is explained by partial volume of "normal" brain and tumor or necrosis.

J. Location 6 (Contralateral Control): Examination 2, Post Brachytherapy (4.5 months)

1. Peak Identification and Quality Assurance: Reading the Spectrum from Right to Left

A good quality PROBE spectrum (Fig. 5.165) shows incomplete water suppression and hence some baseline artifact beyond the mI peak. Diagnostic quality is not compromised.

Reading systematically from right to left, the normal brain protein-lipid profile at 0.9 to 1.33 ppm is present, while lactate is absent.

Peaks are identified for NAA, β,γ-Glx, second and third NAA, Cr, Cho, mI, α-Glx, a broadened second Cr and a possible second mI peak. Apart from a somewhat reduced NAA/Cr, this spectrum is not notably abnormal.

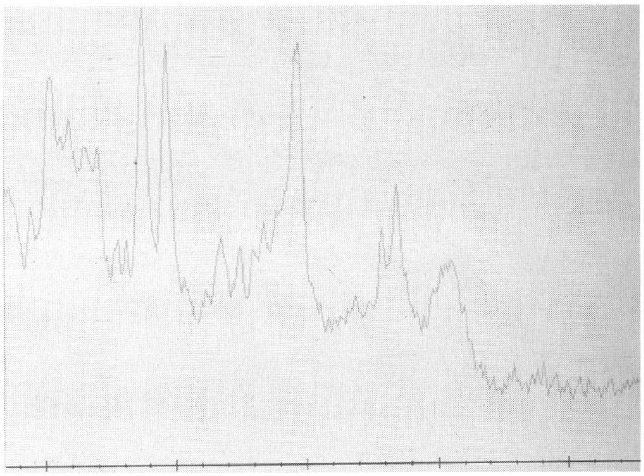

Fig. 5.164 A PROBE spectrum from a location similar, but not identical, to that measured four months previously, shows similar features of active tumor and necrosis. (Courtesy of Dr. D. Hungerford, Mobile Infirmary, Alabama.)

2. Interpretation and Analysis

Peak ratios by inspection are: NAA/Cr = 1.25 (some 10% reduced); Cho/Cr = 0.75 (normal for white matter); mI/Cr = 0.43 (normal or reduced); and β,γ-Glx is not increased. The modest abnormality may be a result of therapy or other secondary effects of the tumor on what is a normal appearing contralateral hemisphere on MRI.

K. Summary of Examination 2, Post Brachytherapy (4.5 months)

4½ months after therapy, there were significant MRS signs of tumor activity and necrosis. A further follow-up is indicated.

L. Location 7 (Frontal Tumor): Examination 3, Post Brachytherapy (6 months)

1. Peak Identification and Quality Assurance: Reading the Spectrum from Right to Left

The automated PROBE spectrum (Fig. 5.166) is of excellent quality, and acquired from a VOI identical to that 1½ months earlier.

Reading the spectrum systematically from right to left, there is lipid at 0.9 ppm, a lactate doublet at 1.33 ppm, and peaks for NAA, β,γ-Glx, second NAA, Cr, Cho, mI and α-Glx, as well as a second Cr peak, but not a second mI, are readily identified. The spectrum is very comparable to that 1½ months previous, and remains pathological.

2. Interpretation and Analysis

Peak ratios by inspection are NAA/Cr = 0.91 (lower than previously); Cho/Cr = 0.85–0.9 (still elevated); and mI/Cr = 0.36 (low). β,γ-Glx is probably not elevated, just better resolved in this spectrum, judging from the relative intensities of the third and second NAA peaks.

M. Summary of Examination 3, Post Brachytherapy (6 months)

There is significant interval change, with further increase in lipid and lactate and a small reduction in NAA, consistent with tumor progression or necrosis.

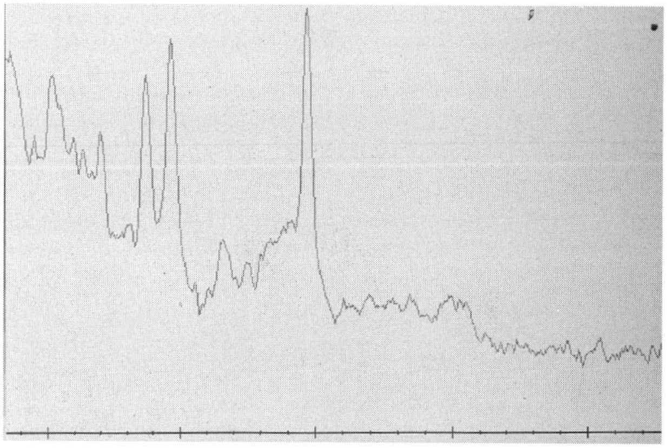

Fig. 5.165 A PROBE spectrum from a corresponding unaffected brain region in the contralateral hemisphere. None of the features of tumor or necrosis are present, despite subtle differences from normal. (Courtesy of Dr. D. Hungerford, Mobile Infirmary, Alabama.)

N. Location 8 (Midparietal Tumor): Examination 4, Post Brachytherapy (8 Months)

1. Peak Identification and Quality Assurance: Reading the Spectrum from Right to Left

There are dramatic differences in this PROBE-Q spectrum (Fig. 5.158) from any of the previous tumor spectra, although the location of the VOI is similar to Location 5, and only approximately that of Fig. 5.154B and Fig. 5.161. Despite "improvement" in MRI appearance, mass effect is more obvious in Fig. 5.157A, and the VOI is almost exclusively within the tumor bed.

Reading the PROBE-Q spectrum systematically from right to left, there is lipid at 0.9 ppm and 1.3 ppm in the proportions of approximately 1:4, no NAA, broad β,γ-Glx, no second NAA, a small Cr and more prominent Cho, mI (at 3.56 ppm), indistinct α-Glx region, but identifiable second Cr are all seen. It should be remembered that while concentrations of all metabolites in this spectrum are arguably low, it is a feature of MRS that peaks are scaled down if, as in this case, a single metabolite dominates. Thus, lipid at 1.3 ppm may represent 30–60 mM concentration, and Cr and Cho might nevertheless lie within the ranges of 5 and 2.5 mM, respectively.

The spectrum is that of necrotic tumor, with persistent or recurrent activity, despite brachytherapy eight months previous.

2. Interpretation and Analysis

Lipid is not quantified, but is present in massive concentration with the anticipated peak ratio between methylene (1.3 ppm) and methyl (0.9 ppm) peaks from long-chain triglyceride. There is no lactate doublet at 1.33 ppm. Peak ratios calculated within automated PROBE-Q are: NAA/Cr = 0.67 (obviously an overestimate, since the computer has failed to recognize the complete absence of any sharp NAA peak at 2.02 ppm); Cho/Cr = 1.52 (twice normal); mI/Cr = 0.57 (normal). β,γ-Glx/Cr and α-Glx/Cr are increased, by inspection, but [Cr] is possibly reduced.

The spectrum is that of necrosis within viable tumor.

O. Location 9 (Occipital Tumor Margin): Examination 8, Post Brachytherapy (8 Months)

PROBE-Q with automated acquisition and peak ratio calculation after baseline correction and integration of peak areas was performed eight months after therapy. Only the numerical result is displayed; however, the PROBE spectrum observed in Fig. 5.159 is no different from routine PROBE, and the baseline correction applied is not displayed.

1. Peak Identification and Quality Assurance: Reading the Spectrum from Right to Left

An excellent and diagnostic tumor spectrum with well-resolved metabolite peaks, all of which are reduced in intensity by "scaling" to the dominant lipid peaks.

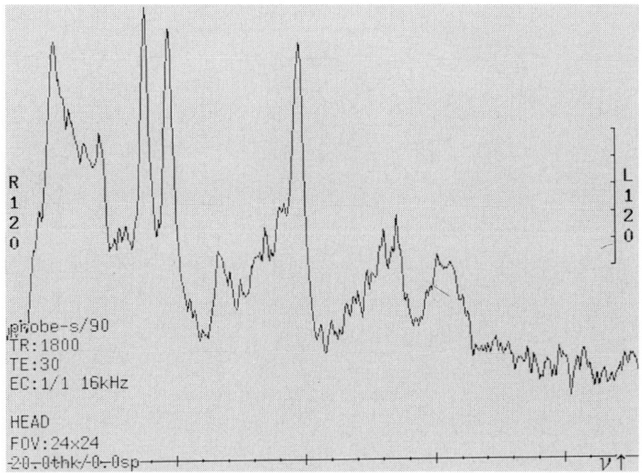

Fig. 5.166 Automated PROBE examination 6 months after treatment. There are no major changes in the 6 weeks since the previous examination. (Courtesy of Dr. D. Hungerford, Mobile Infirmary, Alabama.)

Reading the PROBE-Q spectrum systematically from right to left, the lipid peaks at 0.9 ppm and 1.3 ppm have the characteristic proportions of long-chain triglyceride, with a ratio of approximately 1:2.5 for 0.9 and 1.3 ppm peaks. There is no lactate doublet at 1.33 ppm. NAA at 2.02 ppm, and the second NAA peak are perceptible but essentially absent. β,γ-Glx is prominent but poorly resolved. There are peaks of Cr, and Cho, small peaks of sI or glucose, and mI, as well as peaks of α-Glx and possibly the second Cr.

The spectrum, like that previous, is highly suggestive of residual tumor activity, or recurrence, within a necrotic tumor bed, following brachytherapy.

2. Interpretation and Analysis

Peak ratios acquired automatically by PROBE-Q are excess lipid (not quantified, but six times Cr ~30–50 mM if [Cr] is normal); NAA/Cr = void by computer, or < 0.3 by inspection; Cho/Cr = 2.03 (three times normal); mI/Cr = void by computer, or 0.5 by inspection (normal). β,γ-Glx/Cr, not quantified, but $\leqslant 1.0$ by inspection (increased two- to three-fold).

Signal to noise by PROBE-Q, and by inspection, is less in this PROBE-Q (Fig. 5.159) than the previous PROBE-Q (Fig. 5.158). All these features point to a real reduction in [Cr] in this tumor region.

P. Conclusion

As in the previous spectrum of this examination, a great deal of necrosis has occurred in the interval since brachytherapy, as would be anticipated. However, it is hard to escape the conclusion that recurrent tumor, noted as early at 7½ months earlier, and repeatedly in PROBE examinations, is now a major feature of the spectrum.

Q. Summary of the Patient's History Post-MRS

The patient died 8 months after the last MRS examination from a recurring glioblastoma.

R. Other Comments

MRS of a lesion to differentiate recurrent tumor from necrosis obviously has the advantage over biopsies that surgery is avoided. Another difference that should be noted is that MRS gives an answer for a volume of 1 cc or more, but details from within the chosen volume are not available. A biopsy, on the other hand, is a much smaller sample and may miss a possible recurrence, but the biopsy will give a more detailed answer for the small piece of tissue that is investigated.

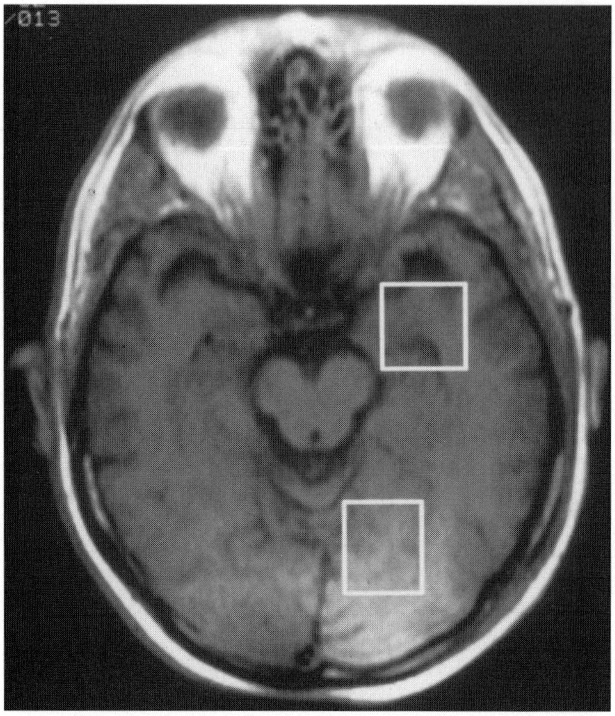

Fig. 5.167 MRI indicating locations examined at 12 months postsurgery by MRS.

5.7.9 TUMOR: POSTSURGICAL RECURRENCE

A. Purpose of MRS

Identify tumor recurrence after surgery and radiotherapy, and rule out contralateral spread.

B. Summary of Patient's History

The patient, a 51-year-old male, was diagnosed with occipitoparietal glioblastoma of left hemisphere, treated by surgical resection and postsurgical radiotherapy. A year later MRI suggested a recurrence. Subsequent history and MRS examinations are described up to the present (5 years after the initial detection of the glioblastoma).

C. MRS Method, Volume of Interest, and Comments about Examinations

Examination 1 (6 months postsurgery):

STEAM (TE = 30 ms, TM = 13.7 ms, TR = 1500 ms) was performed six months postsurgery. The VOI was dictated by the original tumor location (Location 1), and by abnormal MRI appearance, in left inferior occipital cortex (Location 2). To avoid skull and scalp lipid, the VOI was partly anterior to the main MRI pathology. The partial volume of normal-appearing brain in the VOI was increased. As "controls," a contralateral VOI (Location 3) was examined, (Images of Locations 1–3 are not shown).

Examination 2 (12 months postsurgery):

A second MRS (STEAM,TE = 30 ms, TM = 13.7 ms, TR = 1500 ms) was performed approximately six months later (Examination 2, Location 1), and included a remote VOI as a control (Examination 2, Location 2), indicated in Fig. 5.167.

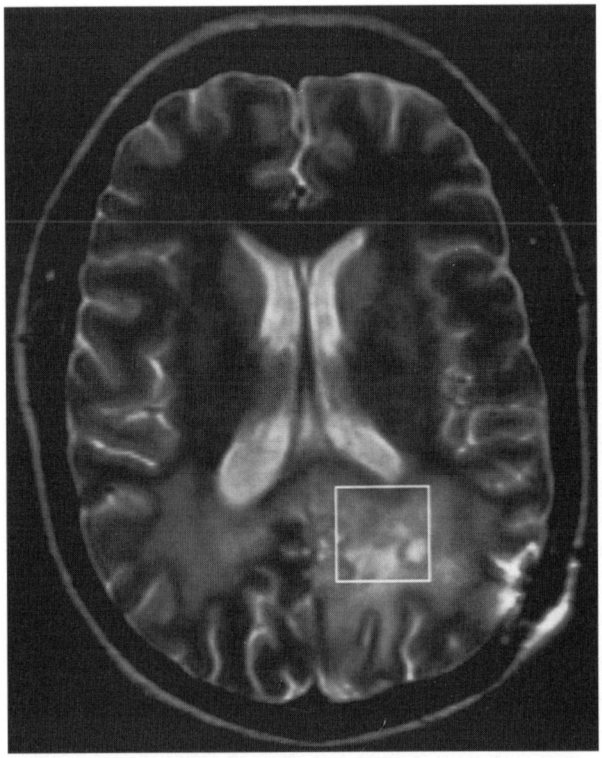

Fig. 5.168 Axial MRI (18 months postsurgery) shows little change from that acquired six months previously. The location of the VOI for a PROBE study is different from those acquired previously and should be interpreted cautiously.

Examination 3 (18 months postsurgery):

A limited MRI/MRS examination was performed 18 months postsurgery in the location indicated (Fig. 5.168) using PROBE-STEAM (TE = 30 ms, TM = 13.7 ms, TR = 1500 ms).

Examination 4 (5 years postsurgery):

Onset of right hemiplegia and detection of gadolinium-enhanced changes on MRI lead to repeat biopsy (glioblastoma) and MRS (STEAM, TE = 30 ms, TM = 13.7 ms, TR = 1500 ms) of the location shown Fig. 5.169.

Examination 5 (5½ years postsurgery):

A three months follow-up examination to evaluate the effect of chemotherapy, (MRI not shown).

D. Examination 1: Location 1 (Tumor Bed Six Months Postsurgery)

1. Peak Identification and Quality Assurance: Reading the Spectrum from Right to Left

Spectral quality is poor, with a large artifact dominating all of the peaks beyond 3.4 ppm, Fig. 5.170. The pathological information is nevertheless obvious. Reading from right to left, two very large lipid peaks at 0.9 and 1.3 ppm are seen, with no evidence of a lactate doublet at 1.33 ppm, NAA is absent, with only a broad "hump" at 2.1 ppm, but narrow peaks of Cr and Cho are present.

2. Interpretation and Analysis

The only relevant peak ratio is Cho/Cr = 1.24, with NAA/Cr ~0. Quantitation showed [Cr] only 30% of normal. The unquantified lipid is nevertheless in great excess. The water content, 80.7%, is high.

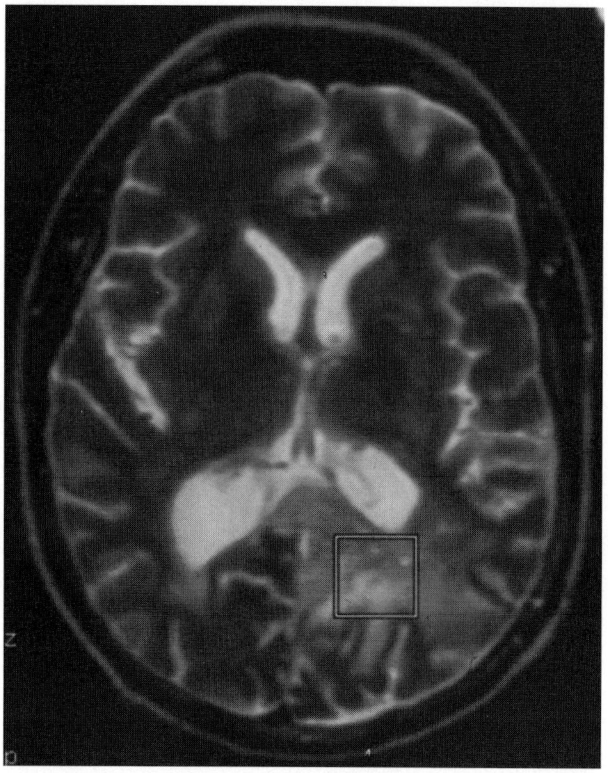

Fig. 5.169 MRI indicating the MRS location examined 5 years postsurgery.

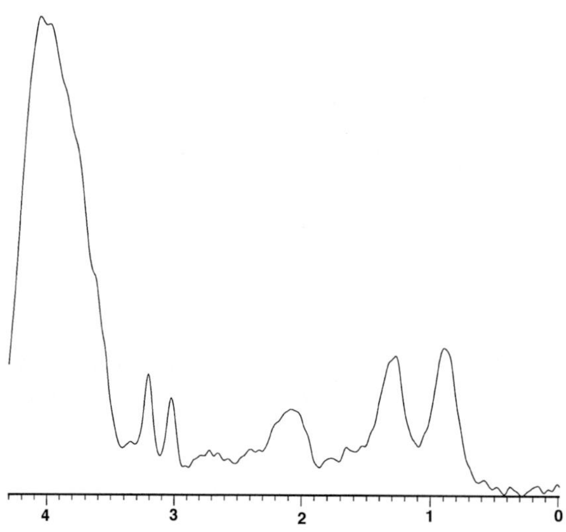

Fig. 5.170 MRS of postsurgical tumor-bed at 6 months. Despite poor technical quality, presence of lipid and excess Cho/Cr, with low or absent NAA is consistent with tumor recurrence.

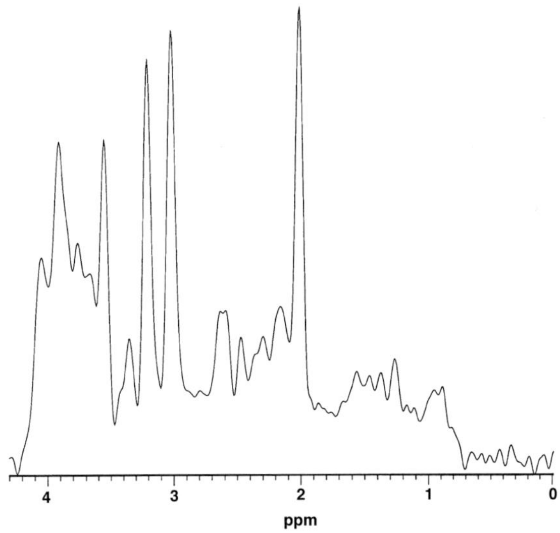

Fig. 5.171 MRS of normal appearing contralateral occipital cortex. NAA/Cr is reduced, and Cho/Cr and mI/Cr are increased. While not conclusive of tumor, this result is consistent with post–radiation changes.

E. Examination 1: Location 2 (Contralateral Hemisphere)

1. Peak Identification and Quality Assurance: Reading the Spectrum from Right to Left

The technical quality of Fig. 5.171 is far superior to that from the surgical bed. Reading the spectrum systematically from right to left, a minor lipid peak is probably present, but no lactate (the "doublet" is questionable, and not at 1.33 ppm). Prominent peaks for NAA, β,γ-Glx, the second NAA, Cr, Cho, sI and glucose, mI, α-Glx, a second Cr and mI are all easily identified. The spectrum is nevertheless abnormal, at this contralateral site adjacent to a surgical tumor bed.

2. Interpretation and Analysis

Reading from right to left, metabolite peak ratios are: NAA/Cr = 1.06 (20% reduced); Cho/Cr = 0.94 (increased, for this largely grey matter location); mI/Cr = 0.74 (~40% increased). [Cr] is 90% of normal, and brain water of 73.7% is normal. The changes are consistent with gliosis, rather than necrosis or tumor.

F. Examination 1: Location 3 (Distant "Normal" White Matter)

1. Peak Identification and Quality Assurance: Reading the Spectrum from Right to Left

This spectrum from white matter, Fig. 5.172, is of excellent quality with essentially normal appearance. Reading systematically from right to left, the lipid-protein peaks at 0.9 ppm and 1.5 ppm are within normal limits. Peaks for NAA, β,γ-Glx, second NAA, Cr, Cho, sI, mI, α-Glx, and second Cr are easily identified.

2. Interpretation and Analysis

Metabolite peak ratios are: NAA/Cr = 1.29 (significantly below mean for age); Cho/Cr = 0.87 (marginally elevated); and mI/Cr = 0.61 (normal). [Cr] is 100% of normal, and brain water 64.4% is typical of normal white matter. β,γ-Glx/Cr = 0.31, and α-Glx/Cr = 0.58 (not significant).

G. Summary of Examination 1

Two features of this examination are noteworthy: the evidence of persistent tumor, within necrosis, and the subtle abnormality of spectra acquired contralateral and even distant from the site of the original tumor. A follow-up examination was indicated, and performed 6 months later.

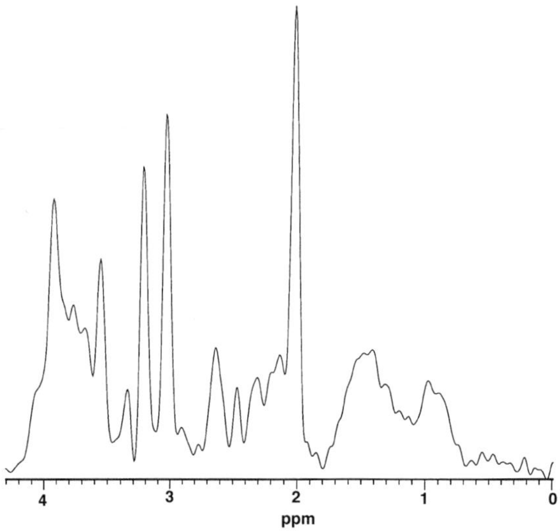

Fig. 5.172 Normal white matter location. Distant from tumor and treatment, this MRS is normal.

H. Examination 2: Location 1 (Tumor Bed 12 Months Postsurgery)

1. Peak Identification and Quality Assurance: Reading the Spectrum from Right to Left

The spectrum, Fig. 5.173, is in all respects identical to that acquired 6 months earlier, including the huge and dominant "peak" beyond 3.4 ppm, previously called "artifact." It may be an unidentified metabolite.

2. Interpretation and Analysis

Lipid, absent NAA, and Cr and Cho peaks are identified. Signal to noise is low overall, consistent with necrosis and the presence of recurrent tumor.

I. Examination 2: Location 2 (Normal Temporal Lobe)

1. Peak Identification and Quality Assurance: Reading the Spectrum from Right to Left

The spectrum from white matter, Fig. 5.174, is of excellent quality apart from a small outer-volume lipid artifact at 1.9 ppm that may influence NAA/Cr ratio.

Reading systematically, normal protein-lipid peaks at 0.9 ppm, and doubtful lactate doublet at 1.33 ppm, are followed by NAA, β,γ-Glx, second NAA, Cr, Cho, sI, mI, α-Glx, and the second peaks of Cr and mI.

2. Interpretation and Analysis

Metabolite peak ratios are NAA/Cr = 1.18 (15% reduced); Cho/Cr = 0.77; mI/Cr = 0.55; β,γ-Glx/Cr = 0.31; α-Glx/Cr = 0.44. [Cr] is 100% and brain water is 75.8%. With the exception of reduced NAA/Cr, the spectrum is normal.

J. Summary of Examination 2

A number of characteristic features of posttreatment MRS are illustrated, in particular the necrosis.

K. Conclusion

Despite the patient's satisfactory clinical course, 1 year postsurgery there is evidence of recurrence at the site of surgery.

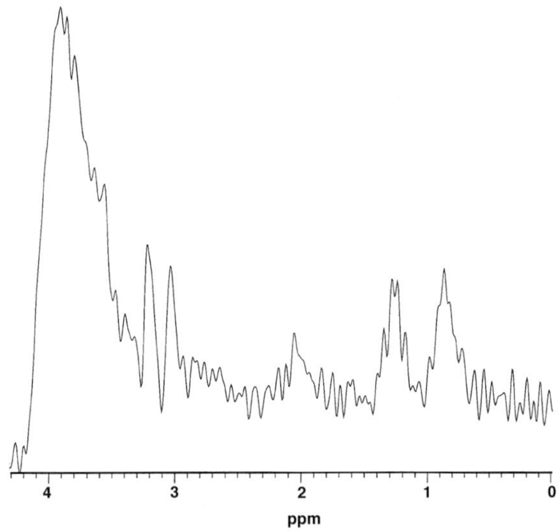

Fig. 5.173 Re-examination 12 months after surgery. MRS is highly suggestive of recurrent tumor. Note the artifact which dominates the spectrum nevertheless permits recognition of elevated Cho/Cr.

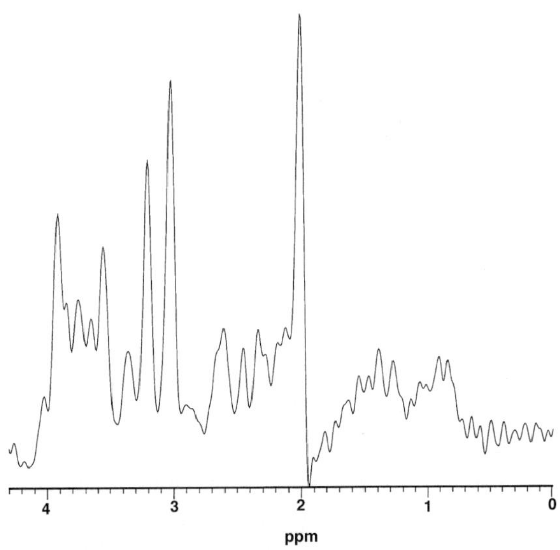

Fig. 5.174 MRS of unaffected anterior temporal lobe at 12 months. MRS is essentially normal for this location and shows no evidence of either tumor or effects of irradiation.

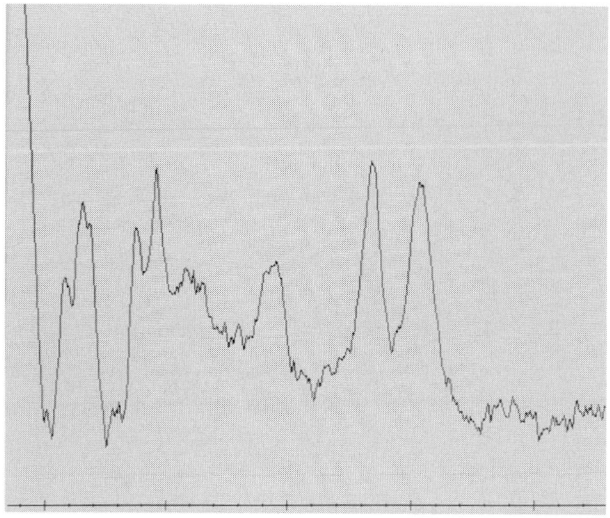

Fig. 5.175 Automated PROBE MRS STEAM. Although quality is compromised by a baseline artifact at 3.2 to 3.4 ppm, most of the details of this spectrum can be seen. Lipid is elevated and NAA/Cr is dramatically reduced. Cho/Cr cannot easily be determined, but is not markedly elevated. This follow-up examination confirms necrosis, but the presence of recurrent tumor is not apparent. This may be the result of inappropriate placement of the VOI.

L. Examination 3 (18 Months Postsurgery)

A limited MRI/MRS examination at 18 months postsurgery is indication of necrosis only. Evidence of recurrence is lacking in this MRS examination. Placement of the VOI in the high intensity region of MRI and superior to the original tumor bed may be questioned. A further examination was not performed at this time. Additional radiotherapy was delivered locally. The spectrum is shown in Fig. 5.175, but not described in detail here.

M. Examination 4 (Five Years Postsurgery)

1. Peak Identification and Quality Assurance: Reading the Spectrum from Right to Left

The spectrum, Fig. 5.176, is that of recurrent glioma. Reading from right to left lipid and lactate is seen, NAA is absent, but β,γ-Glx, Cr, Cho, and mI are observed.

2. Interpretation and Analysis

Lipid and lactate are prominent. NAA is absent, with β,γ-Glx of normal appearance and Cho/Cr is increased.

N. Summary of Examination 4

MRS is supportive of the histological diagnosis of recurrent malignant glioma.

O. Examination 5 (5½ Years Postsurgery, 3 Months Postchemotherapy)

Chemotherapy was instituted on the basis of biopsy and MRS information. The patient improved symptomatically.

1. Peak Identification and Quality Assurance: Reading the Spectrum from Right to Left

Good quality spectrum, Fig. 5.177, of recurrent glioma. Reading from right to left lipid and lactate is seen, NAA is absent, but β,γ-Glx, Cr, Cho and mI are observed.

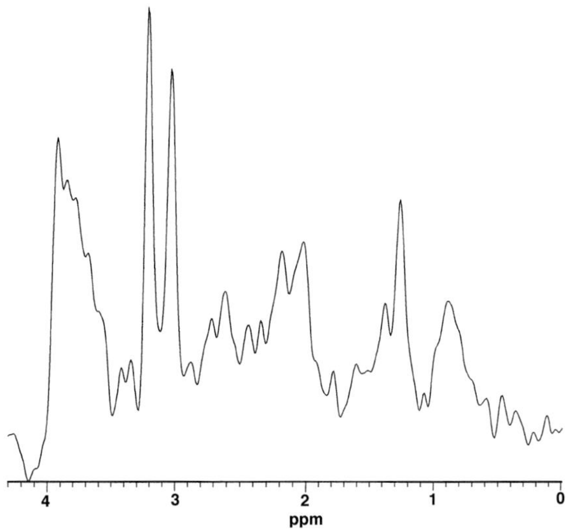

Fig. 5.176 MRS examination 5 years after surgery performed after the onset of right hemiplegia. The MRS is consistent with recurrent glioblastoma.

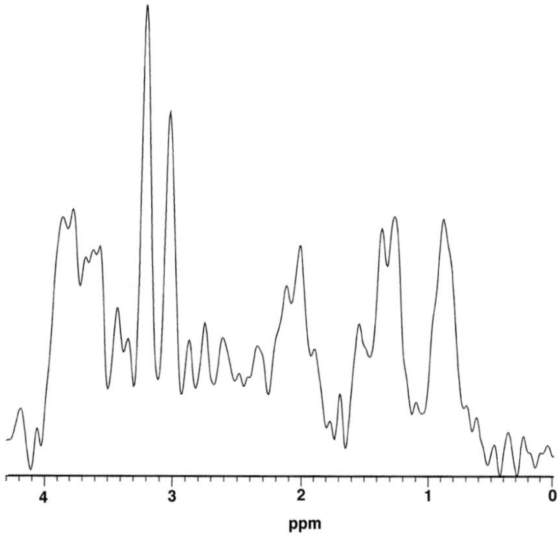

Fig. 5.177 MRS examination 5½ years after surgery, and after 3 months of chemotherapy. The MRS is still consistent with recurrent glioblastoma, and the only evidence of a response to chemotherapy is the increased lipid.

2. Interpretation and Analysis

Compared to Examination 4, lipid is markedly increased. Other spectral details. Reduced NAA/Cr and increased Cho/Cr, low mI/Cr, are essentially unaltered from Fig. 5.176.

P. Summary of Examination 5

The only evidence of possible response to chemotherapy is the marked increase in lipid.

Q. Conclusion

The conclusion of the present series of examinations is that MRS confirms tumor recurrence and indicates the need for further therapy. Spectral quality is very different between the early and later MRS examinations but disease management is significantly improved by these clinical MRS examinations. MRI was inconclusive, and repeated surgical biopsy is costly and carries high risk.

R. Summary of Patient's History Post-MRS

The patient survives, and is currently responding to chemotherapy.

S. Other Comments

MRS is an integral part of clinical evaluation of brain tumors. In this case, repeated MRS lead to altered and improved management.

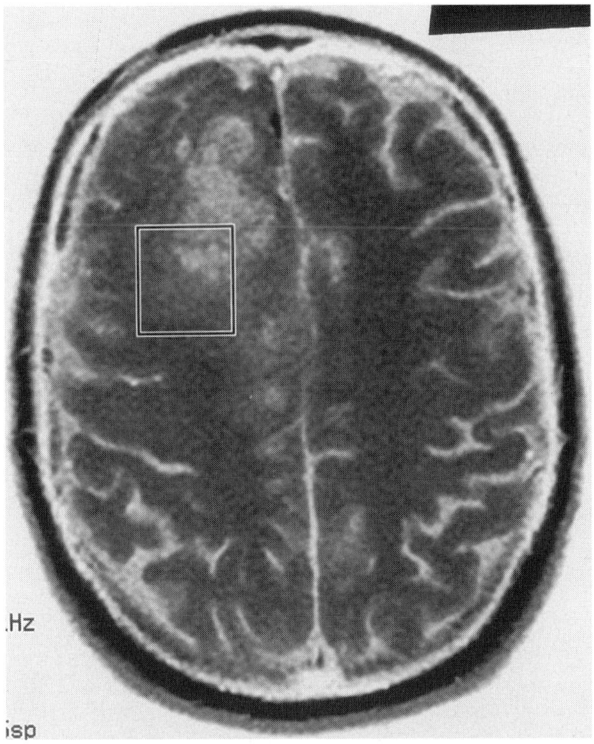

(A)

Fig. 5.178 Axial MRI (A) and (C) and corresponding PROBE-Q MRS (B) and (D) in a patient with undiagnosed brain mass. The PROBE-Q and automated print out define the pathology at the tumor margin (B), and can be contrasted with the nearly normal contralateral findings (D). (Courtesy of Dr. D. Hungerford, Mobile Infirmary, Alabama.)

5.7.10 UNTREATED GLIOMA

A. Purpose of MRS

Differential diagnosis of frontal mass on MRI, biopsy guidance, and treatment planning.

B. Summary of Patient's History

Clinical presentation in 63-year-old male, previously well.

C. MRS Method, Volume of Interest, and Comments about Examination

PROBE-Q STEAM (TE = 30 ms, TM = 13.7 ms, TR = 1000 ms), was performed. A VOI was placed at the tumor margin, to include normal appearing brain (Location 1), and Location 2 was the near symmetrical contralateral (see Fig. 5.178A,C). Note that the different TR has no significant impact on spectral appearance in adults, but adds the benefit of reduced scanning time.

D. Location 1 (Tumor)

1. Peak Identification and Quality Assurance: Reading the Spectrum from Right to Left

The PROBE spectrum (Fig. 5.178B) is of excellent quality, with slight oversuppression of water, but a good baseline.

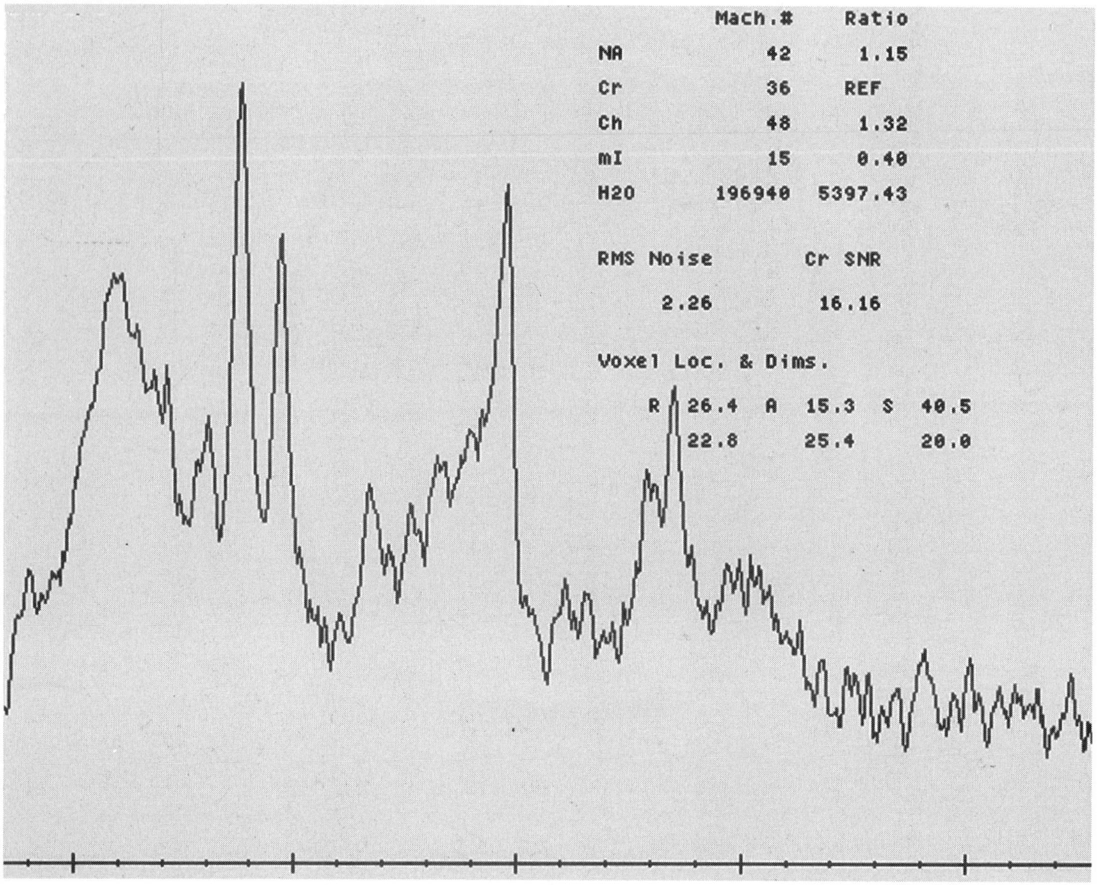

(B)

Fig. 5.178 Continued.

Reading the spectrum from right to left, there is lipid at 0.9 ppm, a lactate doublet at 1.33 ppm, unidentified (possibly lipid) peaks at 1.7 and 1.8 ppm, and a reduced NAA peak is readily identified. β,γ-Glx, prominent in relation to the second NAA peak, Cr, Cho, sI, mI, and α-Glx regions, are-well resolved, but the second Cr and mI peaks are unclear.

2. Interpretation and Analysis

In addition to the abnormal lipid and lactate (well-resolved doublet at 1.33 ppm with 7 Hz splitting, and no alanine doublet detectable), reduced and slightly broadened NAA, elevated Cho/Cr and reduced [Cr] are apparent.

Peak ratios provided by PROBE-Q are: NAA/Cr = 1.15 (probably an over-estimate due in part to underlying β,γ-Glx); Cho/Cr = 1.32 (80% increased, relative to control, and to contralateral location); and mI/Cr = 0.40 (reduced, rather than increased). An estimate of [Cr] from the Cr signal-to-noise ratio is 67% (⅔) of contralateral, and thus significantly reduced. β,γ-Glx/Cr appears increased, but cannot be substantiated by the inadequately resolved α-Glx in this spectrum.

The spectrum is in all respects appropriate for a malignant, primary tumor. The persistence of NAA and of Cr is not surprising given the intentional inclusion of up to 50% of adjacent "normal" brain within the VOI.

E. Location 2 (Contralateral and Uninvolved Hemisphere)

1. Peak Identification and Quality Assurance: Reading the Spectrum from Right to Left

The PROBE-Q spectrum (Fig. 5.178D) is incomplete, but of diagnostic quality. Water suppression is good, with the exception of loss of the resonances beyond mI at 3.56 ppm.

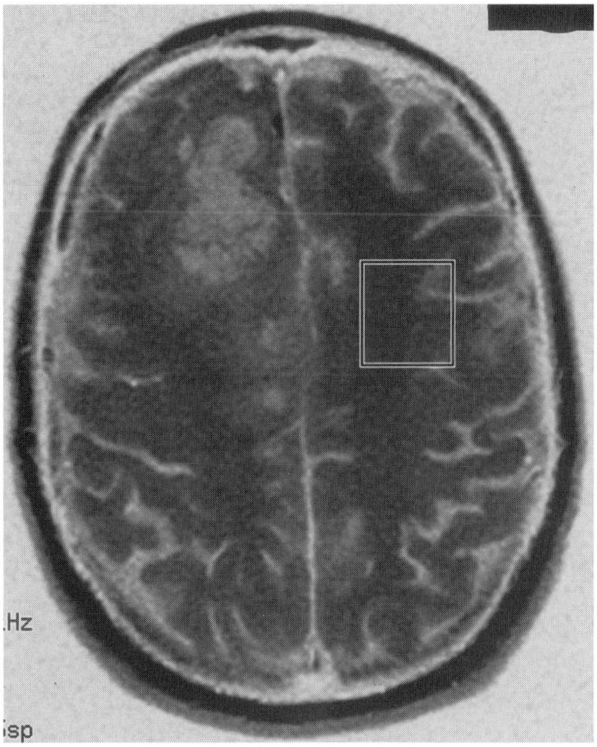

(C)

Reading the spectrum from right to left, there is no lipid or lactate. NAA, β,γ-Glx, second NAA, Cr, Cho, sI and glucose are well seen. *Myo*-inositol is dubious due to the water-suppression artifact described.

2. Interpretation and Analysis

This normal appearing PROBE-Q spectrum provides automated peak ratios as follows: NAA/Cr = 1.26 (significantly reduced from normal); Cho/Cr = 0.73 (normal); mI/Cr should be discounted, although near normal, at 0.63. The absence of lipid and lactate and the presence of normal β,γ-Glx are also noteworthy.

F. Conclusion

A typical malignant tumor spectrum, with modest MRS abnormality in the contralateral hemisphere. The result is adequate to guide stereotactic biopsy and subsequent therapy, if indicated.

G. Summary of Patient's History Post-MRS

Biopsy confirmed glioma. No further history available.

H. Other Comments

Note that the metabolite ratios in PROBE-Q are calculated based on a spectrum from which the residual water has been removed. The spectrum shown next to the numbers is the spectrum as it appears before the elimination of residual water.

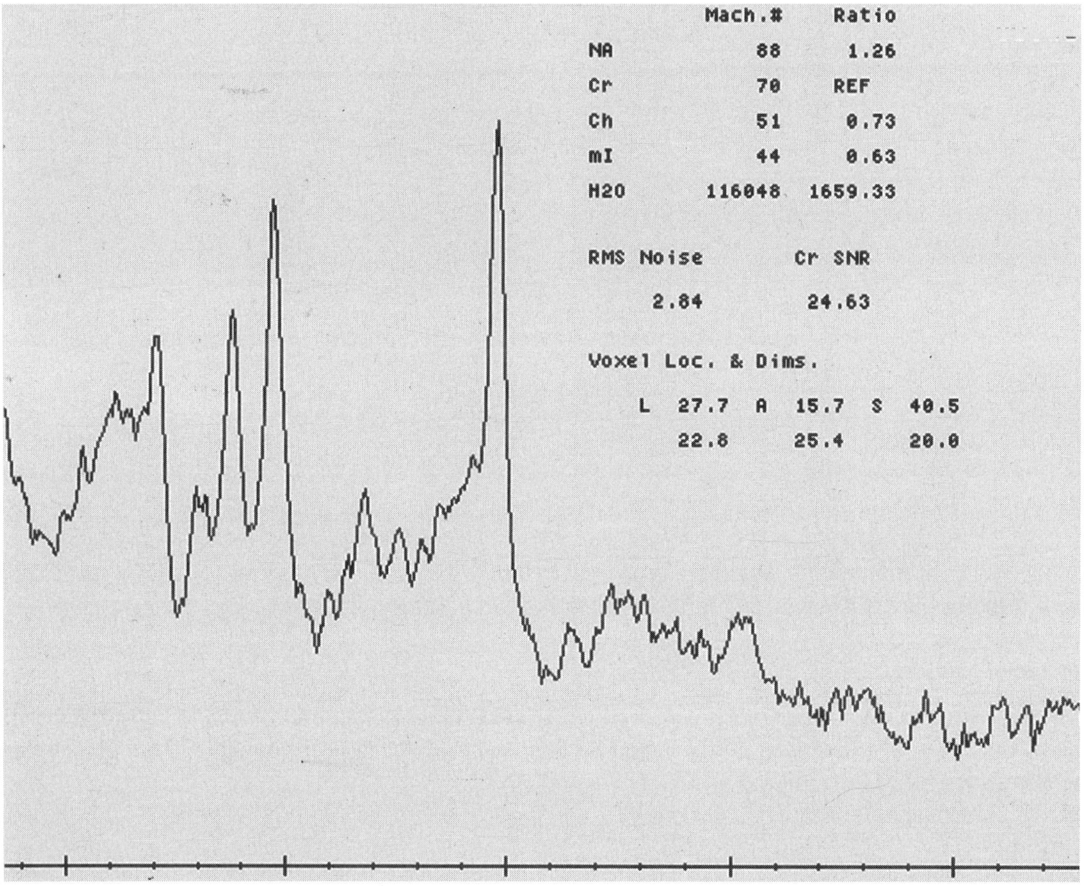

(D)

Fig. 5.178 Continued.

5.7.11 RECURRENT HEMANGIOBLASTOMA: NEGATIVE

A. Purpose of MRS

MRS was performed to rule out recurrence of cerebellar hemangioblastoma removed surgically 14 years previously when patient was 18 years old.

B. Summary of Patient's History

Male 32 years old. Small lesions less than 1 cm^3 were apparent on MRI, four in right cerebellum and two in the left.

C. MRS Method, Volume of Interest, and Comments about Examination

Automated PROBE-Q STEAM (TE = 30 ms, TM = 13.7 ms, TR = 1000 ms) was performed. The bilaterally symmetrical VOIs ($25 \times 25 \times 20$) were selected within the cerebellum to include the MRI-visible lesions, which in each case occupy no more than 20% of the VOI.

D. Location (Right Cerebellum)

1. Peak Identification and Quality Assurance: Reading the Spectrum from Right to Left

PROBE-Q spectrum and numerical analysis (Fig. 5.179) is of excellent technical quality, with the exception of incomplete resolution of Cr from Cho.

Reading the spectrum from right to left, there is abnormal excess of lipid at 0.9 ppm and 1.4 ppm, without evidence of splitting at 1.33 ppm (due to poor spectral resolution throughout). There follow peaks of NAA, β,γ-Glx, second NAA, Cr, Cho (incompletely resolved), sI, mI, α-Glx, a second Cr, and possible second mI. Water suppression is excellent with almost horizontal baseline.

2. Interpretation and Analysis

The lipid is abnormal. Peak ratios established by automated analysis are: NAA/Cr = 1.15 (true normal not available in this patient); Cho/Cr = 0.9 (obviously an overestimate; by inspection, Cho/Cr is about 0.6); mI/Cr = 0.55 (by inspection, 0.4); β,γ-Glx/Cr and α-Glx/Cr appear normal.

There is no evidence of tumor recurrence from this spectrum; lipid alone may be evidence of necrosis or gliosis. The contribution of "normal" cerebellum to the VOI is 80–90%, obscuring any pathology.

E. Location 2 (Left Cerebellum and Lesion)

1. Peak Identification and Quality Assurance: Reading the Spectrum from Right to Left

Automated PROBE-Q and numerical evaluation (Fig. 5.179) is of excellent quality, including good resolution of Cr from Cho.

Reading the spectrum from right to left, lipid peaks at 0.9 ppm and 1.5 ppm are within normal limits, and there is no lactate doublet at 1.33 ppm. NAA, β,γ-Glx, second and third NAA peaks, Cr, Cho, sI, and mI are all clearly identified; α-Glx and the second Cr peak are dubious.

2. Interpretation and Analysis

Lipid is normal. Peak ratios by automated PROBE-Q analysis are: NAA/Cr = 1.06 (possibly low for this location); Cho/Cr = 0.72 (as it is by inspection); mI/Cr = 0.40. All values are within normal limits, although NAA/Cr may be inappropriate, and low. This is the only indication that the MRI-visible lesion contributes significantly to the larger VOI examined.

F. Conclusion

The essentially negative examination cannot be interpreted as ruling out recurrent hemangioblastoma.

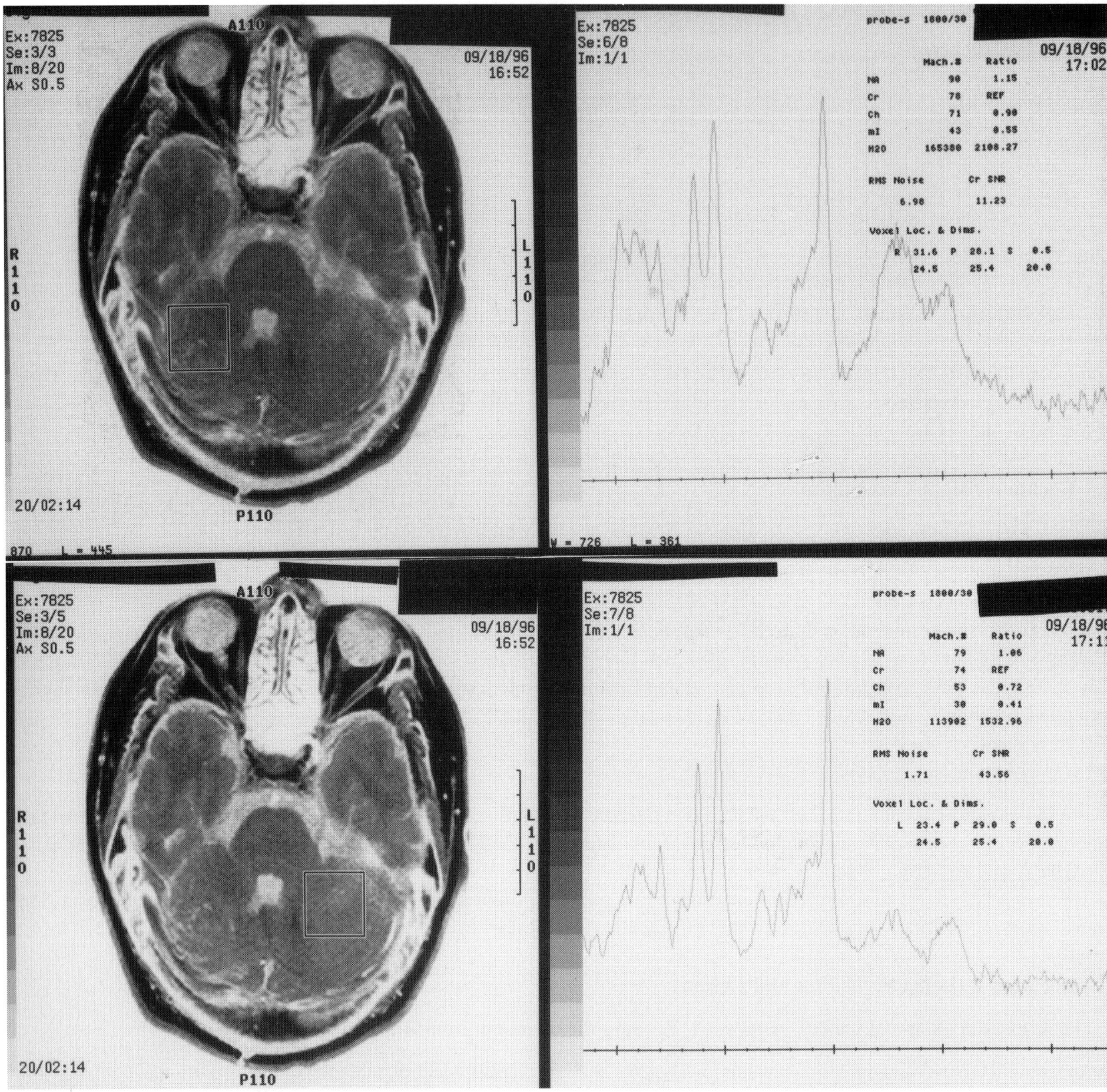

Fig. 5.179 Both VOI's include (poorly) visible MRI lesions, but despite excellent technical quality, cannot be conclusively interpreted as the partial volume of normal cerebellum is in each case too large, (Courtesy of Dr. D. Hungerford, Mobile Infirmary, Alabama).

G. Summary of Patient's History Post-MRS

Independent evidence confirmed tumor recurrence.

H. Other Comments

Despite excellent technical quality, this PROBE-Q examination illustrates the compromise necessary when interrogating lesions which are at, or below, the limits of spatial resolution for ^{1}H MRS.

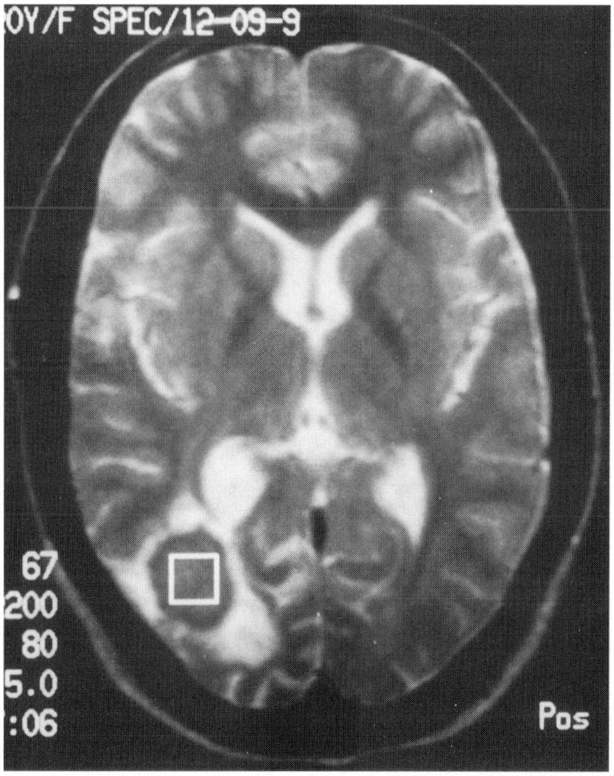

Fig. 5.180 A T_2-weighted MRI showing the location of the VOI totally within the hypointense well-defined mass in the right occipital region. (Courtesy of Dr. R. K. Gupta, Sanjay Gandhi Post-Graduate Institute of Medical Sciences, Lucknow, India.)

5.7.12 TUBERCULOMA VERSUS LYMPHOMA

A. Purpose of MRS

To evaluate a mass approximately $25 \times 24 \times 28$ mm^3, that was hypointense of T_2 weighted images to determine if the mass is a lymphoma or a tuberculoma.

B. Summary of Patient's History

Female 19, who complained of headache and visual symptoms.

C. MRS Method, Volume of Interest, and Comments about Examination

STEAM (TE = 20 ms, TM = 30 ms, TR = 3000 ms) was used to measure in the lesion averaging 128 scans, Fig. 5.180. The patient did not need sedation and was cooperative. Furthermore movement was avoided by fixing the head in the coil. The examination took place in India, where tuberculoma is a likely inclusion in the differential diagnosis.

D. Location in Lesion

1. Peak Identification and Quality Assurance: Reading the Spectrum from Right to Left

The S/N of the spectrum Fig. 5.181 is good for 128 averages. Artifacts are not seen, except possible partly inverted signals in the lipid region. The water suppression is perfect.

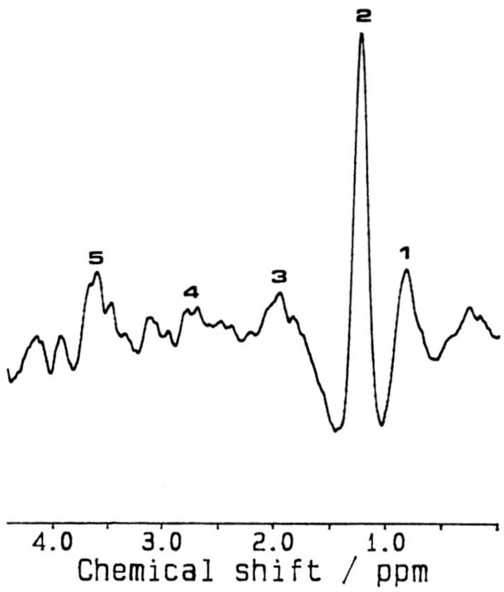

Fig. 5.181 MRS in a tuberculoma. The assignments were based on high resolution in vitro 1 and 2 dimensional NMR spectroscopy of the excised tissue. 1: CH_3; 2: $(CH_2)_n$; 3: $H_2C{=}CH$; 4: $CH{-}CH_2CH{=}$; 5: phosphoserine (for further details see Gupta et al. Magn. Reson. Med. 36: 829–833, 1996). (Courtesy of Dr. R. K. Gupta, Sanjay Gandhi Post-Graduate Institute of Medical Sciences, Lucknow, India).

Reading the spectrum from right to left typical long chain lipids are seen at 0.9 ppm (CH_3) and 1.3 ppm $(CH_2)_n$. There is no NAA, but some signal in the β,γ-Glx region, no second NAA, no Cr or any of the other usually assignable peaks. Small broader resonances are seen between 2 and 4 ppm.

2. Interpretation and Analysis

The metabolite ratios to Cr are meaningless since the Cr is absent, and absolute water or metabolite concentrations are not available. The main finding is lipid. There is no clear evidence of lactate, Cho or any other of the metabolites. The MRS is therefore not consistent with lymphoma or other kinds of active tumor.

The labels 1–5 on Fig. 5.181 indicate assignments which were based on high resolution in vitro 1 and 2 dimensional NMR spectroscopy of the excised tissue. These have retrospectively been recognized in this in vivo spectrum, but a certain assignment of these broad signals from the in vivo spectrum alone would not have been possible. The pattern, the "finger print" if you wish, may however with some experience, and with the addition of clinical information be used to identify tuberculoma.

E. Conclusion

The MRS in the lesion is consistent with tuberculoma, not with lymphoma.

F. Summary of Patient's History Post-MRS

The patient was operated on. The mass was excised and histopathology was consistent with tuberculoma. Post operatively, the patient was kept on antituberculous therapy and showed complete resolution of symptoms.

G. Other Comments

Any assistance offered by in vivo MRS in the diagnosis of tuberculoma is welcomed, because the alternative method of diagnosis, culturing a biopsy requires surgery, have a low yield and take 6–8 weeks. This is too long to wait before starting antituberculous therapy. Dr. R.K. Gupta and his group also suggest using high field NMR spectroscopy on excised tissue to further aid the differentiation between a granuloma with and without tuberculosis, the presence of phenolic lipids represent the biochemical fingerprints of Mycobacterium tuberculosis in a granuloma.

This is yet another example of how MRS can influence the care of patients, allowing drug therapy with potential side effects only to be given to the group of patients who will benefit from it.

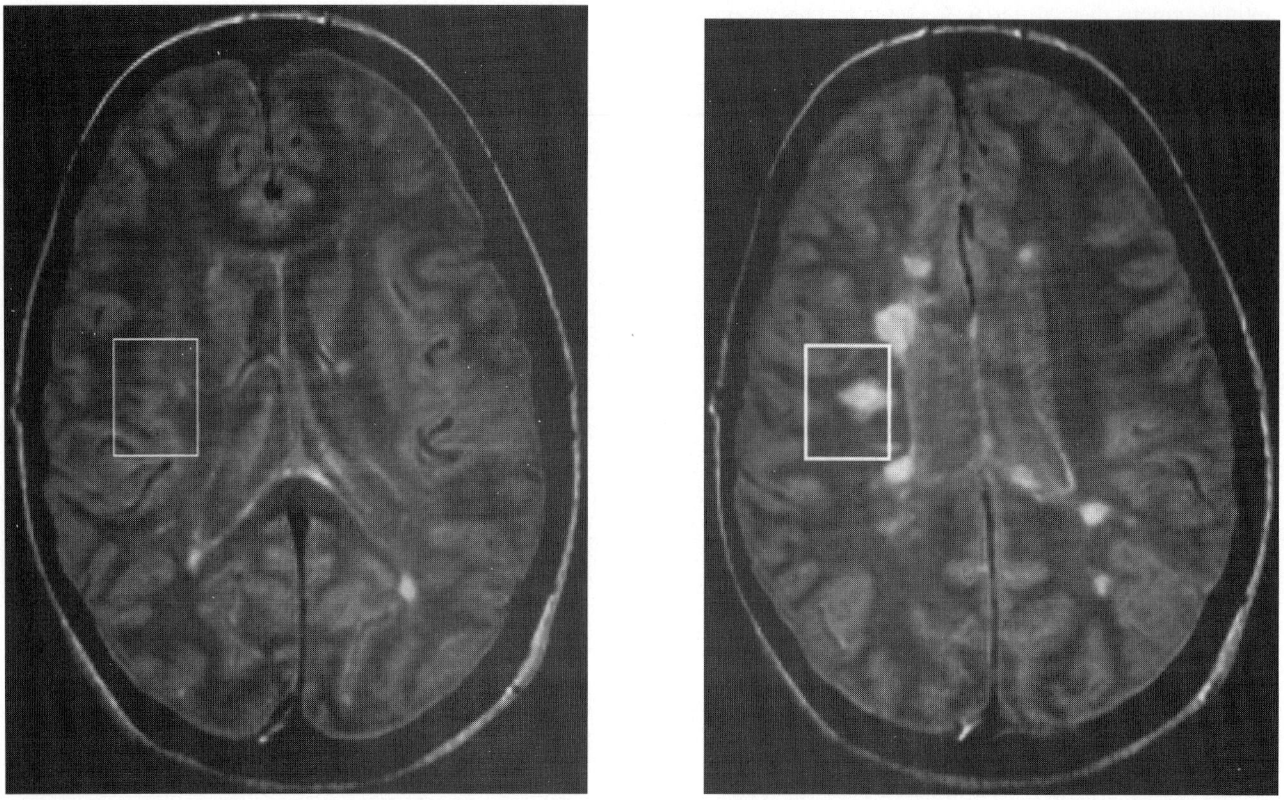

(A) (B)

Fig. 5.182 T_1-weighted axial MRI in 33 year old female with multiple-sclerosis, showing all 3 plaques (A–C) from which PROBE was acquired.

5.8 Multiple Sclerosis and Other Inflammatory Disease

5.8.1 MULTIPLE SCLEROSIS: QUIESCENT PLAQUE

A. Purpose of MRS

Define "activity" in multiple sclerosis plaque.

B. Summary of Patient's History

Female, age 33 years, with long duration of multiple sclerosis. At the time of MRS, she was clinically quiescent, without specific therapy (compare to the following case).

C. MRS Method, Volume of Interest, and Comments about Examination

PROBE-STEAM (TE = 30 ms, TM = 13.7 ms, TR = 1500 ms) was performed in a region of white matter that included a plaque which probably only contributes 10% to the signal of the entire VOI. Thus, the plaque is well visualized in only one of the three 5 mm axial slices displayed, Fig. 5.182.

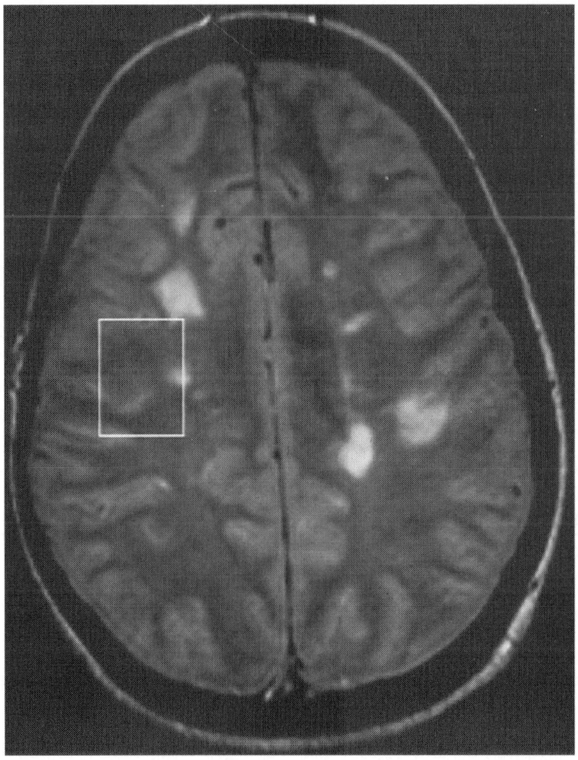

(C)

Fig. 5.182 Continued.

D. White Matter Location

1. Peak Identification and Quality Assurance: Reading the Spectrum from Right to Left

The technical quality of the PROBE spectrum, Fig. 5.183, is excellent, and despite the very small partial volume occupied by the lesion under investigation illustrates some pathological features.

Reading the spectrum systematically from right to left, lipid peaks at 0.9 ppm and 1.48 ppm are pathologically elevated. There is no identifiable lactate doublet at 1.33 ppm, or alanine doublet at 1.48 ppm. Continuing reading, the following peaks are identified: NAA, β,γ-Glx, second and third NAA peaks, Cr, Cho (appropriate for a white matter location), sI, mI and α-Glx, as well as second peaks for Cr and mI. In addition to unusual and elevated lipid resonances, excess of mI and possibly also of Cho, are identified.

2. Interpretation and Analysis

Reading from right to left, metabolite ratios by inspection are: NAA/Cr = 1.58 (probably normal for centrum semi-ovale white matter); Cho/Cr = 0.73 (normal white matter); mI/Cr = 0.69 (elevated 25% compared with normal white matter).

E. Conclusion

An abnormal white matter PROBE examination from a region involved in long-persistent MS. Excess lipid implies some "activity" while elevated mI/Cr, but normal Cho/Cr, and NAA/Cr are characteristic of quiescent plaque. Because the partial volume is so small, the report is necessarily cautious.

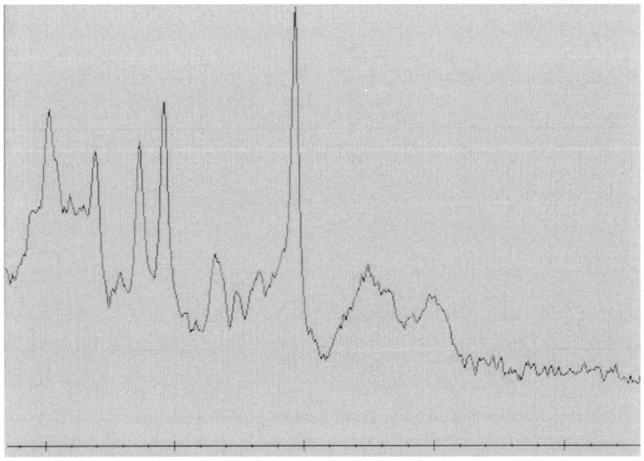

Fig. 5.183 The white matter spectrum from a region of the right centrum semiovale shows abnormalities of chronic, quiescent multiple sclerosis, despite the apparently small proportion of the VOI which is occupied by plaque.

F. Other Comments

The strategy applied to MRS of focal as opposed to generalized, brain lesions is empirical and pragmatic. A substantial VOI in a homogeneous brain region guarantees a readily readable spectrum, while "key-hole" approaches to very small foci in non-homogeneous brain regions are almost doomed to technical failure. Of two acceptable compromises, this case illustrates the more conservative: a VOI is selected to include the entire "lesion" together with a larger partial volume of "normal" brain. In MS, even brain which appears normal on MRI is usefully examined with MRS.

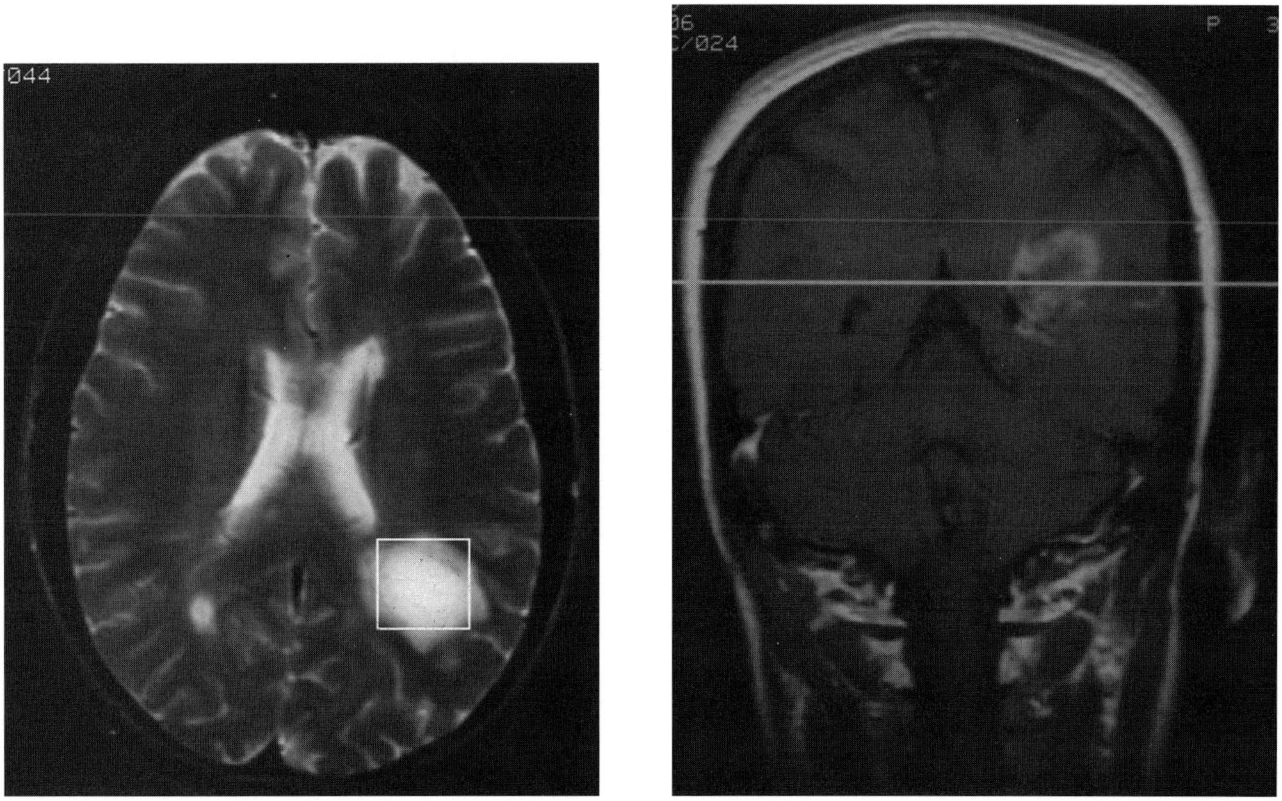

(A) (B)

Fig. 5.184 (A) Axial T_2-weighted MRI at presentation showing symmetrical, probably MS plaques, and PROBE locations measured at presentation, and in three follow-up examinations. (B) A postcontrast coronal MRI of gadolinium enhanced lesions seen in figure above, and examined with PROBE, and subsequently with MRS.

5.8.2 MULTIPLE SCLEROSIS

A. Purpose of MRS

Monitoring of activity, and differential diagnosis of giant multiple sclerosis plaque.

B. Summary of Patient's History

Female, age 39 years, with a ten year history of multiple sclerosis suffered a severe exacerbation of symptoms. MRI was initially misdiagnosed as "tumor."

C. MRS Method, Volume of Interest, and Comments about Examination

PROBE-STEAM (TE = 30 ms, TM = 13.7 ms, TR = 3000 ms) was performed at presentation (Examination 1) in HMRI's parietal white matter location, almost entirely occupied by gadolinium-enhanced lesion (Fig. 5.184).

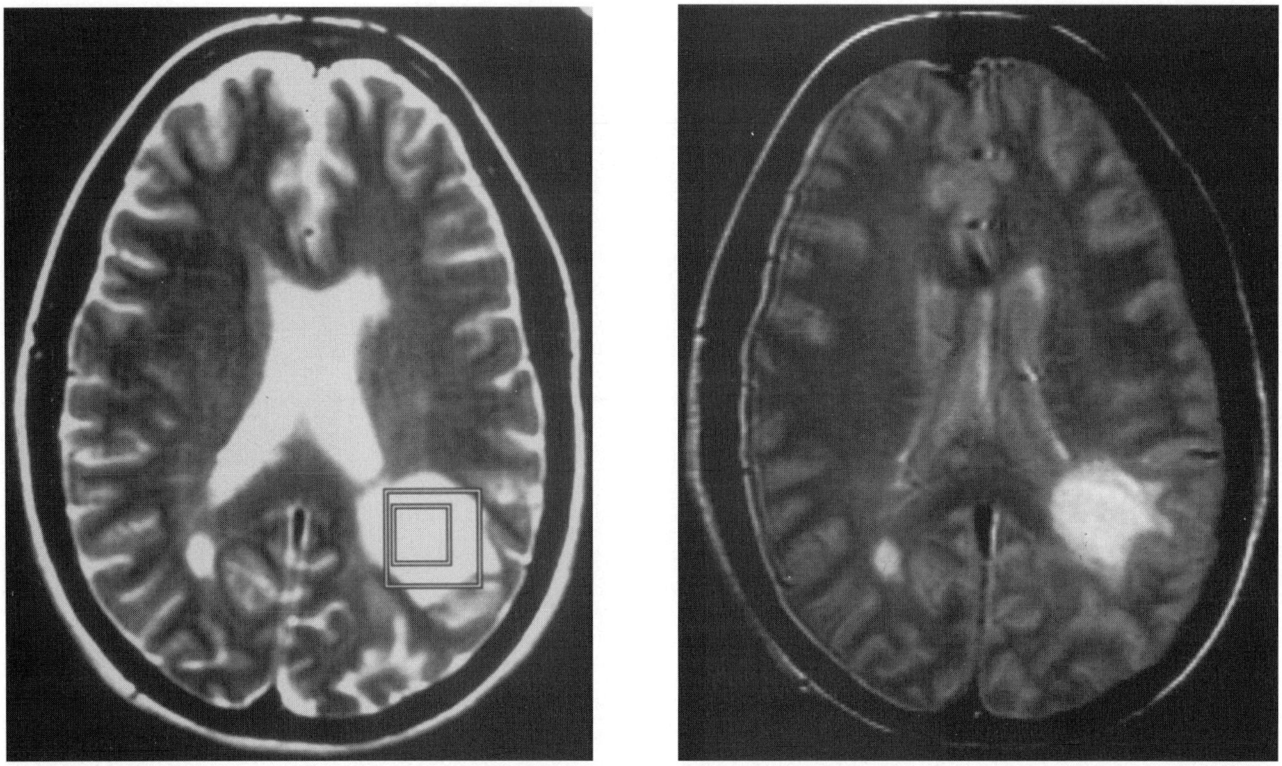

(A) (B)

Fig. 5.185 Progression of MRI appearance in giant MS plaque. Four images to illustrate regression of a highly active MS plaque over a period of 45 days. (A) Day 3: fast spin echo, T_2-weighted image showing, if anything, slightly greater edema (cf. MRS examination 2). (B) Day 38: axial MRI shows no perceptible change in size of the "lesion" but a clearer demarcation from the surrounding edema; gadolinium-enhancement is confined to the periphery of the lesion. (C) Day 45: fast spin echo T_2-weighted MRI showing more clearly the demarcation between "plaque" and surrounding edema, and the locations used for MRS of the active (examination 3, Location 1), and a small inactive, contralateral plaque (examination 3, Location 2).

MRS-STEAM (TE = 30 ms, TM = 13.7 ms, TR = 3000 ms) was performed three days later (Examination 2), (Fig. 5.185A). An imaging study only was performed on day 38, (Fig. 5.185B). MRS was performed 45 days after presentation (Examination 3) in the affected VOI, as well as the contralateral, uninvolved equivalent white matter VOI (Fig. 5.185C).

The first examination revealed a diagnostic MRS pattern, and reversed the MRI impression. Subsequent examinations documented the spontaneous resolution of the intense inflammatory events surrounding the natural history of the giant MS plaque.

C. MRS Method, Volume of Interest, and Comments about Examination

PROBE-STEAM (TE = 30 ms, TM = 13.7 ms, TR = 3000 ms) was performed at presentation (Examination 1) in HMRI's parietal white matter location, almost entirely occupied by gadolinium-enhanced lesion (Fig. 5.184).

MRS-STEAM (TE = 30 ms, TM = 13.7 ms, TR = 3000 ms) was performed three days later (Examination 2), (Fig. 5.185A). An imaging study only was performed on day 38, (Fig. 5.185B). MRS was performed 45 days after presentation (Examination 3) in the affected VOI, as well as the contralateral, uninvolved equivalent white matter VOI (Fig. 5.185C).

The first examination revealed a diagnostic MRS pattern, and reversed the MRI impression. Subsequent examinations documented the spontaneous resolution of the intense inflammatory events surrounding the natural history of the giant MS plaque.

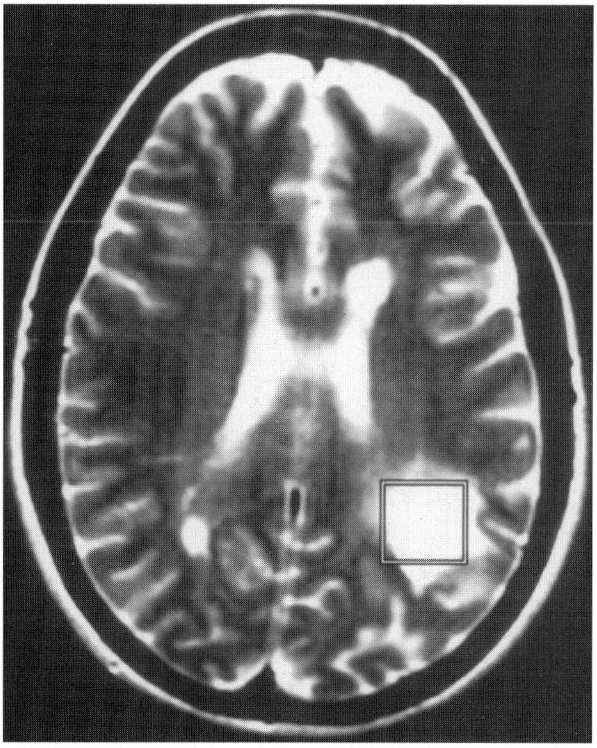

(C)

Fig. 5.185 Continued.

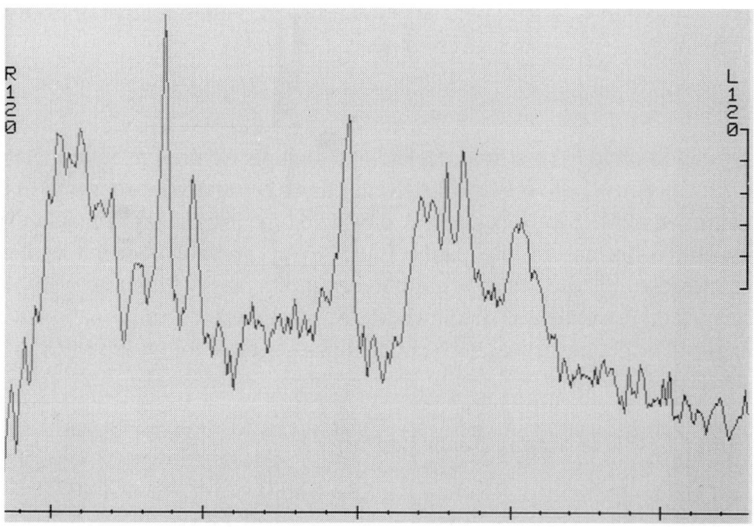

Fig. 5.186 The "white matter" PROBE spectrum of an active MS plaque shows the abnormal metabolite profile of a highly inflammatory lesion, with lipid, lactate, alanine, reduced Cr and excess Cho, but very nearly normal amounts of NAA. The latter is inconsistent with the diagnosis of a primary brain tumor.

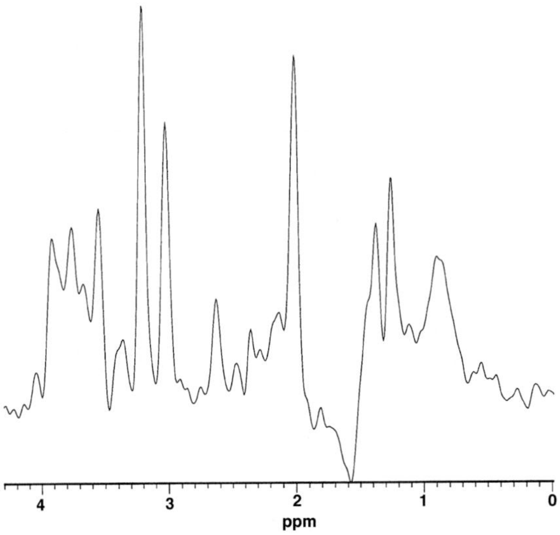

Fig. 5.187 MRS of acute MS plaque after 3 days (VOI = 10.6 cm^3). Some resolution of abnormalities noted in earlier MRS-examinations has occurred over this short interval. Lipid and lactate remain, and Cho/Cr is markedly elevated but the previous alanine doublet is not observed.

D. Examination 1: White Matter Focal Lesion

1. Peak Identification and Quality Assurance: Reading the Spectrum from Right to Left

The technical quality of the automated PROBE spectrum (Fig. 5.186) is excellent, but the metabolite pattern is strikingly abnormal. Water suppression is very slightly excessive ("oversuppressed"), resulting in a baseline that uncharacteristically slopes down from right to left. There are no artifacts within the metabolite spectrum, and diagnostic interpretation is unimpaired.

Reading the spectrum systematically from right to left, lipid at 0.9 ppm is prominent. There is a doublet of lactate at 1.33 ppm, and another of (probably) alanine at 1.48 ppm. Peaks of NAA, β,γ-Glx, a second NAA peak, Cr, and very prominent Cho, sI and/or glucose, mI and a triplet of α-Glx together with glucose (3.8 ppm) are identified. The second Cr and mI peaks are identifiable, but "peaks" at 4.15 and 4.25 ppm may be either artifacts, representative of excessive water suppression, or evidence of the lactate-alanine methylene quartet rarely observed in *in vivo* spectra.

2. Interpretation and Analysis

Reading from right to left, in addition to excess lipid, lactate and alanine are present. The metabolite ratios (adjusting for baseline) are: NAA/Cr = 1.72; Cho/Cr = 1.96; mI/Cr = 1.04. Judging from relatively poor S/N in this PROBE examination, which was subsequently confirmed in MRS performed three days later, the inappropriate metabolite ratios are a reflection of a reduced [Cr], possibly only 60% of normal in white matter. Thus, NAA is probably reduced, while Cho is markedly elevated, and mI is either normal or increased. These findings are entirely in keeping with an actively demyelinating MS plaque.

In summary, an enhancing lesion in a patient with known MS is distinguished from a malignant brain tumor by this short-echo time PROBE examination. The diagnosis was substantiated in a series of follow-up MRS, MRI and MRI with contrast studies over the ensuing 45 days.

E. Examination 2: Location 1 (White Matter Focal Lesion)

1. Peak Identification and Quality Assurance: Reading the Spectrum from Right to Left

The MRS spectrum (Fig. 5.187) was obtained from an identical location to the examination three days previously (voxel size, 10.6 cm^3), and shows general similarities as expected, but a striking change is the virtual disappearance of the alanine doublet (1.48 ppm) previously observed.

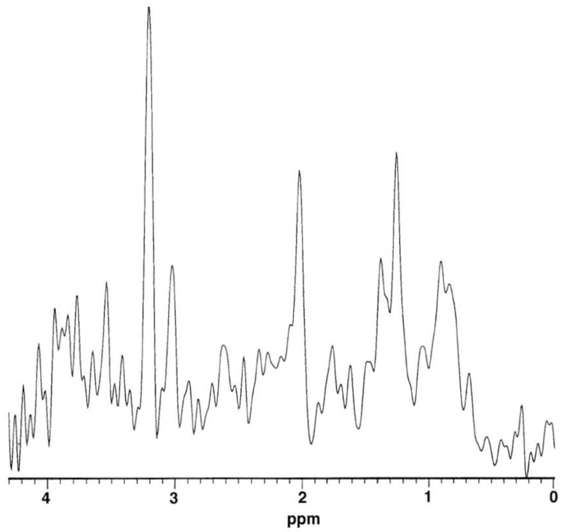

Fig. 5.188 MRS of Small volume (1.7 cm^3) within the previous figure's VOI. Notably NAA is not reduced, while lactate and Cho are increased, indicating that NAA is present at the center of the MS lesion.

Reading the spectrum from right to left, there is lipid, and a very similar intensity of lactate (doublet at 1.33 ppm) with a small shoulder at approximately 1.4 ppm. A baseline artifact, possibly due to lipid within the lesion, but outside the VOI, distorts this region of the otherwise excellent spectrum.

2. Interpretation and Analysis

The metabolite ratios are: NAA/Cr = 1.23; Cho/Cr = 1.41; mI/Cr = 0.70; β,γ-Glx is normal, but $glucose_1$/Cr and $glucose_2$/Cr are clearly increased. Absolute [Cr] is reduced to 71% of normal. In the non-quantitative PROBE examination performed at presentation, it was also surmised that [Cr] might be below normal.

F. Examination 2: Location 2 (Within White Matter Focal Lesion)

1. Peak Identification and Quality Assurance: Reading the Spectrum from Right to Left

Further quantitative analysis of the lesion's center was provided by a VOI equal to 1.7 cm^3 (Fig. 5.188), located entirely within the center of the VOI described above.

Reading the spectrum from right to left, the S/N is necessarily much poorer, but still provides well-resolved resonances for lipid (0.9 ppm), lactate (doublet 1.33 ppm), NAA, Cr, Cho, and possibly mI (at 3.56 ppm).

2. Interpretation and Analysis

In addition to lipid and lactate, which clearly derive from the central portion of this enhancing lesion, the metabolite ratios for NAA/Cr and Cho/Cr are 1.50 and 2.53, respectively (both are elevated compared to the previous spectrum). Absolute [Cr] is 62%, [NAA] is 76%, and [Cho] is 180% of normal. The findings indicate NAA is present in the center of the lesion, rather than a contribution of partial volume from surrounding "normal" brain. The center is thus unlikely to represent glioma, or necrosis.

G. Examination 3: Location 1 (White Matter Focal Lesion)

1. Peak Identification and Quality Assurance: Reading the Spectrum from Right to Left

A good quality MRS spectrum (Fig. 5.189) with abnormal lipid (0.9 ppm), and reduced lactate (1.33 ppm), compared to prior examinations. NAA is more prominent, and β,γ-Glx is better resolved. Cr, Cho, mI, α-Glx, and second peaks for Cr and mI are all identified.

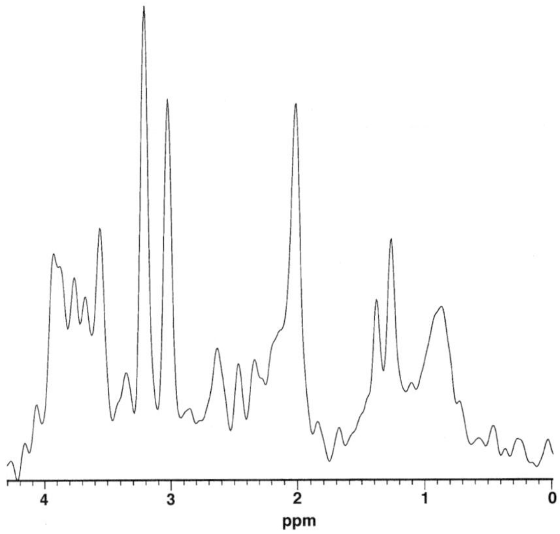

Fig. 5.189 MRS after 45 days. The principal abnormalities of this MS plaque are altered very little, but intensities of lipid and lactate have fallen while Cr and NAA are both increased compared to the MRS from the larger VOI measured at Day 3.

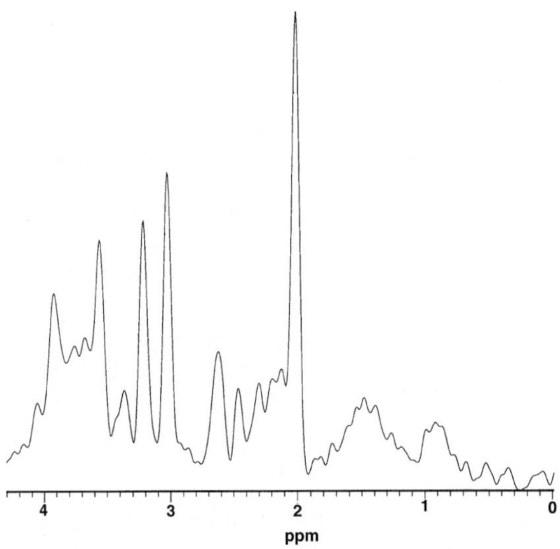

Fig. 5.190 MRS of normal-appearing white matter in this patient with MS, after 45 days. The mI/Cr is abnormal, in this otherwise normal spectrum.

2. Interpretation and Analysis

Lipid and lactate are elevated, and the metabolite peak ratios are: NAA/Cr = 0.99; Cho/Cr = 1.26; mI/Cr = 0.65; β,γ-Glx/Cr and α-Glx/Cr are normal, as are the two glucose resonances.

The spectrum shows improved S/N, and absolute [Cr] has increased to 89% of normal. Therefore, NAA has partly recovered, while the excess Cho is less marked. All of these changes are consistent with recovery and/or re-myelination of the area examined.

H. Examination 3: Location 2 (Contralateral Control)

1. Peak Identification and Quality Assurance: Reading the Spectrum from Right to Left

An excellent quality white matter spectrum of near-normal appearance serves as a control to the foregoing (Fig. 5.190). When examined in detail, only mI/Cr (elevated from 0.61 to 0.78) represents a significant abnormal finding.

I. Conclusion

A series of MRS examinations confirm the presence of an MS plaque rather than a neoplasm. In particular, NAA persists in the center of the lesion and recovers over time. This is consistent with demyelination, and re-myelination.

Finally, automated PROBE reproduces in all details the diagnostic details of MRS. The sole change, disappearance of alanine, illustrates how rapidly metabolic markers of inflammation may appear, or disappear, even without obvious changes in MRI.

J. Summary of Patient's History Post-MRS

Patient recovered from the present episode, and continued well with occasional relapses of MS.

K. Other Comments

Despite the overlap of MRS features for many focal brain lesions, careful attention to detail often reveals diagnostic differences.

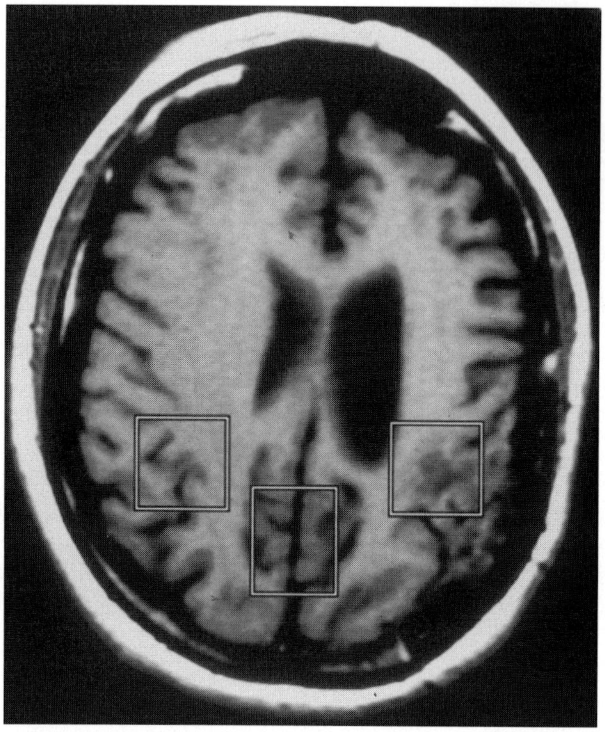

Fig. 5.191 Image indicating the pathological region in this patient with Rasmussen's encephalitis, and the locations of three VOIs used to examine it.

5.8.3 RASMUSSEN'S ENCEPHALITIS (ADULT)

A. Purpose of MRS

Define the extent of encephalitis.

B. Summary of Patient's History

Female, age 46 years, with little history available, apart from MRI evidence of long-standing inflammatory change, and of seizures now well controlled by medication.

C. MRS Method, Volume of Interest, and Comments about Examination

STEAM (TE = 30 ms, TM = 13.7 ms, TR = 3000 ms) was performed in three locations measuring approximately 2 × 2 × 2 cm. Location 1 defines the lesion, Location 2 is from a contralateral, normal-appearing region, and Location 3 is from HMRI's occipital grey matter, also visibly involved in the pathological process, Fig. 5.191.

D. Location 1 (White Matter within Lesion)

1. Peak Identification and Quality Assurance: Reading the Spectrum from Right to Left

The technical quality of spectrum, Fig. 5.192A, is good. There is a small, possibly nonsignificant lipid peak at 0.9 ppm, and a definite lactate doublet at 1.33 ppm. Reading the spectrum from right to left, NAA, β,γ-Glx, second and third NAA peaks, Cr, Cho, sI, mI and α-Glx peaks are all identified, as are second peaks for Cr and mI.

(a)

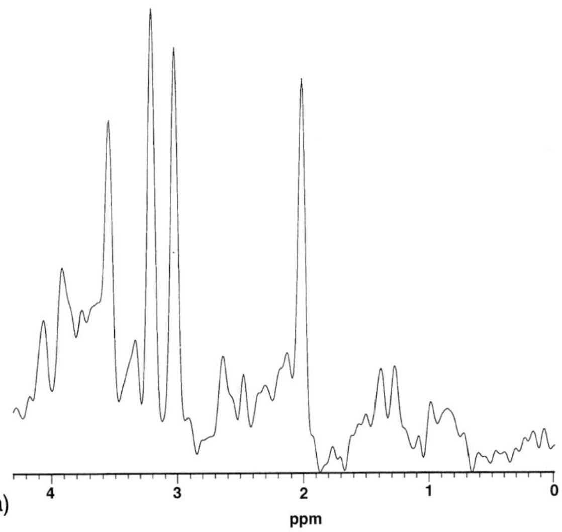

(b)

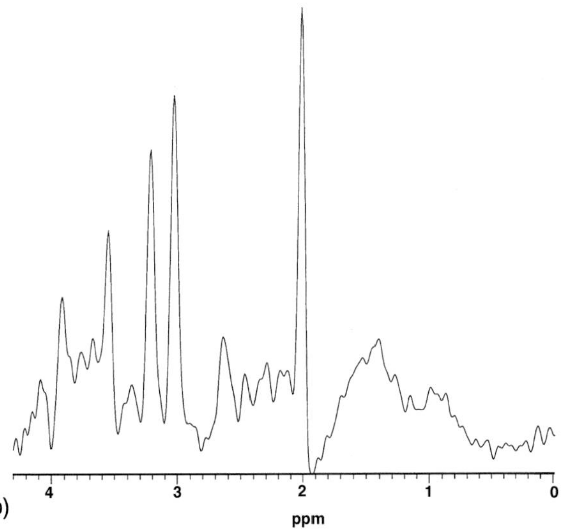

Fig. 5.192 Quantitative MRS from the three brain regions. The most severely affected spectrum (white matter within lesion) is characteristic of a chronic encephalitis, but not specific to Rasmussen's encephalitis (a). Even the apparently normal contralateral white matter is affected in this patient; principally with axonal-neuronal loss (b). Grey matter location shows abnormalities which are a reflection of a partial volume of MRI abnormal tissue included within the VOI, from the first location examined above (c).

2. Interpretation and Analysis

The spectrum is abnormal in several details. In addition to excess lactate, and possibly also lipid, metabolite ratios are: NAA/Cr = 0.92 (34% reduced); Cho/Cr = 1.10 (33% increased); mI/Cr = 0.82 (34% increased); β,γ-Glx/Cr = 0.19 (reduced); α-Glx/Cr = 0.44. Brain water is 76%, high for white matter, but the CSF volume within the VOI is 17.9%, which is increased 5%, and is a measure of cortical atrophy. Absolute [Cr] is 105% of normal, so that the abnormal ratios reflect actual concentration changes.

E. Location 2 (Contralateral White Matter)

1. Peak Identification and Quality Assurance: Reading the Spectrum from Right to Left

A spectrum of good quality, Fig. 5.192B, was acquired from a region of brain which appears normal on MRI, Fig. 5.191.

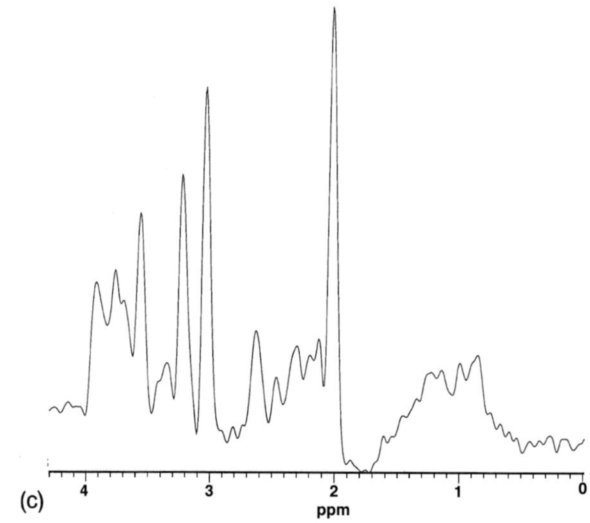

Fig. 5.192 Continued.

Reading from right to left, there is some outer-volume lipid artifact, resulting in a "dip" at 1.95 ppm. Protein-lipid peaks in phase (upright) at 0.9 ppm and 1.4 ppm are probably normal. Peaks are identified as NAA, β,γ-Glx, second NAA, Cr, Cho, sI, glucose, mI and α-Glx peaks, with a second Cr and possibly second mI peak.

2. Interpretation and Analysis

The metabolite ratios are: NAA/Cr = 1.26 (10% below normal for this location); Cho/Cr = 0.85 (normal); mI/Cr = 0.61 (normal); β,γ-Glx/Cr = 0.27 (normal) and α-Glx/Cr = 0.34. [Cr] is normal at 103%, so that the reduction of NAA is real. Brain water (68.3%) and CSF contribution (3.6%), are both normal for white matter and for this location.

F. Location 3 (Grey Matter)

1. Peak Identification and Quality Assurance: Reading the Spectrum from Right to Left

Spectrum, Fig. 5.192c, from mainly normal appearing grey matter, but for which the VOI also includes some tissue of MRI-appearance similar to, Fig. 5.192A. The spectrum is good quality without significant artifacts.

Reading from right to left, a lipid peak at 0.9 ppm, but no convincing doublet for lactate at 1.33 ppm. Peaks identified are NAA, β,γ-Glx, second and third NAA, Cr, Cho, sI, glucose, mI, α-Glx (unusual appearance), second Cr, but no second mI.

2. Interpretation and Analysis

The metabolite ratios are: NAA/Cr = 1.23 (reduced 9%); Cho/Cr = 0.74 (25% above normal grey matter); mI/Cr = 0.64 (slightly, but significantly elevated). [Cr] is 86% (slightly reduced), so NAA is further reduced. Brain water is 70.5% (essentially normal for grey matter) and CSF comprises 11.9% of VOI (no significant cortical atrophy within this VOI); β,γ-Glx/Cr = 0.27 (normal) and α-Glx/Cr = 0.56, normal for Glx, but α-Glx/Cr is possibly artificially increased due to elevated glucose ($glucose_1$/Cr = 0.58, and $glucose_2$/Cr = 0.15).

G. Conclusion

Within the region of cortex most obviously affected by Rasmussen's encephalitis, MRS is highly suggestive of an active inflammatory process (lactate, lipid, Cho) with axonal loss and gliosis. The remaining regions of brain examined are also abnormal.

H. Summary of the Patient's History Post-MRS

The seizures continued.

I. Other Comments

Acute viral encephalitis (herpes, PML in AIDS, and HIV encephalitis) is characterized by much lipid and lactate, and high Cho/Cr. In this case, a more chronic picture, with neuronal/axonal loss and gliosis is likely, but not proven.

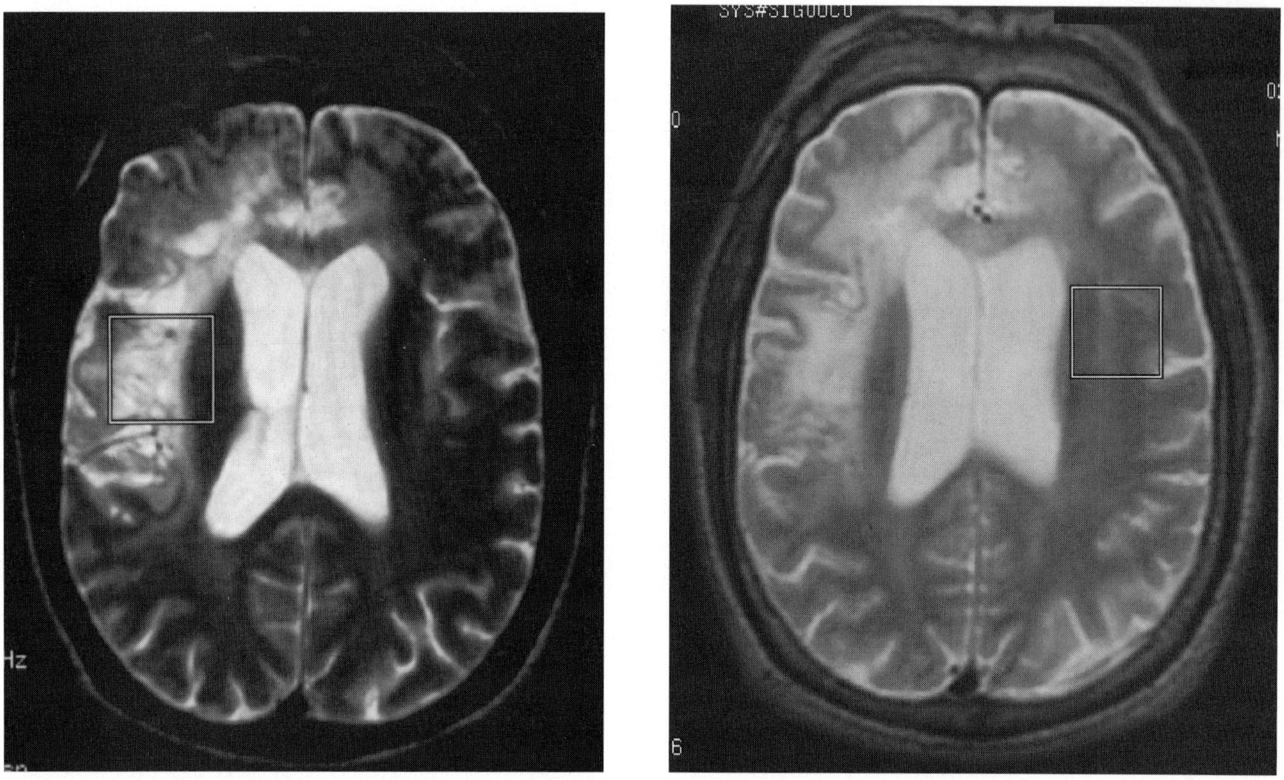

(A) (B)

Fig. 5.193 The position of Location 1 in the right hand side lesion is shown on this T_2-weighted MRI (A). The position of the contralateral VOI (B). (Courtesy of Dr. Wang, Rancho Los Amigos Hospital, Downey, California.)

5.8.4 HERPES ENCEPHALITIS

A. Purpose of MRS

To evaluate a patient who was diagnosed with herpes encephalitis and a multitude of complicating factors, including alcohol abuse. The latter led to the request of ruling out hepatic encephalopathy.

B. Summary of Patient's History

Male, age 49 years, diagnosed with herpes encephalitis six months prior to the MRS/MRI examination. He has a long history of i.v. drug abuse, has tuberculosis, is confused and agitated, and suffers from memory loss. There is no hypertension, no diabetes, and no renal failure.

C. MRS Method, Volume of Interest, and Comments about Examination

Automated PROBE-STEAM (TE = 30 ms, TM = 13.7 ms, TR = 1500 ms) was performed in two locations. The first was a right hand side, parietal white matter VOI placed in the affected area as seen on MRI (Fig. 5.193), and therefore more frontal than HMRI's standard parietal white matter location. Location 2 was the corresponding left hand side in mostly normal appearing white matter. The patient received 7 mg i.v. Valium prior to the examination. MRI contrast (Magnevist®) was used, but it was not administered before the MRS measurements had been finalized.

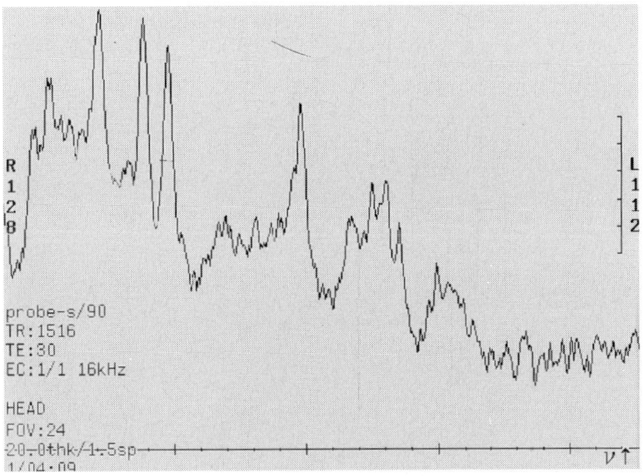

Fig. 5.194 An abnormal white matter spectrum in a MRI visible lesion in a patient with herpes encephalitis. (Courtesy of Dr. Wang, Rancho Los Amigos Hospital, Downey, California.)

D. Location 1 (Right White Matter in Lesion)

1. Peak Identification and Quality Assurance: Reading the Spectrum from Right to Left

The technical quality of the spectrum (Fig. 5.194), is excellent with acceptable S/N, narrow peaks, and no artifacts. The baseline slopes smoothly up towards the left due to residual water signal; this is not a problem, but nevertheless has to be accounted for when estimating the peak ratios.

Reading the spectrum from right to left, lipids and/or macromolecules are seen at 0.9 ppm and at 1.2–1.8 ppm underlying one or two doublets. The lactate doublet is seen at 1.33 ppm, possibly partially overlapping with an alanine doublet centered at 1.48 ppm. A certain assignment of alanine can however not be made as S/N of these resonances do not allow it, and the possibility of it only being "lipids" and lactate cannot be ruled out. Then follows a low NAA, and insignificant β,γ-Glx, and second NAA. The Cr, Cho, mI and the second Cr are all clear peaks. The second mI is also seen.

2. Interpretation and Analysis

The S/N is lower than in the spectrum from the contralateral VOI (Location 2), which was acquired under identical conditions. This is probably due to a reduced fraction of healthy remaining brain tissue in the VOI of Location 1. The [Cr] is not available from the automated PROBE sequence. [Cr] (the normalized brain tissue concentration) cannot be assumed normal in this patient, but the spectrum provides valuable information in the form of ratios and abnormal resonances. Excess lipids and lactate are consistent with inflammation understood in the context of membrane breakdown and poor oxygenation. The significance of the possible finding of alanine is unknown, it has been seen in meningioma and in multiple sclerosis (see cases 5.7.6 and 5.8.2).

The metabolite ratios are estimated and compared to the standard white matter ratios: NAA/Cr = 0.7 (markedly decreased), Cho/Cr = 1.0 (increased), and mI/Cr = 0.9 (increased); note that the baseline has been accounted for. Unless [Cr] differs drastically from normal it is safe to say that NAA is markedly decreased, and mI is markedly increased. Whether Cho is increased, or normal, depends a little more on [Cr], but it is unlikely that Cho is decreased.

Decreased NAA/Cr, increased Cho/Cr and mI/Cr and in some cases excess lipid/lactate have been found in other (mostly white matter) lesions where inflammation or other types of destruction is taking place (examples: multiple sclerosis, progressive multifocal leukoencephalopathy [PML], HIV encephalitis, adrenoleukodystrophy [ALD], and in bacterial abscesses). The spectrum is therefore not a surprising one for herpes encephalitis.

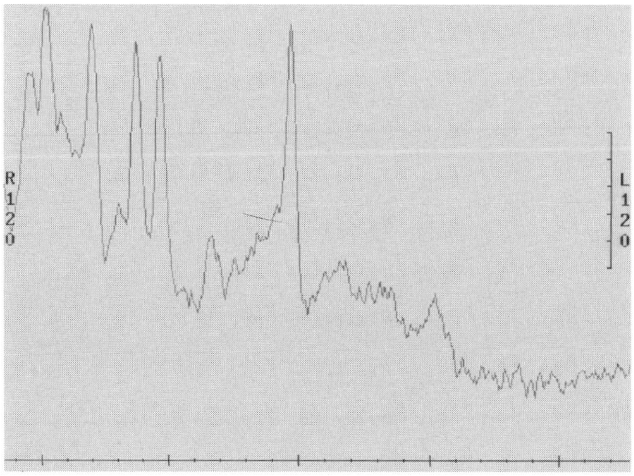

Fig. 5.195 The mostly normal appearing contralateral VOI gave a similarly abnormal spectrum, but missing the excess lipid/lactate resonances. Neither of the spectra are consistent with hepatic encephalopathy, another suspected complication in this patient, (Courtesy of Dr. Wang, Rancho Los Amigos Hospital, Downey, California).

E. Location 2 (Left White Matter)

1. Peak Identification and Quality Assurance: Reading the Spectrum from Right to Left

The quality of the spectrum (Fig. 5.195) is very good with excellent S/N, narrow peaks, and no artifacts. A nonproblematic slope of the baseline due to residual water is again seen.

Reading the spectrum from right to left, NAA, β,γ-Glx, the second NAA, Cr, Cho, sI, mI, α-Glx, and the second Cr and mI are identified.

2. Interpretation and Analysis

Again, only ratios are available, and they are estimated to be: NAA/Cr = 1.2 (decreased), Cho/Cr = 1.0 (increased), and mI/Cr = 1.0 (increased). This is a quite abnormal white matter spectrum, considering the mostly normal MRI. Partial volume of a lesion at the bottom edge (not shown) of the VOI may contribute, at the most, 10% of the VOI, and can therefore not explain the extent of the abnormalities.

F. Conclusion

The PROBE spectra are not consistent with hepatic encephalopathy. Both spectra are abnormal with changes typically seen in degenerated, or inflamed brain tissue, and are therefore not surprising for herpes encephalitis. The changes are worst in the MRI visible lesion, but also present contralaterally. An occipital location would have been helpful to confirm the most likely global nature of these changes. Extended or global loss of NAA can certainly explain the memory loss and behavioral abnormalities in this patient.

G. Other Comments

A part of the lesion in the temporal lobes looked even worse, but it is difficult to obtain good quality spectra from the temporal lobes, so time was not wasted on an attempt to do this.

A VOI further away from the lesion, for example in occipital grey matter, would have been useful to get a better impression of how globally the brain was affected.

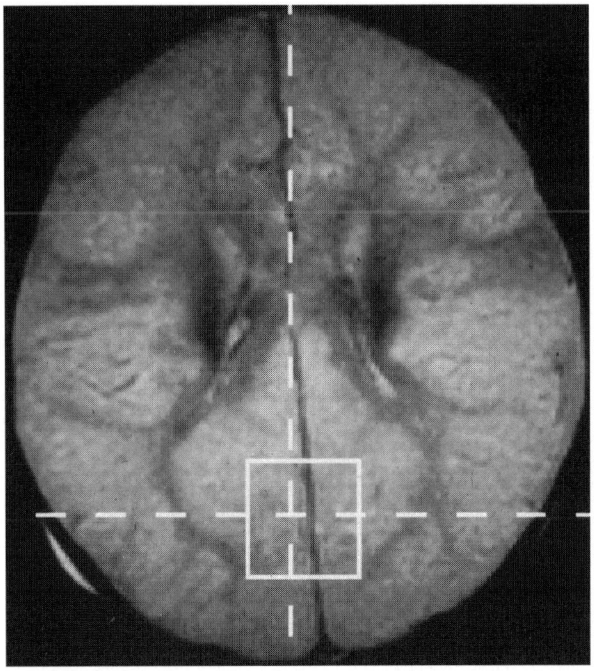

Fig. 5.196 Axial image showing the location of the VOI. Multiple abnormalities were found on the MRIs, including diffuse cerebral edema. (Courtesy of Drs. Ashwal and Holzhouser, Loma Linda Medical Center, California.)

5.9 Pediatrics: Abnormal Development, Reyes Syndrome, Inborn Errors of Metabolism

5.9.1 HYPERGLYCINEMIA

A. Purpose of MRS

MRS (and MRI) was requested to evaluate the brain of an infant with severe metabolic acidosis of unknown etiology.

B. Summary of Patient's History

A one-month-old female (term infant) was found gasping and unresponsive in her crib. The infant was intubated, and a CT of the head showed severe cerebral edema. There was severe metabolic acidosis of unknown etiology, and abnormal liver enzymes.

C. MRS Method, Volume of Interest, and Comments about Examination

STEAM (TM = 30 ms, TR = 3000 ms) was performed in the occipital grey matter (see image, Fig. 5.196) at TE 20 ms (Fig. 5.197) and TE 270 ms (Fig. 5.198).

D. Grey Matter Location (TE 20 ms)

1. Peak Identification and Quality Assurance: Reading the Spectrum from Right to Left

The first impression is a severely abnormal spectrum (Fig. 5.197), but the resolution and S/N are good. In comparing the quality of this MRS and the majority of other cases presented, it should be noted that the spectra of this case are acquired with a scanner with non-shielded gradients, and that post-processing, such as line-broadening and filtering of the water signal, are

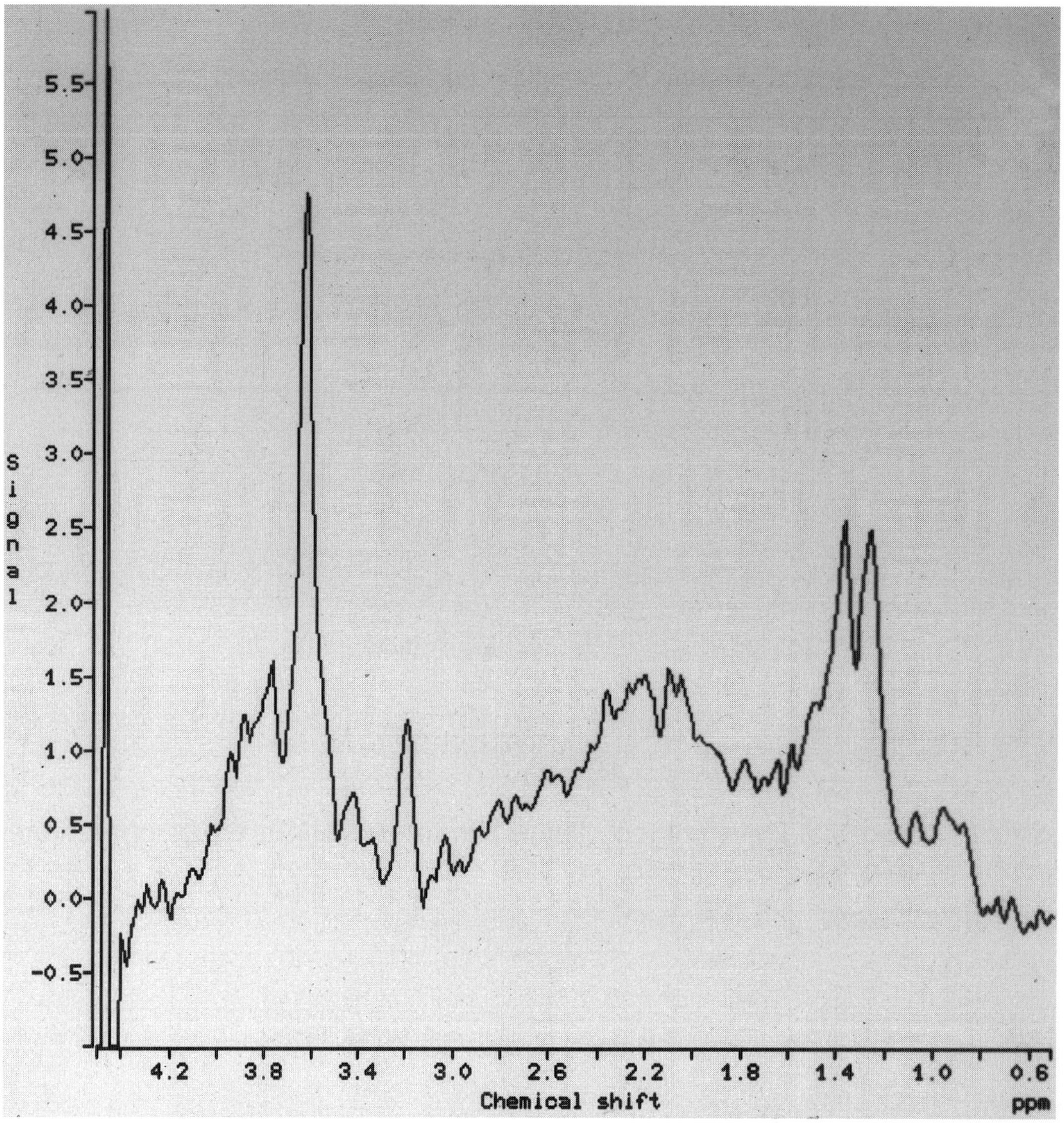

Fig. 5.197 The TE 20 ms STEAM spectrum shows severe abnormality. An unusual peak is found at 3.6 ppm. (Courtesy of Drs. Ashwal and Holzhouser, Loma Linda Medical Center, California.)

not identical. A different appearance of peaks and noise between peaks, some residual water signal, and even a less perfect baseline should be anticipated.

Reading from right to left, a very small amount of lipid is seen at 0.9 ppm, and a clear lactate doublet at 1.33 ppm, probably with some underlying lipid. NAA is absent, but Glx is seen. The baseline drops towards 3.0 ppm (probably an effect of eddy currents). Cr is absent, but Cho is seen at 3.2 ppm. The next large peak at 3.6 $\pm$ 0.02 ppm might at first sight be assigned to mI, but that would be a mistake. Although we cannot exclude minor amounts of overlapping mI, the resonance is 0.04 ppm from mI and must be assigned to something else (see below).

2. Interpretation and Analysis

Ratios are meaningless because Cr is absent. Features of the spectrum are prominent lactate, missing NAA and Cr, some Cho and the peak at 3.6 ppm. In normal brain it is well established that the 3.56 ppm peak consists mainly of mI, and that

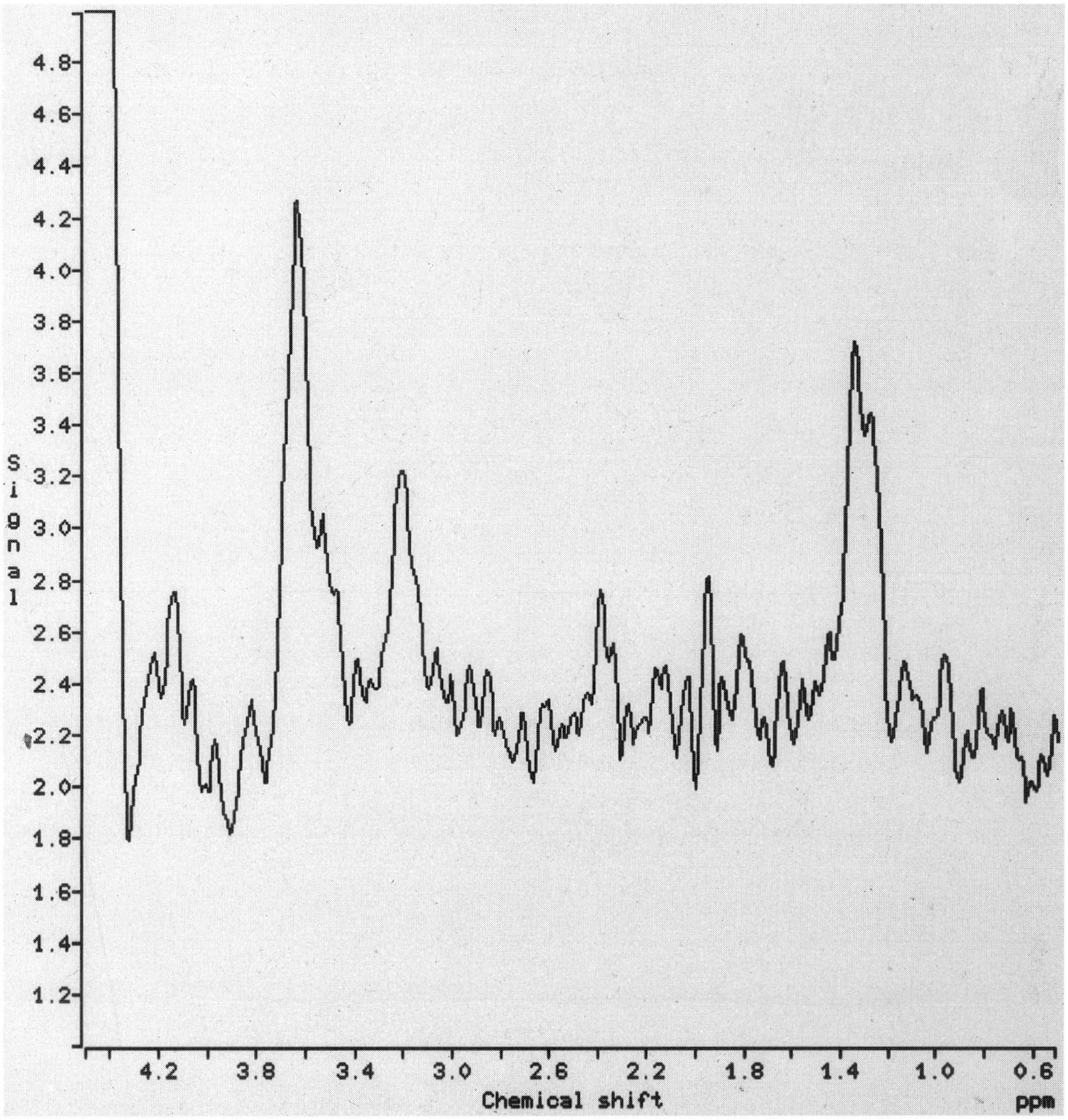

Fig. 5.198 The TE 270 ms STEAM spectrum still shows a peak at 3.6 ppm. This confirms the probable singlet nature (apparent long T_2) of the resonance. (Courtesy of Drs. Ashwal and Holzhouser, Loma Linda Medical Center, California.)

underlying glycine (also resonating at 3.56 ppm) contributes a maximum of 10%. Glycine is a singlet, which differs from the mI in that the apparent T_2 (no coupling) is much longer, so glycine will, just as NAA, Cr and Cho, remain at long echo times, whereas mI will vanish. In order to determine if the 3.6 ppm peak also has properties of a singlet, a long-echo time spectrum was acquired.

E. Grey Matter Location (TE 270 ms)

1. Peak Identification and Quality Assurance: Reading the Spectrum from Right to Left

This long-echo spectrum (Fig. 5.198) has a flatter baseline than the short TE 20 ms spectrum, which is not surprising; outer volume water or lipid signals are more attenuated, eddy currents are less of a problem, and true underlying short T_2 resonances (humps) have disappeared at TE 270 ms. However, the resolution is not as good as in the first spectrum. Some asymmetry

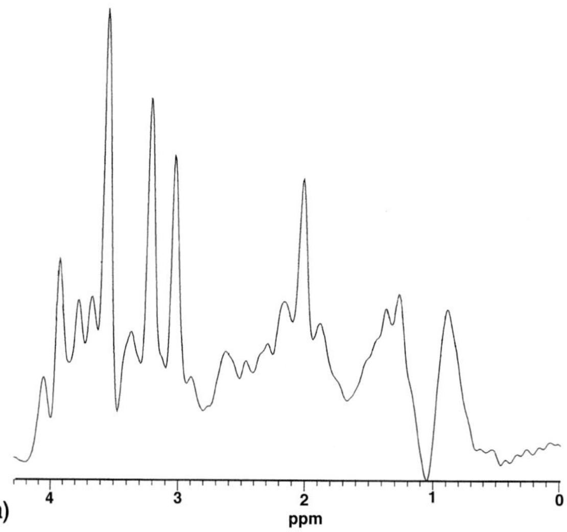

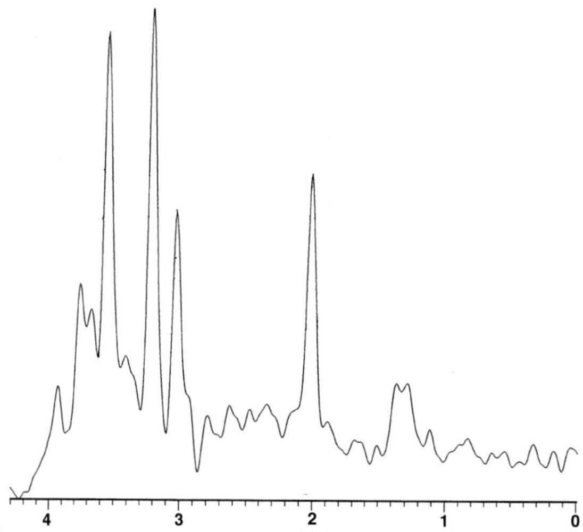

Fig. 5.199 Grey matter spectrum TE = 30 ms (a) and TE = 270 ms (b) from a patient with non-ketotic hyperglycinemia. Part of the peak at 3.56 ppm seen in (a) is assigned to glycine, because it persists at TE = 270 ms (b).

is noted in the 3.2 and 3.6 ppm peaks, probably indicating movement during the acquisition. But the purpose of the measurement, to investigate the 3.6 ppm peak, is not jeopardized by the broader peaks. The peak of 3.6 ppm has not vanished, so it is likely to be a singlet.

Reading from right to left we see lactate, which is hardly a doublet due to the poor resolution, but at long-echo time this peak cannot be due to lipids, so the TE 20 ms finding of lactate is confirmed. A small "spike" is seen at 2.0 ppm (this is not a resonance). A resonance, very close to the noise level, is seen at 2.4 ppm (see below). Cho is confirmed, and the peak at 3.6 ppm is still present at TE 270 ms, which signifies that it is a long T_2 metabolite. The 3.6 ppm peak differs by 0.04 ppm in chemical shift from that of glycine. However, a speculative possibility is that, it may still be glycine that has shifted due to an abnormal pH value. Exchange of the neighboring chemical group, NH_3^+ and COO^-, means that glycine's chemical shift is probably more susceptible to pH than most other metabolites seen in proton brain spectra. Creatine is absent, and there is a huge lactate peak, both of which are compatible with unusual pH. But before the 3.6 ppm peak can be assigned to glycine, the alternate possibilities need to be investigated in chemical solutions, and the pH in the baby's brain needs to be established, for example by ^{31}P MRS.

2. Interpretation and Analysis

The findings of the short TE spectrum are confirmed. There is prominent lactate, no NAA, no Cr, but some Cho is present. The large peak at 3.6 ppm is difficult to assign, but might be glycine. The small peak at 2.4 ppm is most likely real, but since the S/N and the resolution could be better, another examination is recommended before anything certain can be said about this peak. Possibilities are that this is a very little signal, which drowns in the broader resonances of the TE 20 ms spectrum, but shows up in the TE 270 ms spectrum because it has a long T_2, probably a singlet peak. Succinate, which is a singlet that resonates at 2.4 ppm, can be considered among candidates if the peak is reconfirmed in a later examination.

F. Conclusion

Excess lactate, and absent NAA and Cr, indicate poor outcome. A dominant peak at 3.6 ppm needs further investigation before it can be assigned.

G. Summary of Patient's History Post-MRS

Elevated blood levels of glycine was found, and the patient is believed to have suffered from nonketotic hyperglycinemia, although a certain diagnosis was not reached.

H. Other Comments

The assignment of the 3.6 ppm peak is difficult. One possibility which was considered was that the display of the chemical shift reference was chosen incorrectly. In other word the doublet is not lactate, the peak called "Cho" is not Cho, and the unknown peak's chemical shift is correspondingly shifted from 3.6 ppm. But the consideration of other possible doublets and singlets appearing with the given chemical shift differences was not fruitful (Doublet at "X" ppm, singlet at "X" + 1.89 ppm, and singlet at "X" + 2.27 ppm).

Grey matter spectra from a confirmed case, another patient, are shown in Fig. 5.199 for comparison. This patient's gestational age was 42 week and plasma glycine was elevated (1725 μmol/L, normal range 224–514 μmol/L), and the diagnosis was non-ketotic hyperglycinemia.

The spectra were measured at TE = 30 ms and TR = 1500 ms (A), and TE = 270 ms and TR = 3200 ms (B). The peaks at 3.56 and 4.06 ppm in the TE = 30 ms spectrum are first assigned to mI, but part of the 3.56 ppm peak originates from glycine which persists at TE = 270 ms.

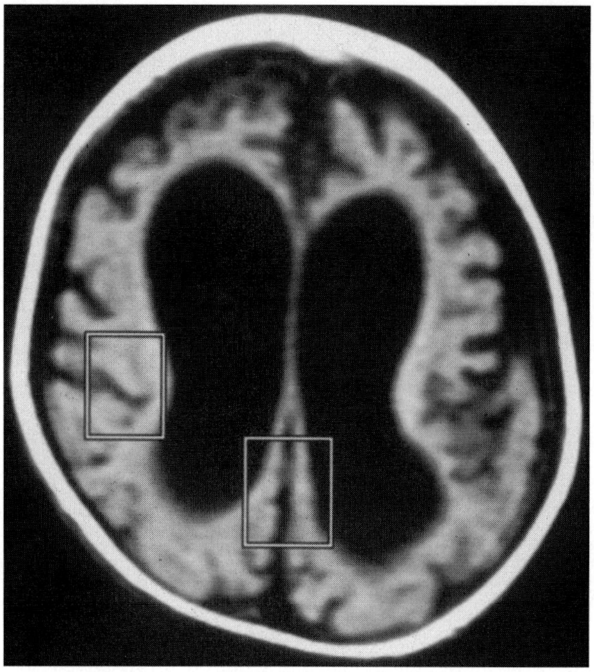

Fig. 5.200 T_1-weighted axial MRI illustrating the locations for MRS. The images are striking for massive symmetrical ventricular dilatation, cortical thinning, and absence of normal sulci, secondary to congenital CMV and hypoxic-ischemic injury.

5.9.2 CYTOMEGALOVIRUS (CMV) AND PERINATAL HYPOXIA: LATE SEQUELAE

A. Purpose of MRS

To determine cause of seizures, and aid in prognosis.

B. Summary of Patient's History

Female, age 18 months at the time of MRS; lay in fetal position and appeared decorticate. Patient was born with severe congenital cytomegalovirus encephalitis, and thereafter probable perinatal asphyxia and superimposed hypoxic-ischemic encephalopathy.

C. MRS Method, Volume of Interest, and Comments about Examination

Localized STEAM (TE = 30 ms, TM = 13.7 ms, TR 1500 ms) was performed in HMRI's occipital grey matter (Location 1), and parietal white matter (Location 2). VOIs were measured three days after admission (Fig. 5.200).

D. Location 1 (Grey Matter)

1. Peak Identification and Quality Assurance: Reading the Spectrum from Right to Left

The technical quality of the spectrum (Fig. 5.201) is sufficient, but it includes inverted lipid and artifacts of water suppression. Nevertheless, abnormalities in the principal metabolites can be reliably defined.

Reading from right to left, the lipid/lactate region is totally distorted by outer-volume lipid (the VOI is too close to the scalp and skull). NAA is totally absent; β,γ-Glx is present; no second NAA peak is present; Cr, Cho, sI, glucose and mI (3.56 ppm) are well defined; α-Glx, second Cr and mI are not observed. Further "peaks" are artifacts.

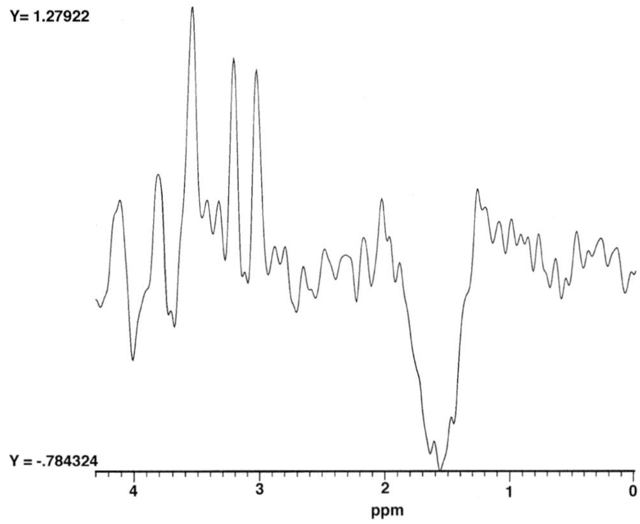

Fig. 5.201 The grey matter spectrum acquired at age 18 months, after uterine CMV and perinatal asphyxia, is consistent with severe hypoxic injury, cortical atrophy with gliosis, and neuronal loss (note outer-volume lipid artifact obscures presence of absence of lactate).

2. Interpretation and Analysis

Reading from right to left, the metabolite ratios are: NAA/Cr = 0 (reduced); Cho/Cr = 1.06 (elevated 100% for 18 month infant); mI/Cr = 2.27 (elevated 300% for age). [Cr] is only 42% of normal, so metabolite ratios are over-estimated. β,γ-Glx/Cr = 0.18 (normal). $Glucose_2$/Cr is elevated (0.39 compared to 0.12 for normal); elevation is contributed by 39% CSF in VOI. Brain water is normal (73%).

E. Location 2 (White Matter)

1. Peak Identification and Quality Assurance: Reading the Spectrum from Right to Left

Reading from right to left, the lipid peak is "in phase" and represents abnormal excess due to necrosis (Fig. 5.202).

2. Interpretation and Analysis

Reading from right to left, the metabolite ratios are: NAA/Cr = 0.89 (reduced 45%); Cho/Cr = 1.08 (increased); mI/Cr = 0.96 (increased two-fold); β,γ- Glx/Cr = 0.39, and α-Glx/Cr = 0.85 (both elevated).

F. Conclusion

Both VOIs indicate severe reduction or absence of NAA, low [Cr], and a real excess of Cho, consistent with old, severe hypoxic ischemic encephalopathy, or severe viral encephalitis, with much reduced neuronal density in the severely atrophied cerebral cortex. Prognosis is essentially persistent vegetative state. Elevated mI/Cr and Cho/Cr probably represent gliosis in surviving cortex.

G. Summary of the Patient's History Post-MRS

Brain stem dysfunction; patient died.

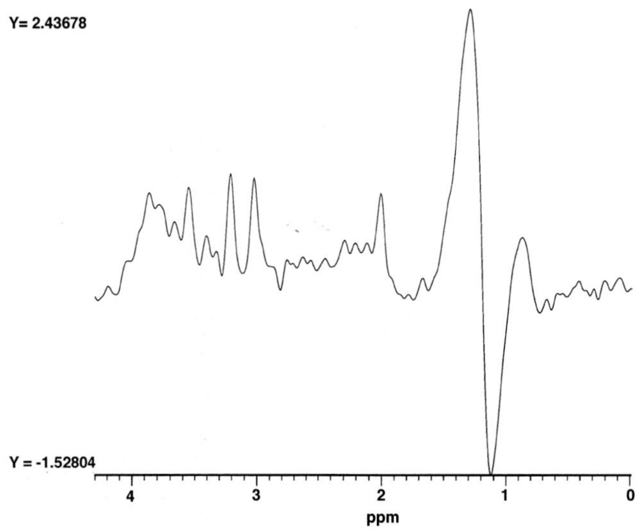

Fig. 5.202 The white matter spectrum contrasts with that from grey matter in this patient. Note the "scaling" of all metabolite peaks is drastically reduced by the presence of a large lipid peak, which itself is most probably outer-volume lipid artifact. Only 128 scans were acquired. NAA, Cr, Cho and mI are all present, but at below normal concentrations. Nevertheless, this provides evidence of survival of some neuronal and axonal tissue in white matter.

H. Other Comments

NAA is absent, or at least below the detectable concentration in grey matter, however, there is a very small signal at 2.0 ppm. A normal NAA peak consists of ~95% NAA, but also has some underlying signal, including some small resonance of glutamate and glutamine.

MRS is indicated *early* in CMV and hypoxic encephalopathy. When performed later, as in this case, the etiology of the severely abnormal MRS can rarely be precisely established. However, MRS is a very valuable planning tool providing more accurate outcome prediction than MRI.

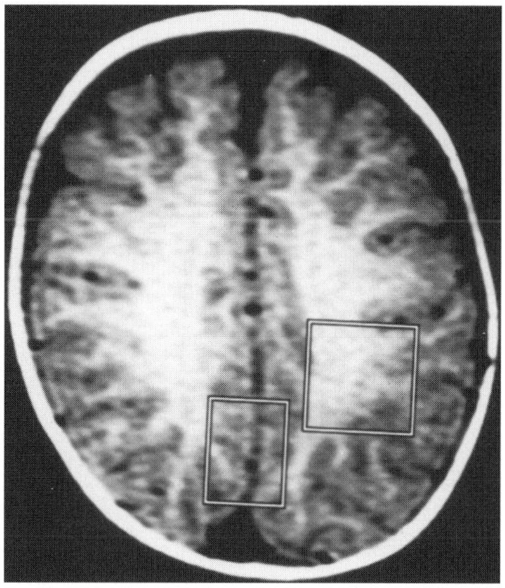

Fig. 5.203 The 2 locations used for MRS are shown on this MRI.

5.9.3 VIRAL ENCEPHALITIS: REYES SYNDROME

A. Purpose of MRS

Investigation of comatose child suspected of having Reyes syndrome (diagnosis and prognosis).

B. Summary of Patient's History

Male, age 9 months, was admitted to the hospital suffering from acute onset of coma, and hyperammonemia (NH_3 = 248 μmoles/l; normal <100 μmoles/l) following exposure to chicken pox. Aspirin, a known contributor to Reyes syndrome, was detected in the blood as well.

C. MRS Method, Volume of Interest, and Comments about Examination

STEAM (TE = 30 ms, TM = 13.7 ms, TR = 1500 ms) was performed in HMRI's parietal white matter and occipital grey matter locations, as shown in Fig. 5.203. The MRS was performed on the second day post admittance to the hospital; the patient was on ventilator support.

D. White Matter Location (Location 1)

1. Peak Identification and Quality Assurance: Reading the Spectrum from Right to Left

The technical quality of the spectrum is acceptable, showing huge abnormalities (Fig. 5.204).

Reading from right to left, there is a lactate doublet at 1.33 ppm. Continuing from right to left, the following metabolite peaks are identified: NAA; β,γ-Glx (increased); second NAA; Cr; Cho; mI (decreased); α-Glx (increased).

2. Interpretation and Analysis

Reading from right to left, lactate is detected, and the metabolite ratios are: NAA/Cr = 1.40 (normal for age); Cho/Cr = 0.99 (normal for age); and mI/Cr = 0.76 (normal for age); [Cr] is 0.77 (decreased). The CSF fraction in the VOI is 5%.

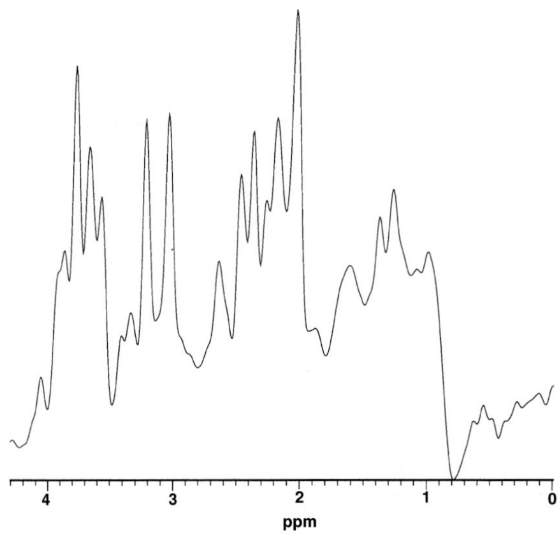

Fig. 5.204 Severely abnormal MRS from the white matter, Location 1. The main feature is increased Glx, which can be proven truly to be glutamine, when compared to a spectrum of a glutamine solution. The metabolite ratios NAA/Cr, Cho/Cr and mI/Cr all appear normal for age, but one should be aware of the huge underlying Glx signals, which contributes to the NAA and mI signals. Therefore the true mI peak is a smaller peak found "riding" on top of the "shoulder" of the α-Glx. Likewise, a "shoulder" of β,γ-Glx is found under the NAA peak.

The major feature of the spectrum is the vastly elevated Glx, which was assigned to glutamine by difference spectroscopy (subtraction of a spectrum of a test sample containing glutamine from the brain spectrum). The glutamine concentration is probably 20 mM or more (normal = 4–6 mM).

E. Grey Matter Location (Location 2)

1. Peak Identification and Quality Assurance: Reading the Spectrum from Right to Left

This very abnormal spectrum, Fig. 5.205, is comparable to that obtained from white matter. There is elevation of the baseline which may exaggerate the lactate region. This is in part attributable to oversuppression of water in the preparation for ^{1}H MRS. Nevertheless, the spectrum has exceptional resolution and is easily interpreted.

Reading from right to left: Lactate, NAA major excesses of all components of β,γ-Glx, the second NAA peak (at 2.6 ppm), a small unassigned peak at 2.9 ppm, Cr, Cho, sI, glucose, mI and two peaks of greatly exaggerated α-Glx are all recognized. The second Cr is present but the second mI at 4.06 ppm cannot be defined with certainty.

2. Interpretation and Analysis

There is a lactate doublet at 1.33 ppm, (confused by baseline artifact) and adjacent peaks at 1.6–1.8 ppm which may be lipid but cannot be so assigned. Peak ratios are NAA/Cr = 1.03 (low); Cho/Cr = 0.64 (normal for location); and mI/Cr = 0.67 (normal). β,γ-Glx/Cr = 0.70 (dramatically elevated) and α-Glx/Cr is also apparently elevated, but not accurately quantified.

It is important to "correct" NAA/Cr for the underlying excess of Glx; a true estimate may be NAA/Cr ~ 0.6–0.7, a very significant reduction.

Absolute [Cr] = 0.92, not abnormal. Brain water 80.9% is probably appropriate for age. The spectrum is a striking confirmation of the widespread metabolite changes seen in parietal cortex.

F. Conclusion

Glutamine is extremely elevated suggesting that the severe encephalopathy is induced by hyperammonemia. The mI and NAA appear normal. Absolute [Cr] is decreased. This abnormal MRS is consistent with acute Reyes syndrome. The elevated glutamine, the presence of lactate, and the low Cr leads to a guarded prognosis.

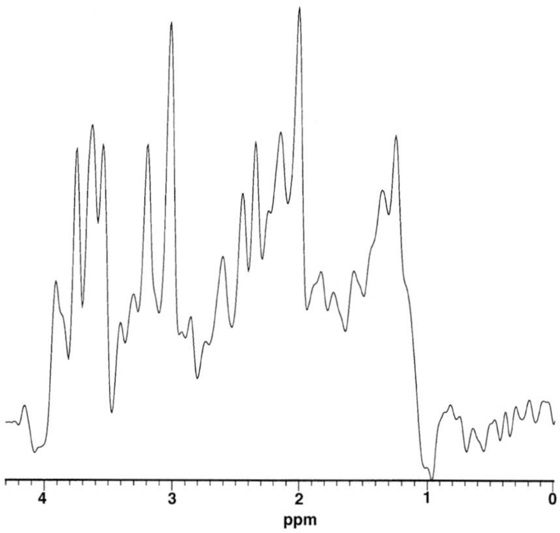

Fig. 5.205 Location 2, grey matter. Similar features to those described above for white matter. There is a close similarity in the major features of the spectra from two different areas of the brain in this, and most other global encephalopathies.

G. Summary of Patient's History Post-MRS

The diagnosis of Reyes syndrome was confirmed. The patient had received aspirin through a "cold-drug" administered without the advice from a physician. The child was discharged from the hospital with an optimistic prognosis, but has since been lost to follow-up. In 50% of Reyes syndrome cases, patients die, while the remaining 50% show neurological sequelae.

MRS was also performed on day 4 and 8 post admittance. Lactate persisted. Glutamine decreased. NAA decreased to 60% of normal, and mI fell to less than half of normal, and [Cr] remained at only 60% of normal.

H. Other Comments

Common pitfalls related to Glx can be illustrated by this case: The complex Glx signal contributes a minor signal underlying NAA, usually not enough to be of concern, thus NAA is not corrected for this underlying Glx. But when Glx is elevated, NAA/Cr is misleading. A correction based upon β,γ-Glx is proposed whereby ($NAA/Cr_{observed} - \beta,\gamma\text{-}Glx/Cr_{excess}$) = "corrected" NAA/Cr. The mI/Cr is also strongly affected by alteration in α-Glx. The same approximate correction (viz. true mI/Cr = $mI/Cr_{obs} - \alpha\text{-}Glx/Cr_{excess}$) would have identified early and severe mI depletion in this patient.

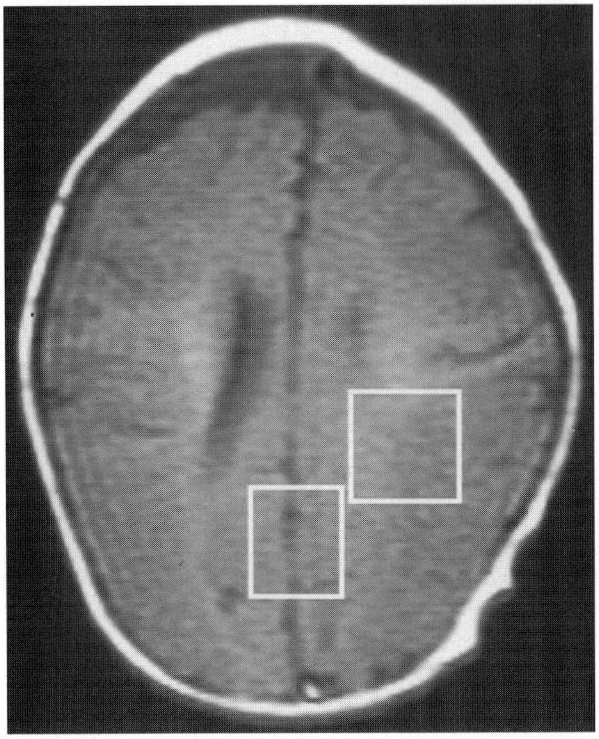

Fig. 5.206 The T_1-weighted MRI showing the two MRS locations.

5.9.4 MYOPATHY: RULE OUT MITOCHONDRIAL ENCEPHALOPATHY

A. Purpose of MRS

Floppy baby (myopathy), brain MRS to rule out mitochondrial encephalo-myopathy and Leigh's syndrome.

B. Summary of Patient's History

Male, age ~4 months (term; GA 57 weeks), with myopathy since birth, and a neurological diagnosis of "moderate encephalopathy" consistent with white matter disease. However, previous MRI was normal.

C. MRS Method, Volume of Interest, and Comments about Examination

STEAM (TE = 30 ms, TM = 13.7 ms, TR = 1500 ms) was performed in HMRI's parietal white (Location 1) and occipital grey (Location 2) matter, Fig. 5.206.

D. Location 1 (White Matter)

1. Peak Identification and Quality Assurance: Reading the Spectrum from Right to Left

The technical quality of the spectrum Fig. 5.207 is excellent, and to the trained eye, appropriate for an infant of this gestational age.

Reading the spectrum from right to left, possible excess lipid at 0.9 and 1.3 ppm, but no lactate identified, by doublet at 1.33 ppm. Continuing reading, the following resonances are identified: NAA, β,γ-Glx, second NAA, Cr, Cho, glucose and/or sI, mI (at 3.56 ppm), α-Glx, and second Cr. NAA is *not* decreased, and Cho is *not* increased when appropriate age correction is applied.

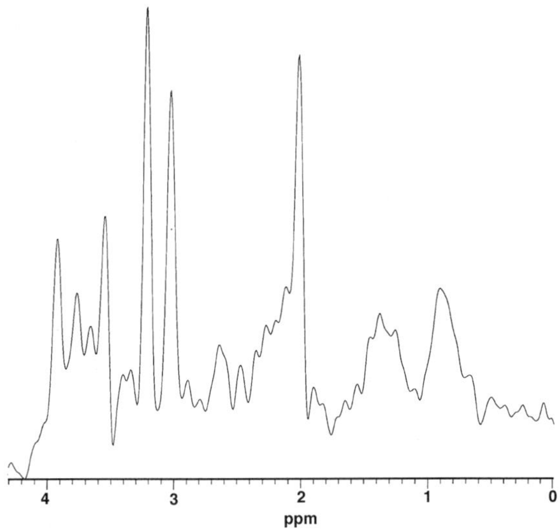

Fig. 5.207 A near normal MRS from parietal white matter in a floppy baby. There is no evidence of mitochondrial encephalopathy.

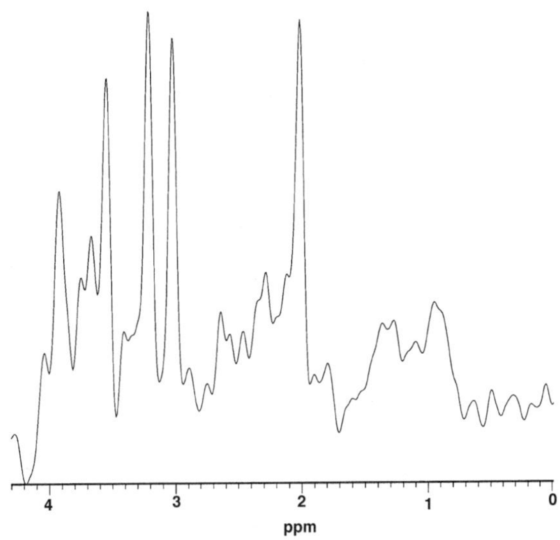

Fig. 5.208 A near normal MRS from occipital grey matter. Again, as in white matter, there is no evidence of mitochondrial encephalopathy, or other metabolic disease.

2. Interpretation and Analysis

There is no lactate detected, and the metabolite ratios are: NAA/Cr = 1.1 (appropriate); Cho/Cr = 1.26 (at 57 weeks GA normal Cho/Cr = 1.05); mI/Cr = 0.65 (appropriate); [Cr] is 100%; β,γ-Glx/Cr = 0.27, and α-Glx/Cr = 0.44 (normal). Brain water is 85.4%, which is appropriate for age.

E. Location 2 (Grey Matter)

1. Peak Identification and Quality Assurance: Reading the Spectrum from Right to Left

The grey matter spectrum (Fig. 5.208) is also of excellent quality, and similar in all details to that from grey matter.

Reading the spectrum systematically from right to left, a similar lipid region at 0.9 and 1.3 ppm, and no lactate doublet identified at 1.33 ppm. Continuing reading, the following resonances are identified: NAA, β,γ-Glx (slightly unusual pattern), second NAA (not clearly resolved), Cr, Cho, sI and glucose, mI (at 3.56 ppm), α-Glx, and second peaks of Cr and mI.

2. Interpretation and Analysis

Reading systematically from right to left, there is no excess lactate detected (as would be expected if Leigh's syndrome, or mitochondrial defect were present in this infant). The metabolite ratios are: NAA/Cr = 1.05 (appropriate); Cho/Cr = 1.04 (age appropriate); mI/Cr = 0.89 (appropriate). [Cr] is 87%, which is possibly normal for age; β,γ-Glx/Cr = 0.38 and α-Glx/Cr = 0.44 (both normal).

F. Conclusion

Neither spectrum shows any notable abnormality, although white matter in particular may be one month behind its correct MRS milestones. There is no evidence of Leigh's disease or encephalopathy. The examination is therefore non-contributory.

G. Summary of the Patient's History Post-MRS

The child died two months after this MRS examination. Despite a full autopsy, no definite diagnosis was achieved.

H. Other Comments

MRS is a useful screening examination in difficult neonatal encephalopathies with a very high success rate. Not so in this case.

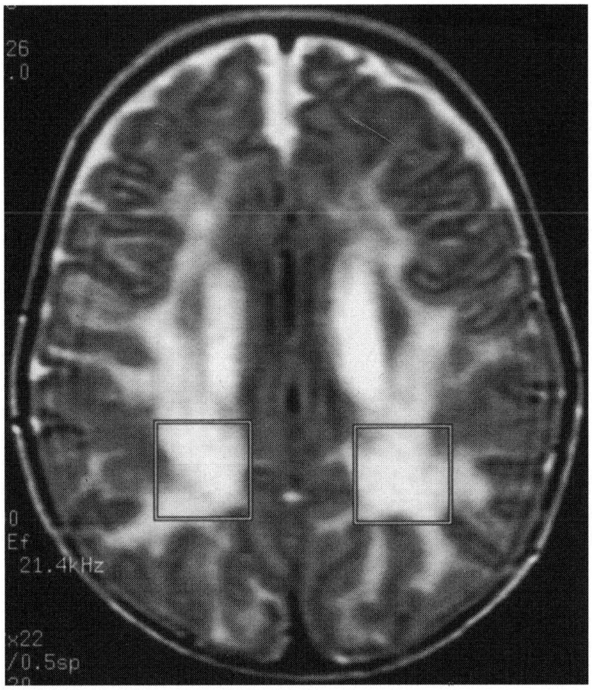

Fig. 5.209 ^{1}H MRS of white matter was acquired from white matter as indicated on the MRI in a 6 month old infant with Canavan's disease.

5.9.5 WHITE MATTER DISEASE? RULE OUT CANAVAN'S DISEASE

A. Purpose of MRS

The primary purpose was to rule out Canavan's disease in an infant with a large head and, who is developmentally delayed. If the patient did not have Canavan's disease, then the secondary purpose was to investigate other possible causes such as one of the white matter diseases seen in infants. A possible differential diagnosis was Alexander's disease. MRS findings in Canavan's disease are abnormal and very specific (increased NAA), in Alexander's disease, and in many other infantile white matter diseases and inborn errors, the MRS findings are also abnormal, but not as specific for each disease (see chapter 6). Therefore MRS is the perfect modality to rule out Canavan's disease.

The purpose of the second and third examination was to follow up on the results of the first examination, to monitor possible progression, to confirm the unusual findings of severely increased brain water, and to monitor if medical treatments had any positive effect.

B. Summary of Patient's History

Female 6 months old (first examination 6 months, second examination 9½ months, and third 13 months old), born term with no complications. Now presenting with developmental delay. Urine NAA excretion was measured, a finding which is considered diagnostic of NAA-deacylase deficiency (i.e. Canavan's disease).

C. MRS Method, Volume of Interest, and Comments about Examination

Examination 1: STEAM (TE = 30 ms, TM = 13.7 ms, TR = 1500 ms, 256 averages; in one measurement TE was 270 ms to compare to earlier published Canavan's cases, and to evaluate possible metabolite T_2 effects; in another measurement TR was

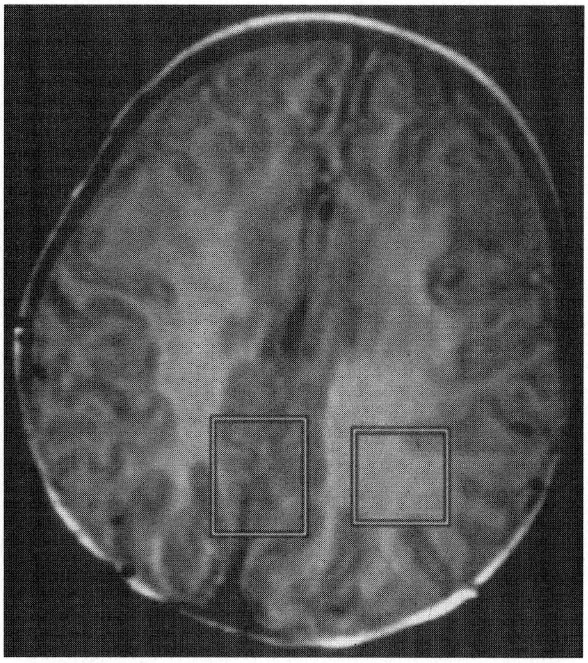

Fig. 5.210 MRI showing the grey and white matter locations measured at 9½ months.

5000 ms with 100 averages to evaluate possible metabolite T_1 effects) was performed in HMRI's parietal white matter location, both right and left hand sides, Fig. 5.209.

Examination 2: STEAM (TE = 30 ms, TM = 13.7 ms, TR = 1500 ms) in the occipital grey matter and parietal white matter locations, Fig. 5.210.

Examination 3: STEAM (TE = 30 ms, TM = 13.7 ms, TR = 1500 ms) in the parietal white matter location, Fig. 5.211.

The infant was sedated with chloral hydrate (50 mg/kg body weight) and patient movement was avoided in all three examinations.

D. Examination 1 (Aged 6 Months): Location 1 (Left Hand Side White Matter, TE = 30 ms, TR = 1500 ms, 256 Averages)

1. Peak Identification and Quality Assurance: Reading the Spectrum from Right to Left

The resolution, the baseline and the S/N are all perfect and there are no artifacts, (Fig. 5.212). There is some minor signal of lipid and/or macro molecules at 0.8–0.9 ppm and 1.3 ppm, this is probably not significant as similar amounts are sometimes seen in normals. There is *no* lactate. The following peaks are identified: NAA, β,γ-Glx, second NAA, Cr, Cho, overlapping sI and glucose at 3.36 and 3.43 ppm, mI, α-Glx, and the second Cr and mI, both in correct proportions compared to their corresponding first peaks.

The relative heights of the peaks are clearly abnormal, and even more abnormal when one is reminded of the appearance of the normal infant MRS, (see cases 5.1.1 and 5.1.2): NAA/Cr is about double the normal. Already at this point, based on the NAA/Cr ratio alone, it can be concluded that the MRS is consistent with Canavan's disease.

2. Interpretation and Analysis

All metabolite ratios, absolute metabolite concentrations, and water concentration will be compared to the corresponding normal values for the appropriate age, 66 weeks gestational age, see Chapter 2. The metabolite ratios are NAA/Cr = 2.20, (1.20 is normal for age); Cho/Cr = 0.79, (1.05 is normal for age); and mI/Cr = 1.28, (0.67 is normal for age). [Cr] = 80% of the normal

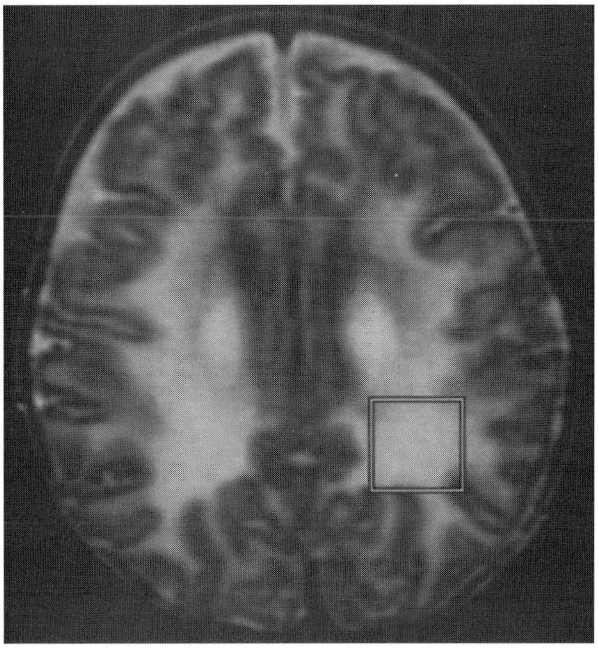

Fig. 5.211 The white matter location measured at 13 months.

adult [Cr], (normal [Cr] at this age is about 90% of the normal adult value). [NAA] is 124% of the normal adult value*, and normal [NAA] at this age is about 85% of the normal adult value, so the patient's [NAA] is about 45% increased for age (1.24/0.85 = 146%).

The water content is very high: 97% (of "the concentration of pure water"), but inspection of all technical aspects, fit of the seven water measurements, calibration, absence of patient movement etc., lead to the conclusion that this value is real. For comparison, the highest normal water content, 90%, is found in neonates, and at the age 6 months normal water content is 85–90%. Normal adult water content in this VOI is 69%.

E. Examination 1: Location 1 (Left Hand Side White Matter, TE = 270 ms, TR = 3000 ms, 256 Averages)

1. Peak Identification and Quality Assurance: Reading the Spectrum from Right to Left

Same location as above, Fig. 5.213, measured with a long TE of 270 ms reveals another high quality spectrum similar to the previous except that the metabolites with short apparent T_2 are absent or very much reduced. Metabolites with short *apparent* T_2 includes both the metabolites with true short T_2 and those influenced by strong coupling. Therefore the following metabolites are not seen: lipids, Glx, the second NAA, and glucose at 3.43 ppm. Both of the mI peaks are very much reduced, and only visible because they were originally very dominating.

*The critical reader, might want to recalculate the [NAA] based on the NAA/Cr ratio and the absolute Cr. But, note that the metabolite ratios are calculated based on fitted peak heights, (peak adjusted to have the same width), whereas the [Cr] and [NAA] are calculated based on a number of peak area fitting routines (see Kreis et al., Magnetic Resonance in Medicine 30:424–437 (1993)). The two methods of processing will of course not give exact "matching" results, since relaxation time corrections and corrections for underlying broad resonances are included in the second processing scheme. An attempt to recalculate the patients [NAA] from NAA/Cr = 2.20 and [Cr] = 80% of adult value, will not match exactly the [NAA] = 124% of adult value found by the peak area fitting routines including relaxation corrections etc, but it would be 2.20*80% = 176%. For this patient we use [NAA] as obtained by the peak area fitting routines, and just apply the correction for age using values from the plot Fig. 2.12, Chapter 2. In fact, the values [NAA] = 124% and "recalculated [NAA]" = NAA/Cr * [Cr] = 176% match each other quite well when corrections for underlying broad resonances are considered.

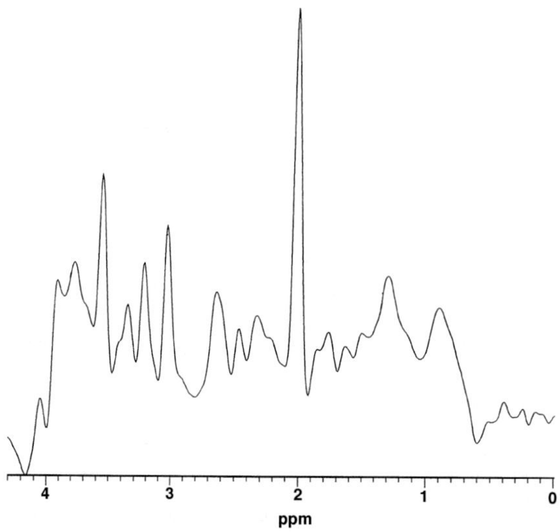

Fig. 5.212 ^{1}H MRS of white matter (Examination 1, Location 1, TE = 30 ms) in a child aged 6 months with Canavan's disease. There is a striking increase in NAA resonances at 2.02 and 2.6 ppm which defines this disease. Scyllo-inositol and mI are also prominent relative to Cr.

Reading from right to left: A trace of lactate is seen followed by NAA, Cr, Cho, sI, and mI. Some remaining α-Glx, second Cr and second mI is also seen. Notice how sI is a well defined peak at 3.36 ppm, because the overlapping glucose at 3.43 ppm has vanished at long TE.

2. Interpretation and Analysis

The ratios are NAA/Cr = 3.41, Cho/Cr = 1.05, and mI/Cr = 0.75. The differences from the TE = 30 ms ratios reflect the fact that the T_2 of the metabolites differ from each other. This is also seen in normal brain, where Cho/Cr is reversed (>1) at TE = 270 ms, even in adults. In this patient NAA is of special interest: The ratio of the NAA peak sizes at 30 and 270 ms is approximately 1.5, which is in good agreement with T_2 of NAA being 524 ± 84 ms, which is the normal for babies however aged 43 ± 3 weeks (see Kreis et al., Magnetic Resonance in Medicine 30:424–437 (1993)). T_2 of NAA is longer in babies than in adults. Similarly, comparing peak sizes at 30 and 270 ms for Cr and Cho, shows that the T_2 values are within normal range for age.

Note that this evaluation of metabolite T_2 values is performed, only to make sure that T_2 effects do not alter the calculation of metabolite concentrations in Canavan's disease, a disease in which this had not previously been proven. At TE = 30 ms, the spectrum is not very much affected by T_2 decay, so T_2 must vary by several 100 ms to have any significant effect on the calculated concentrations. Therefore it is concluded that, based on this one patient, altered T_2 values are not a concern in Canavan's disease. (The T_1 effect of the longer TR = 3000 ms is insignificant as will be shown below). The appearance of the spectrum correspond well with earlier published cases of Canavan's disease measure with long TE.

F. Examination 1: Location 2 (Right Hand Side White Matter, TE = 30 ms, TR = 1500 ms, 256 Averages)

1. Peak Identification and Quality Assurance: Reading the Spectrum from Right to Left

Very similar to the left hand white matter spectrum measured with the same measurement parameters: High quality spectrum, with no artifacts, perhaps with the exception of the baseline sloping slightly down to the left, Fig. 5.214. Lipids are probably not significant, there is no lactate, NAA, β,γ-Glx, second NAA, Cr, Cho, overlapping sI, and glucose at 3.36 and 3.43 ppm, mI, α-Glx, and the second Cr and mI are identified.

2. Interpretation and Analysis

The metabolite ratios are NAA/Cr = 2.37, (1.20 is normal for age); Cho/Cr = 0.73, (1.05 is normal for age); and mI/Cr = 1.35, (0.67 is normal for age). [Cr] is 95% of adult normal or about 100% of the normal age related [Cr]. [NAA] is 152% of the

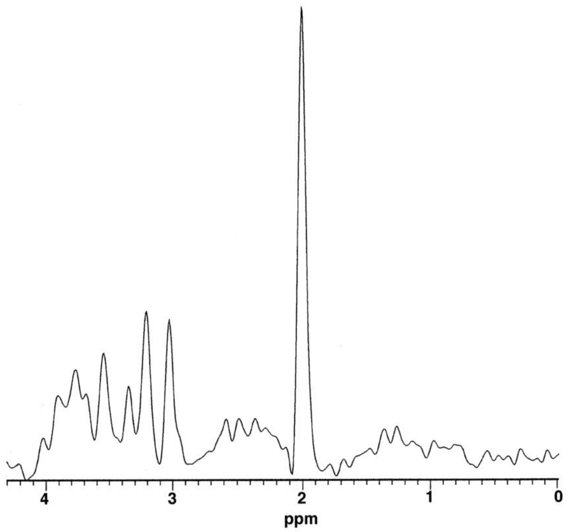

Fig. 5.213 Long-echo time ^{1}H MRS in Canavan's disease, (Examination 1, Location 1, TE = 270 ms). This spectrum is directly comparable to published spectra in the extreme elevation of NAA/Cr. Note that the second NAA peak at 2.6 ppm is much attenuated at long TE. Lactate (probably derived from CSF) is more clearly identified because overlying lipid signals are also lost at TE = 270 ms.

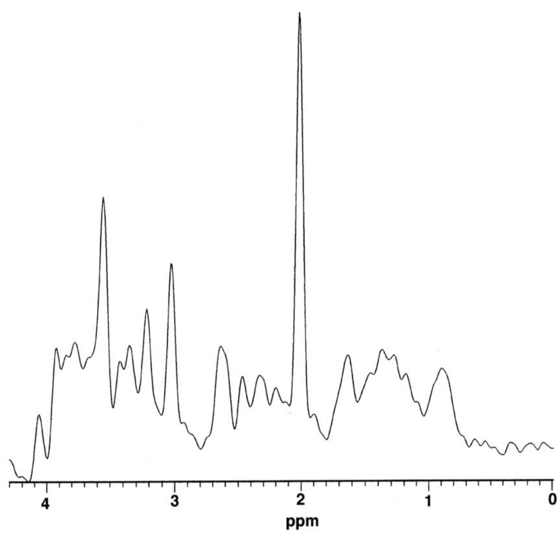

Fig. 5.214 ^{1}H MRS of white matter of contralateral hemisphere (Examination 1, Location 2, TE = 30 ms). The abnormalities are identical to those described above for Location 1.

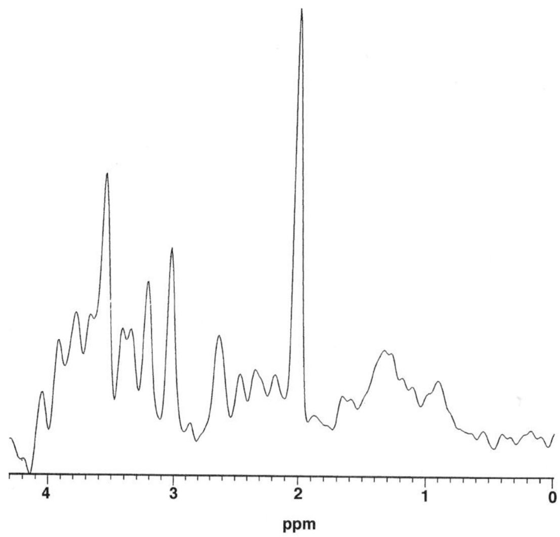

Fig. 5.215 Effect of TR on spectral appearance in Canavan's disease, (Examination 1, Location 2, TR = 5000). Although signal to noise is superior (compare 100 scans in present MRS with 256 in pervious figure), there is no significant difference in spectral appearance; this indicates metabolite T_1's to be similar in Canavan's and normal brain.

normal adult, and normal [NAA] at this age is about 85% of the normal adult value. Thus, the patient's [NAA] is about 80% increased for age (152%/0.85 = 179%).

The water content is again very high: 97% (of "the concentration of pure water"), confirming the result from the left cerebral hemisphere.

G. Examination 1: Location 2 (Right Hand Side White Matter, TE = 30 ms, TR = 5000 ms, 100 Averages)

1. Peak Identification and Quality Assurance: Reading the Spectrum from Right to Left

Notice that with only 100 averages the S/N is comparable to what was obtained with 256 averages in the previous examination. But with this exception the spectrum, Fig. 5.215 is very similar to the previous, lipids are not significant, there is no lactate, NAA, β,γ-Glx, second NAA, Cr, Cho, overlapping sI and Glucose at 3.36 and 3.43 ppm, mI, α-Glx, and the second Cr and mI are identified.

2. Interpretation and Analysis

The ratios are NAA/Cr = 2.30, Cho/Cr = 0.82, and mI/Cr = 1.41. These values and the appearance of the spectrum do not differ very much from the TR = 1500 ms ratios, therefore it is concluded that also for Canavan's disease T_1 relaxation does not influence metabolite ratios significantly, even for TR as short as 1500 ms. Similarly it can be shown that individual metabolite T_1 values for NAA, Cr, Cho and mI are within the normal range for age.

A general observation is that TR = 1500 and the close to fully relaxed TR = 5000 ms give similar spectra. The S/N per single scan is better with 5000 ms, but timewise it does not pay, as the total scan time to get the same S/N is longer with TR = 5000 ms. It can be shown that, given the typical metabolite T_1 values, an optimum TR to get the best S/N per spent time unit, is TR = 1500 ms.

H. Conclusion (Examination 1)

The excess of NAA in both hemispheres is consistent with diagnosis of Canavan's disease. The excess brain water is also in agreement with a diagnosis of Canavan's disease. The excess of mI and possibly lipid/macro molecules may be related to demyelination or delayed myelination. T_1 and T_2 for the metabolites do not differ from normal, and are therefore not a matter of further concern in Canavan's disease.

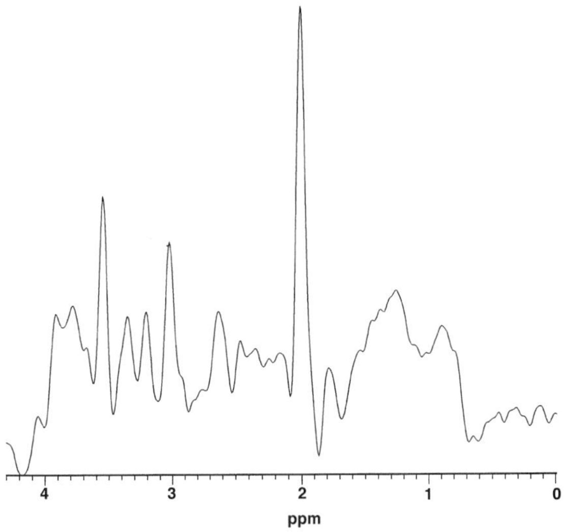

Fig. 5.216 ^{1}H MRS of white matter at 9½ months in Canavan's disease. NAA/Cr is still markedly elevated and other features are unchanged since Examination 1, at 6 months age. Note: An artifact in the lipid region contributes to some apparent reduction in NAA peak intensity.

I. Examination 2 (Aged 9½ Months): Location 1 (Left Hand Side White Matter, TE = 30 ms, TR = 1500 ms, 256 Averages)

1. Peak Identification and Quality Assurance: Reading the Spectrum from Right to Left

Resolution, baseline and S/N are all perfect, Fig. 5.216. A minor artifact in the lipid region near NAA, may contribute to an apparent reduction in the NAA peak intensity. Reading from right to left: Probably insignificant lipids, no lactate, dominant NAA, β,γ-Glx, second NAA, Cr, Cho, significant sI with some overlapping Glucose at 3.36 and 3.43 ppm, mI, α-Glx, and the second Cr and mI are identified.

2. Interpretation and Analysis

All metabolite ratios, absolute metabolite concentrations, and water concentration will be compared to the corresponding normal values appropriate for 81 weeks gestational age. The ratios are NAA/Cr = 2.37, (1.27 is normal for age); Cho/Cr = 0.60, (1.03 is normal for age); and mI/Cr = 1.27 (0.62 is normal for age). [Cr] = 80% of the normal adult [Cr], or 85% of the age related [Cr], as normal [Cr] at this age is about 95% of the adult value. [NAA] = 151% of normal adult corresponding to about 176% of the [NAA] normal for age.

The water content is still very high at 96% of "the concentration of pure water".

J. Examination 2: Location 2 (Grey Matter, TE = 30 ms, TR = 1500 ms, 256 Averages)

1. Peak Identification and Quality Assurance: Reading the Spectrum from Right to Left

Resolution, baseline and S/N are all perfect, and there are no artifacts, Fig. 5.217. Reading from right to left: Insignificant lipids, no lactate, dominant NAA, β,γ-Glx, second NAA, Cr, Cho, significant sI with a shoulder of Glucose, mI, α-Glx, and the second Cr and mI are identified.

2. Interpretation and Analysis

All metabolite ratios, absolute metabolite concentrations, and water concentration will be compared to the corresponding normal values for the age 81 week gestational age in the occipital location. The ratios are NAA/Cr = 1.93, (1.23 is normal for age); Cho/Cr = 0.55, (0.72 is normal for age); and mI/Cr = 0.95, (0.60 is normal for age). [Cr] = 98% of the normal adult [Cr], which is basically normal for age. [NAA] = 126% of normal adult corresponding to about 147% of the [NAA] normal for age.

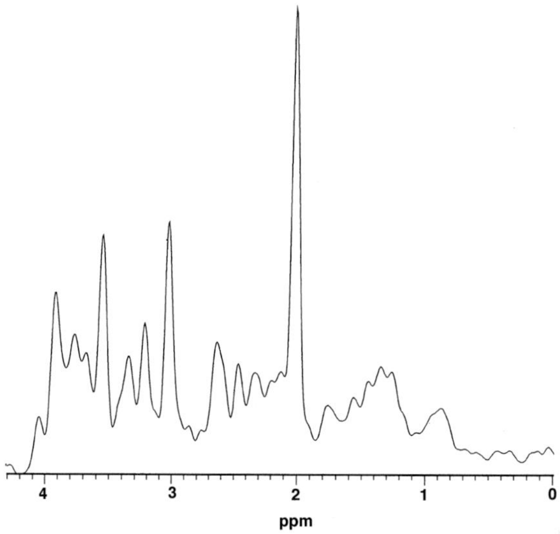

Fig. 5.217 ^{1}H MRS of grey matter at 9½ months in Canavan's disease. NAA/Cr is also increased in grey matter.

The water content in grey matter is 86% of "the concentration of pure water", which is approximately normal for age. The CSF fraction (atrophy index) in this mid occipital grey matter VOI is 14% (increased).

K. Conclusion (Examination 2)

In white matter there are no obvious changes since the first examination, both NAA, mI and water are still increased, and Cho is still decreased. The sI is significant, probably increased in parallel with mI. Grey matter reflects NAA accumulation but not the increase in brain water noted in white matter.

L. Examination 3 (Aged 13 Months): (White Matter, TE = 30 ms, TR = 1500 ms, 256 Averages)

1. Peak Identification and Quality Assurance: Reading the Spectrum from Right to Left

Resolution, baseline and S/N are all perfect. Reading from right to left, Fig. 5.218. Probably insignificant lipids, no lactate, a dip before NAA from out of phase lipid signal, dominant NAA, β,γ-Glx, second NAA, Cr, Cho, significant sI with a shoulder of Glucose, mI, α-Glx, and the second Cr and mI are identified.

2. Interpretation and Analysis

All metabolite ratios, absolute metabolite concentrations, and water concentration will be compared to the corresponding normal values for the age 96 weeks gestational age in the parietal location. The ratios are NAA/Cr = 2.74, (1.31 is normal for age); Cho/Cr = 0.67, (1.01 is normal for age); and mI/Cr = 1.67, (0.62 is normal for age). [Cr] is normal. [NAA] is 187% of adult normal, or about 210% of age corrected normal [NAA], (normal [NAA] at 96 weeks gestational age is about 90% of the adult normal [NAA]).

The water content is persistently elevated at 98% of the concentration of pure water.

M. Conclusion (Examination 3)

The increased water content persists and excesses of NAA, mI, and sI has become more marked. The MRS shows further deterioration in this patient with Canavan's disease.

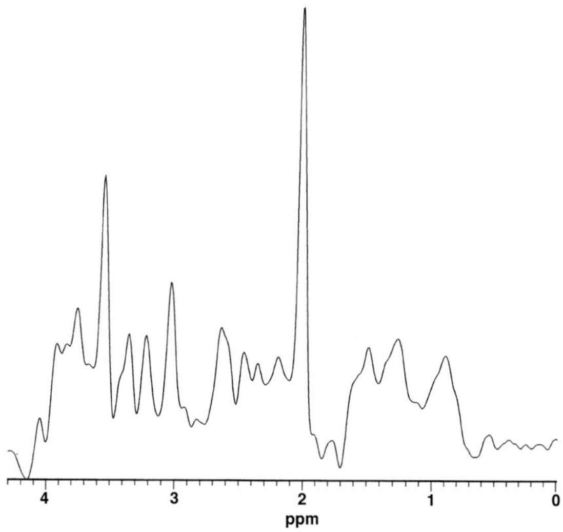

Fig. 5.218 ^{1}H MRS of white matter at 13 months in Canavan's disease. There has been a further increase in NAA/Cr and in [NAA]. The mI and sI are also more prominent and Cho/Cr has fallen.

N. Summary of Patient's History Post-MRS

Between examinations 1 and 3 conservative medical treatment was attempted which obviously did not improve the patient's biochemistry. There has been no clinical deterioration over the 6 month interval.

O. Other Comments

These thorough examinations demonstrate how quantitation needs to be performed carefully, in particular in babies and in patients with diseases not yet studied by MRS. The possibility of abnormal relaxation effects was excluded. Most importantly the water compartments were measured rigorously, showing that brain water was unusually high. If the water had not been quantified, and the water had been assumed normal (normal for age or perhaps even using the normal adult value), and had been used as an internal concentration standard, the metabolite concentrations would have been 40% underestimated.

NOTE IN PROOF

Gene therapy, now being undertaken for treatment of Canavan's disease, is effectively monitored with MRS (unpublished studies: Seymour and Blüml).

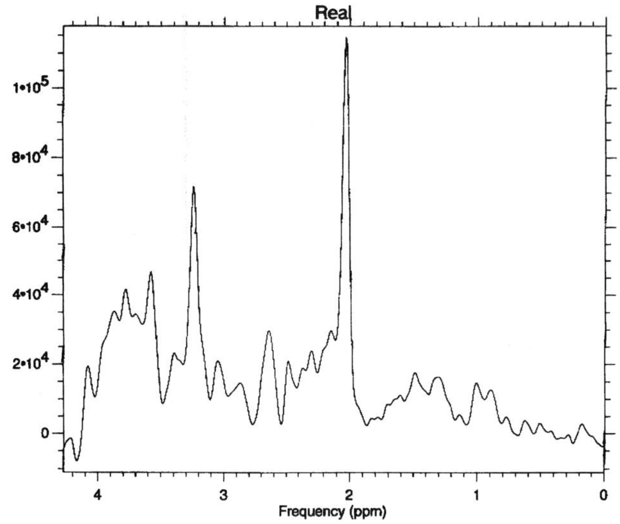

Fig. 5.219 The white matter spectrum in a patient with psychomotor development delay, mild muscle hypotonia and inability to speak. The main finding is that Cr is virtually absent. (Courtesy Dr. Bianchi, IRCCS Stella Maris, Pisa, Italy).

5.9.6 METABOLIC ERROR, ABSENT CREATINE/PHOSPHOCREATINE

A. Purpose of MRS

The purpose of the MRS examination was to (a) Evaluate a child with delayed mental development and muscle weakness: rule out mitochondrial encephalopathy, (b) confirm a previously obtained result, missing Cr, (c) to evaluate the MRS for other abnormalities, and (d) to get a baseline measurement before treating the patient with arginine and/or creatine.

B. Summary of Patient's History

Female toddler (details not available). She is one of two sisters who have psychomotor developmental delay, mild muscle hypotonia and are unable to speak. MR imaging was negative in the previous examination, and is still negative in this examination. MRI and MRS results are similar in the sister.

C. MRS Method, Volume of Interest, and Comments about Examination

STEAM (assumed measurement parameters are: TE = 30 ms, TM = 13.7 ms, TR = 1500 ms) was performed in white matter, inter-hemispheric grey matter, and cerebellum, (images not shown).

D. Location 1 (White Matter)

1. Peak Identification and Quality Assurance: Reading the Spectrum from Right to Left

The resolution, baseline and S/N of the spectrum, Fig. 5.219, is very good and there are no artifacts. Thus, reading this high quality spectrum from right to left; the region from 0 to 2 ppm is unremarkable, (there is no abnormal lipid or lactate), NAA at 2.02 ppm is followed by β,γ-Glx, and the second NAA. Cr is absent (or less than 5% of normal relative to Cho, and mI). Cho, mI, a clear triplet of α-Glx, and the second mI are observed, but notice that the second Cr is also absent.

2. Interpretation and Analysis

This and the following spectra were sampled at 1.5 T with the same STEAM pulse sequence as most other spectra in this handbook, but the post processing routines for this patient were not identical, which may be noted in the line shape of the peaks.

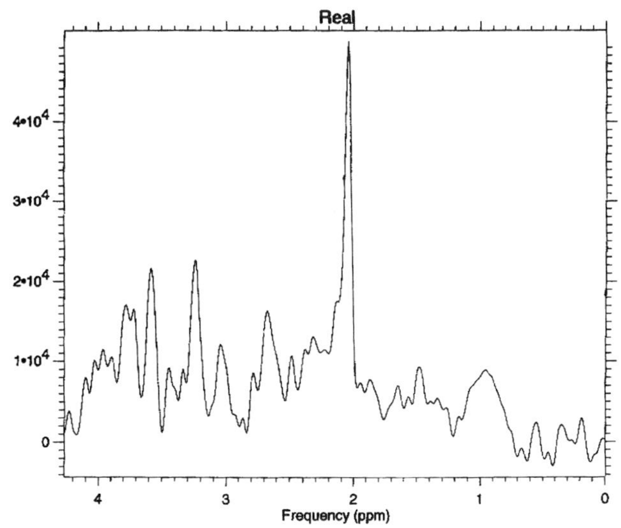

Fig. 5.220 The grey matter spectrum in a patient with psychomotor development delay, mild muscle hypotonia and inability to speak. The spectrum is severely abnormal with a Cr peak that is less than 20% of the normal size. (Courtesy Dr. Bianchi, IRCCS Stella Maris, Pisa, Italy).

Absolute metabolite concentrations are not available, and since Cr is absent, ratios to Cr are meaningless. NAA, Cho and mI cannot be assumed normal. But as a best approximation Cho is assumed normal in the following: The NAA/Cho ratio is estimated to be 1.7, and the normal NAA/Cho ratio is about 1.9, so NAA is reduced about 10%.

Even if we abandon the assumption of normal Cho, we can say that mI, Cho and NAA are within 10–15% of normal, (or that all three are increased or decreased by the same amount, which is unlikely).

E. Location 2 (Grey Matter)

1. Peak Identification and Quality Assurance: Reading the Spectrum from Right to Left

The resolution and baseline are good, but the spectrum, Fig. 5.220, seems a little more noisy than the previous spectrum of white matter. The first significant peak is the NAA peak at 2.02 ppm. After that follows the β,γ-Glx, the second NAA, and a very reduced Cr. Relatively normal Cho and mI are followed by a somewhat unusual α-Glx region, possibly obscured by a mild water suppression artifact. No further peaks are identified.

2. Interpretation and Analysis

Ratios to Cr do not make sense because Cr obviously is reduced. Assuming that Cho is normal, we find that Cr is about 20% of normal, and that NAA is decreased at the most 10%.

F. Location 3 (Cerebellum)

1. Peak Identification and Quality Assurance: Reading the Spectrum from Right to Left

This is a high quality spectrum, Fig. 5.221, with good resolution, S/N, and baseline. A small "dip" at 1.7–1.8 ppm is probably an artifact. Reading from right to left the first significant resonance is NAA at 2.02 ppm, then β,γ-Glx and the second NAA, severely reduced Cr, followed by relatively normal Cho, mI, α-Glx, and the second mI. The second Cr is barely seen at 3.9 ppm.

2. Interpretation and Analysis

Cr is again very much reduced. NAA, Cho, and mI again probably within 10–15% of normal, (note that normal metabolite ratios are different in the cerebellum).

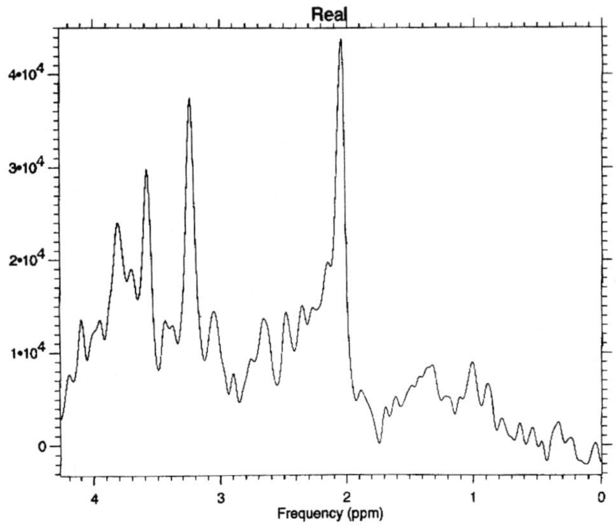

Fig. 5.221 The spectrum from cerebellum in a patient with psychomotor development delay, mild muscle hypotonia and inability to speak. The spectrum is similar to the grey matter spectrum with respect to the severely reduced Cr. (Courtesy Dr. Bianchi, IRCCS Stella Maris, Pisa, Italy).

G. Conclusion

Cr is absent or very much reduced, white matter is affected the most. NAA, Cho and mI are affected at the most 15%. No unusual metabolites are observed.

The virtual absence of total Cr must indicate almost complete absence (or membrane binding) of the high energy metabolite, phosphocreatine. In the absence of lactate, however, we might infer that a second cerebral high energy phosphate, (ATP), is *not* abnormal.

H. Summary of Patient's History Post-MRS

Not available.

I. Other Comments

In a case like this, one should remember that the peaks are results of any metabolites resonating at the given chemical shift, see chapter 4. Thus, also the Cr peak is not a 100% "pure" creatine plus phosphocreatine peak. There is a signal from macro molecules as a very small shoulder on the right hand side underneath the peak, and a coupled resonance of γ-amino-butyric-acid (GABA) also resonates at 3 ppm. These and other resonances are usually small (less* than a few percent) relative to Cr, but when Cr is absent or very small, they need to be considered. The small contributions from "Cr" in this patient may not all be Cr; part of the signal may be from GABA, or other underlying resonances.

A similar case has previously been described by Dr. J. Frahm, Göttingen, Germany: The patient had no Cr in the brain spectrum, the Cr peak was restored after feeding the patient creatine, but the improvement was not permanent. Another difference was that MRI in the patient from Göttingen was abnormal, see chapter 3.

*The background correction factor for Cr, β, was found to be 0.985 by Dr. R. Kreis, see bibliography to chapter 2. This corrects for the residual short T_2 resonances beneath the sharp Cr peak. It is not possible to distinguish between two or more metabolites of *similar* (short) T_2 by this means.

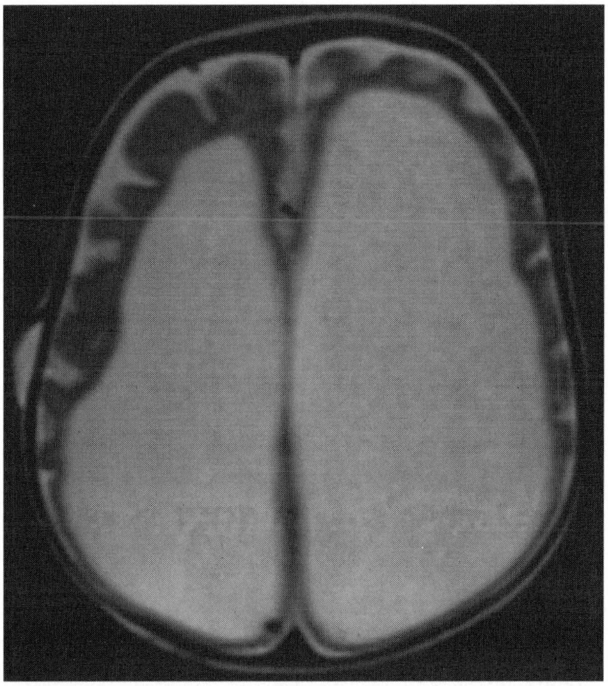

Fig. 5.222 The axial plane where the HMRI standard locations are placed leaves no chance to place a VOI because the cortex is too thin.

5.9.7 HYDROCEPHALUS VERSUS METABOLIC DISEASE

A. Purpose of MRS

To evaluate a patient with severely dilated ventricles before performing shunt surgery to reverse the apparent hydrocephalus. Most metabolic diseases, inborn errors, or other causes of degeneration and atrophy can be ruled out by MRS.

B. Summary of Patient's History

Female, 5 months old at the time of the first MRS examination, who failed to thrive and did not meet developmental milestones. An early CT indicated hydrocephalus, but the head circumference was not enlarged. No acidotic episodes were recorded. A shunt was put in at the age of 4 months and revised two weeks later, as it did not seem to work. During recovery there was acidosis. The need for a repeated shunt surgery or revision was considered at the time of MRS.

C. MRS Method, Volume of Interest, and Comments about Examination

STEAM (TE = 30 ms, TM = 13.7 ms, TR = 1500 ms). The VOI was difficult to place as the cortex was very thin, Fig. 5.222, but a VOI in thallamus and adjacent white matter was chosen as illustrated on Fig. 5.223. The evaluations and calculations were as the best approximation performed using the occipito-parietal white matter as reference. A second MRS was performed in the same location at the age of 1 year and 9 months.

D. Examination 1, Age 5 Months (Thallamus and Adjacent White Matter)

1. Peak Identification and Quality Assurance: Reading the Spectrum from Right to Left

The spectral quality is good, except for a very small dip at 1.1 ppm, which has no impact on the interpretation of the spectrum, Fig. 5.224. Reading from right to left a small and probably insignificant lipid peak is seen at 0.9 ppm. A prominent lactate

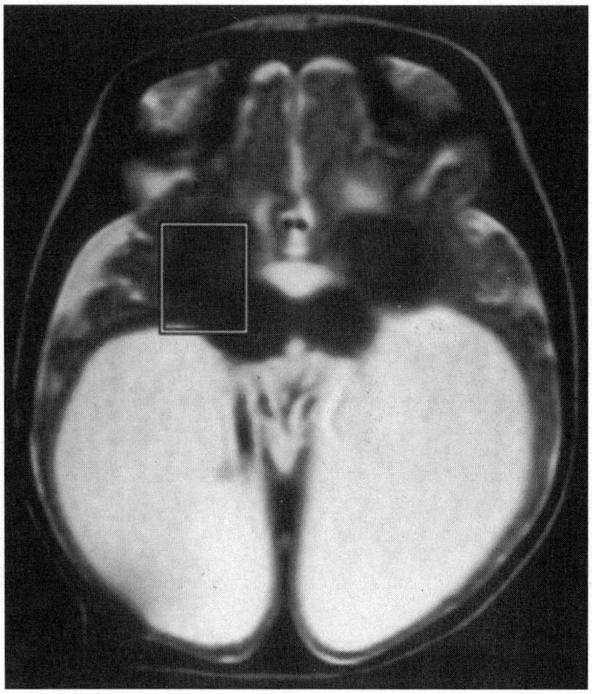

Fig. 5.223 The VOI was placed in thallamus and adjacent white matter as illustrated on this T_2-weighted image.

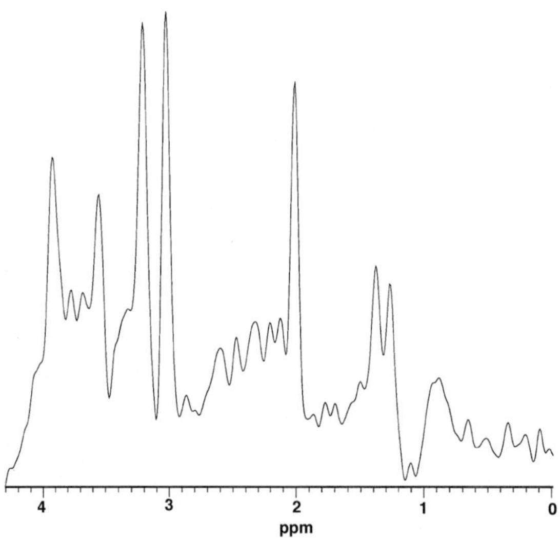

Fig. 5.224 The MR spectrum is from a 5 month old female. Excess lactate and about 30% reduced NAA indicates that shunt surgery will not be successful.

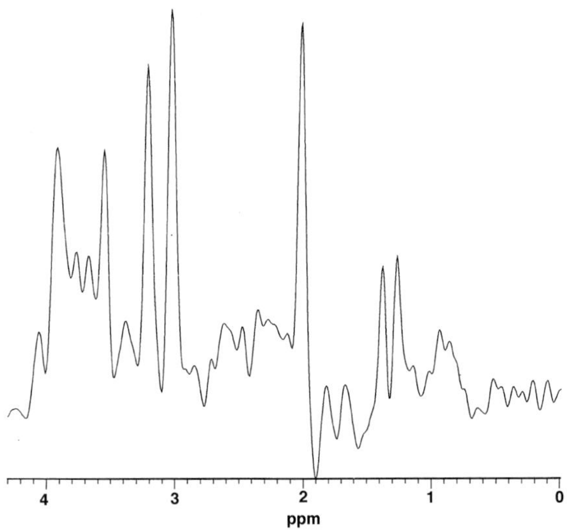

Fig. 5.225 The MR spectrum acquired at age 1 year and 9 months shows that no progression for better or worse has taken place.

doublet is seen at 1.33 ppm followed by NAA. The β,γ-Glx and second NAA is somewhat unusual, but there is not enough evidence to make any special assignments. After Cr follows Cho and mI, there is a shoulder on the left hand side of Cho. α-Glx and the second Cr appear normal. The second mI is not visible.

2. Interpretation and Analysis

The gestational age is 60 weeks. [Cr] which was found to be close to the normal adult value is therefore also close to normal for age, as [Cr] does not vary much with age. The CSF fraction of the VOI was measured to be 16%. There is lactate in CSF, but 16% CSF is far from enough to explain the dominant lactate seen in this spectrum; compare to some of the dementia or liver disease cases, where the CSF fractions are similar due to atrophy, but no lactate is seen. NAA/Cr = 0.84 (normal for age is 1.17), Cho/Cr = 0.95 (normal for age is 1.07), and mI/Cr = 0.59 (normal for age is 0.70). The normal values were all obtained in a different location, occipito-parietal white matter, so the only certain abnormal ratio is NAA/Cr which is about 30% reduced.

E. Conclusion

The MRS with excess lactate and 30% reduced NAA is not normal. It is compatible with a neurological disease of unknown etiology. The enlarged ventricles are most likely caused by neuronal death (or non-development) accompanied by atrophy, and shunt surgery can therefore not be expected to normalize the situation.

F. Examination 2, Age 1 Year and 9 Months (Thallamus and Adjacent White Matter)

1. Peak Identification and Quality Assurance: Reading the Spectrum from Right to Left

The spectrum is again of good quality except some small dips (artefacts) at 1.5–1.9 ppm and possibly also in the β,γ-Glx region, Fig. 5.225. The NAA may be slightly underestimated due to the dip at 1.9 ppm. The spectrum is similar to the previous, showing some lipid at 0.9 ppm, excess lactate, reduced NAA, an unusual β,γ-Glx region, Cr, Cho, possibly sI and glucose, mI, α-Glx, and the second Cr and mI peaks.

2. Interpretation and Analysis

At the gestational age of 131 weeks the metabolite ratios and concentration are very close to the normal adult values, except NAA, which is still increasing. The values from the occipito-parietal white matter will again be used as the best reference available. [Cr] is normal, and the CSF fraction was this time measured to be 19%, no significant change from the first examination. NAA/Cr = 0.96 (still about 30% reduced for age), Cho/Cr = 0.84 (normal), and mI/Cr = 0.64 (normal).

G. Conclusion

No change since the first examination. Excess lactate and although NAA has increased, paralleling the age curve, NAA is still at a level 30% below the normal.

H. Summary of Patient's History Post-MRS

Neurological impairment. Unspecified mitochondrial disease is suspected. Even without a "mitochondrial" diagnosis, treatments aimed at activating electron transport, and removing lactate were attempted.

I. Other Comments

This is a case illustrating how MRS can identify patients who will not benefit from shunt surgery. If the MRS on the other hand is found to be normal then shunt surgery is likely to be successful, as this is consistent with true, or arrested, hydrocephalus.

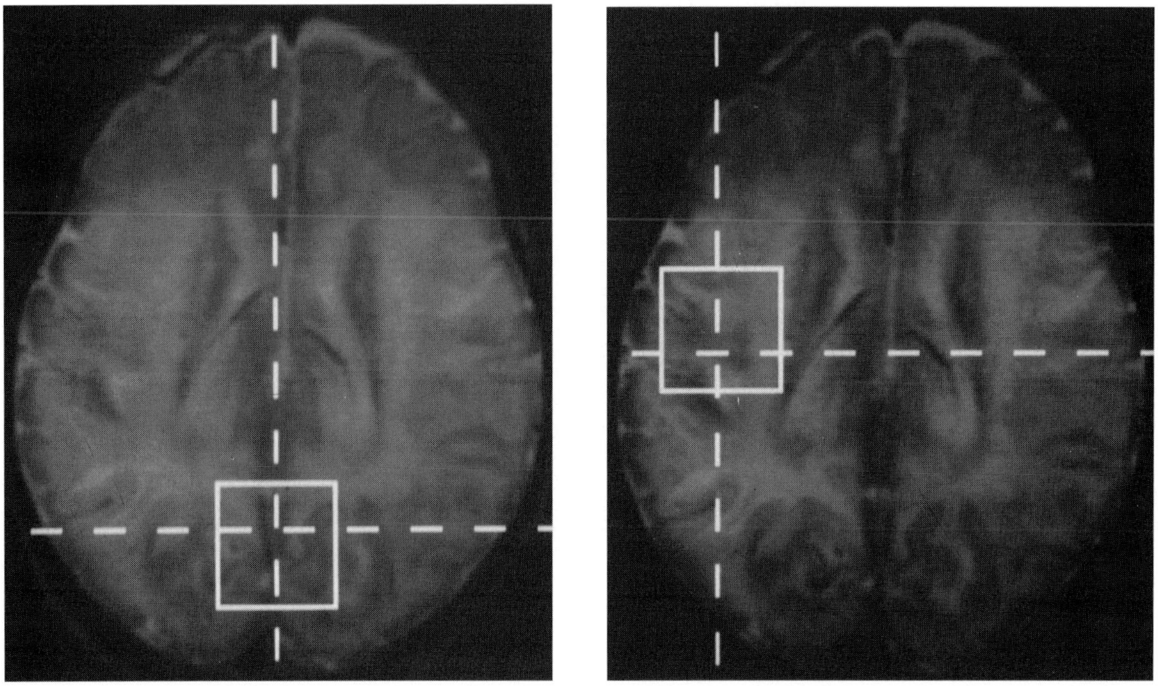

(A) (B)

Fig. 5.226 Axial MR images showing the location of the occipital VOI (A) and the parietal VOI (B). (Courtesy of Drs. Ashwal and Holzhouser, Loma Linda Medical Center, California.)

5.9.8 METABOLIC ACIDOSIS

A. Purpose of MRS

To evaluate a comatose infant for clues of metabolic or other diseases, and to give a prognosis for her future.

B. Summary of Patient's History

Female, one month old at time of MRS, in coma and ventilator dependent. As an infant she was exposed to hepatitis B virus and became colicky and shaky. She was diagnosed with upper respiratory tract infection. While traveling abroad she was given scopolamine and cisapride. Upon returning she was admitted to a local hospital where she had an apneic episode, and was then transferred to a regional hospital.

C. MRS Method, Volume of Interest, and Comments about Examination

STEAM (TE = 20 ms, TM = 30 ms, TR = 3000 ms) was performed in the occipital grey matter (Location 1), and the parietal white matter (Location 2) as shown in Fig. 5.226.

D. Location 1 (Grey Matter)

1. Peak Identification and Quality Assurance: Reading the Spectrum from Right to Left

The spectral quality is good, except for over-suppression of the water leading to a steep slope from 3.8 ppm (Fig. 5.227). The diagnostic value of the spectrum is not in any way affected by the oversuppressed water.

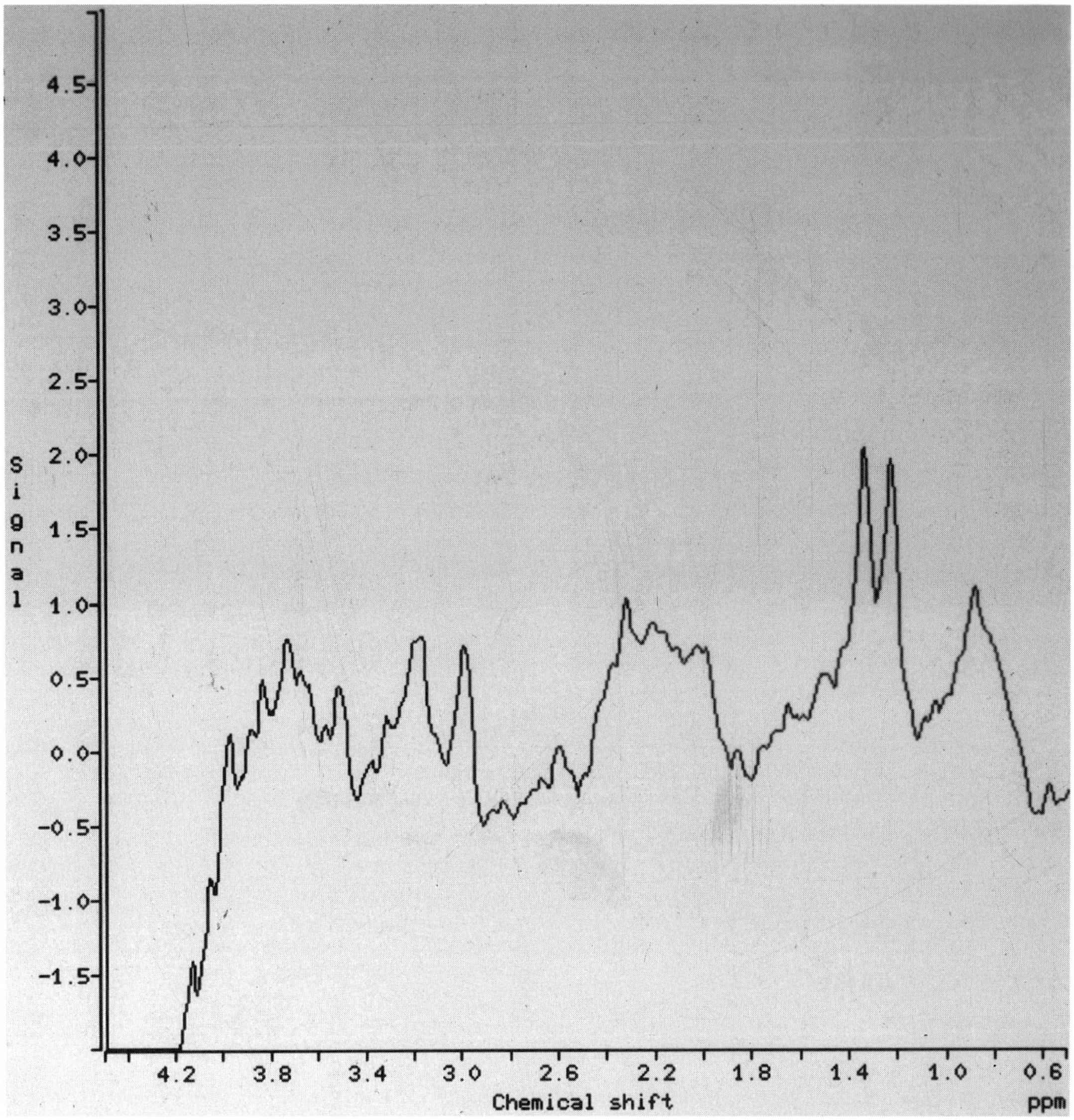

Fig. 5.227 The MRS from the occipital grey matter location shows excess lactate and absent NAA, consistent with a very bad outlook for the 1 month old patient. (Courtesy of Drs. Ashwal and Holzhouser, Loma Linda Medical Center, California.)

Reading from right to left, significant lipid is seen at 0.9 ppm, and a prominent lactate doublet is seen at 1.33 ppm. NAA is virtually absent. Note that the pattern of the β,γ-Glx region cannot be directly compared to the majority of spectra in this book as TE, and in particular TM, is different, but it appears to be increased. After the β,γ-Glx, there is a small peak 2.6 ppm which cannot be the second NAA because the first NAA is missing; it may be aspartate. After Cr follows an unusual broad Cho, and a small mI. The α-Glx is seen, but beyond α-Glx the second Cr and mI are too affected by the water suppression to be noted.

2. Interpretation and Analysis

The gestational age is 44 weeks. [Cr] is not available, but cannot be assumed normal considering the severely abnormal spectrum with lactate as the most dominant resonance. NAA is absent, Glx may be increased, and [Cr] is most likely reduced, leaving both Cho and mI also lower than normal for age. The most important findings, absent NAA and excess lactate, lead to a very poor prognosis. The measurement in the second VOI was performed to confirm this.

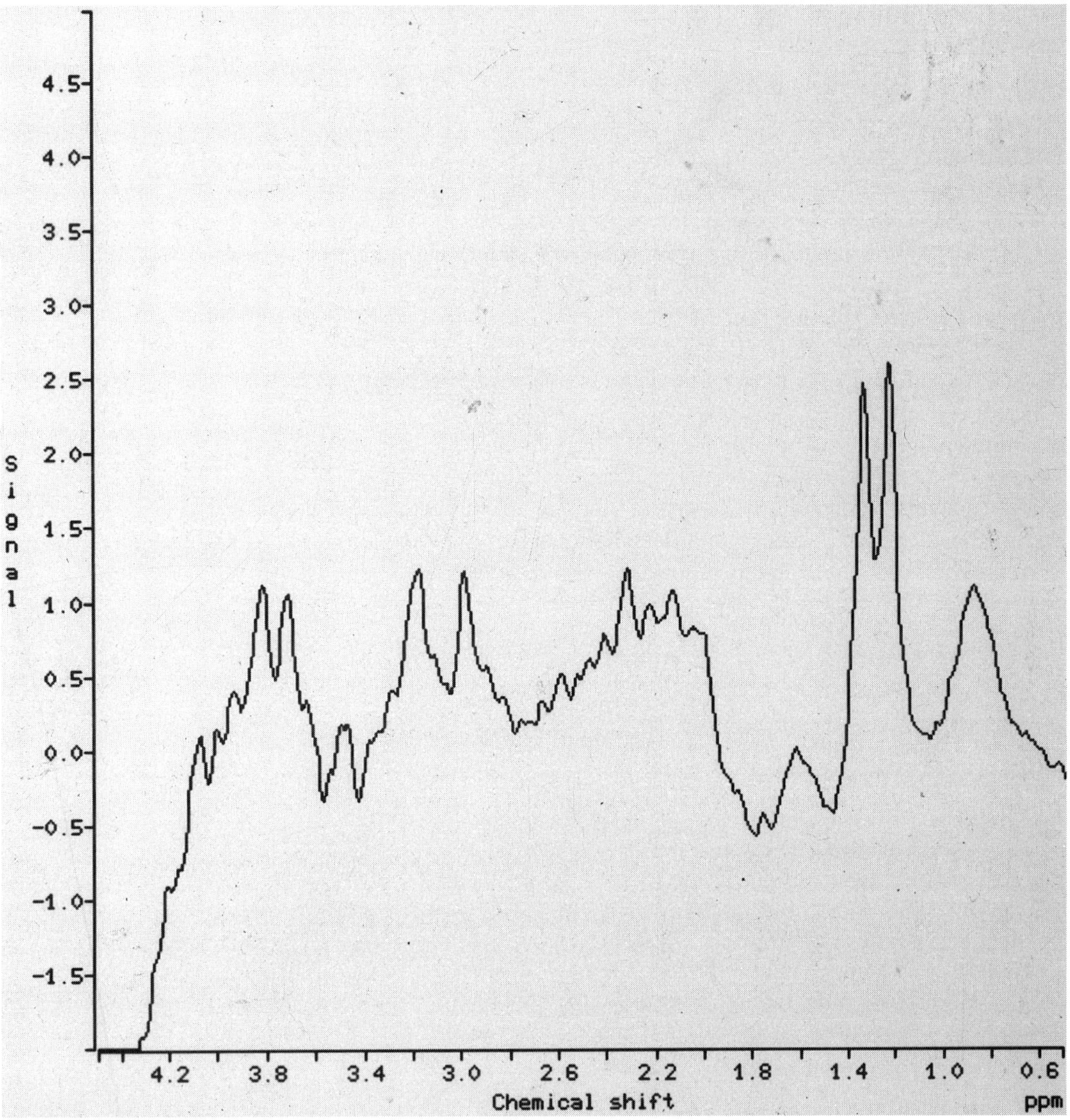

Fig. 5.228 The MRS from the parietal white matter location confirms the result obtained in the occipital grey matter location. (Courtesy of Drs. Ashwal and Holzhouser, Loma Linda Medical Center, California.)

E. Location 2 (White Matter)

1. Peak Identification and Quality Assurance: Reading the Spectrum from Right to Left

The spectral quality is similar to the previous spectrum (Fig. 5.228). The water suppression, and probably the eddy currents, seem to affect the spectrum from 3.3 ppm and beyond, but again the remaining spectrum with good resolution and S/N provides sufficient information.

Reading from right to left, significant lipids are seen at 0.9 ppm (possibly also at 1.3 ppm, underlying the doublet) and at 1.6 ppm. The lactate doublet at 1.33 ppm is huge. NAA is again absent. The β,γ-Glx region is very similar to the previous spectrum, but the small peak at 2.6 ppm is missing in this spectrum. Then follows Cr and Cho. Reading further is not feasible, but also not vital.

2. Interpretation and Analysis

The spectrum from this second VOI is very similar to that of the first, confirming most of the findings, and hereby indicating that the metabolic abnormality is global.

F. Conclusion

Absent NAA and a huge excess lactate, leads to a very poor prognosis. The spectrum is that of extremely severe hypoxia, likely to be secondary to mitochondrial encephalopathy with lactic acidemia.

G. Summary of Patient's History Post-MRS

She did not survive, and at this point no tissue diagnosis has been rendered.

H. Other Comments

Although specific diagnosis cannot always be given, MRS can in most cases provide a prognosis in infants with unexplained failure to thrive, apneic episodes and/or coma. The main indicators are excess/absent lactate and decreased/normal NAA. Note that correction for gestational age is necessary when judging the NAA.

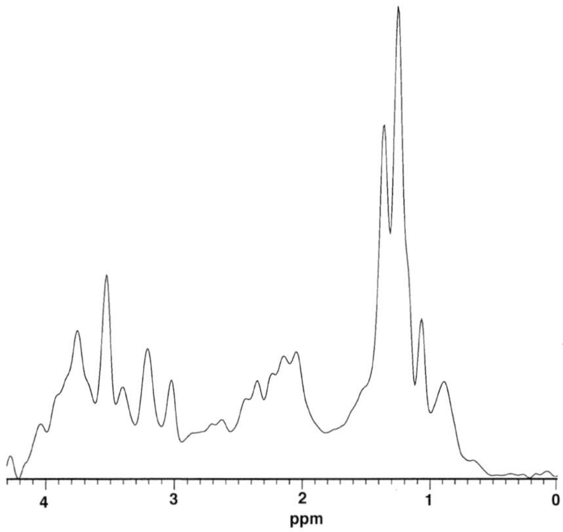

Fig. 5.229 The MRIs showed loss of the sulci in posterior brain (not shown). The locations of the two MRS VOIs were standard. The MRS from the occipito-parietal location is consistent with severe hypoxic encephalopathy. The propylene glycol doublet, shouldering the huge lactate doublet, is a commonly used solvent of pediatric drugs.

5.10 Xenobiotics: Ethanol, Propylene Glycol, Mannitol

5.10.1 PROPYLENE GLYCOL IN THE BRAIN

A. Purpose of MRS

MRS was performed to evaluate prospects for survival in newborn with severe perinatal asphyxia.

B. Summary of Patient's History

A baby girl, three days old (term; 40 weeks gestation). The mother had suffered massive nonspecific vaginal bleeding starting a few hours before admission for emergency Cesarean section. The Apgar scores were 0 both at 1 and 5 minutes, but the baby was resuscitated 14 minutes after delivery. The baby's weight and size was normal, and the maternal history included no unusual circumstances; gravida 2 with good prenatal care and no history of substance abuse.

C. MRS Method, Volume of Interest, and Comments about Examination

STEAM (TE = 30 ms, TM = 13.7 ms, TR = 1500 ms) was performed in HMRI's parietal white matter (Location 1), and in the occipital grey matter (Location 2). The sizes of the VOIs were modified to fit within the small head: white matter was $17 \times 17 \times 16$ mm^3; grey matter was $15 \times 18 \times 16$ mm^3 (image not shown).

The baby was on a ventilator, and the examination was performed under severe time constraints. The measurements of water content and absolute concentrations were therefore omitted.

D. Location 1 (White Matter)

1. Peak Identification and Quality Assurance: Reading the Spectrum from Right to Left

The quality of the spectrum is good, Fig. 5.229. Reading the spectrum from right to left, a very little lipid signal is seen at 0.9 ppm, then follow three peaks from 1.1 to 1.45 ppm. At first view, the two major of these peaks can be assigned to lactate because the doublet is centered at 1.33 ppm with the correct peak separation; however, the smaller peak at 1.1 ppm still needs

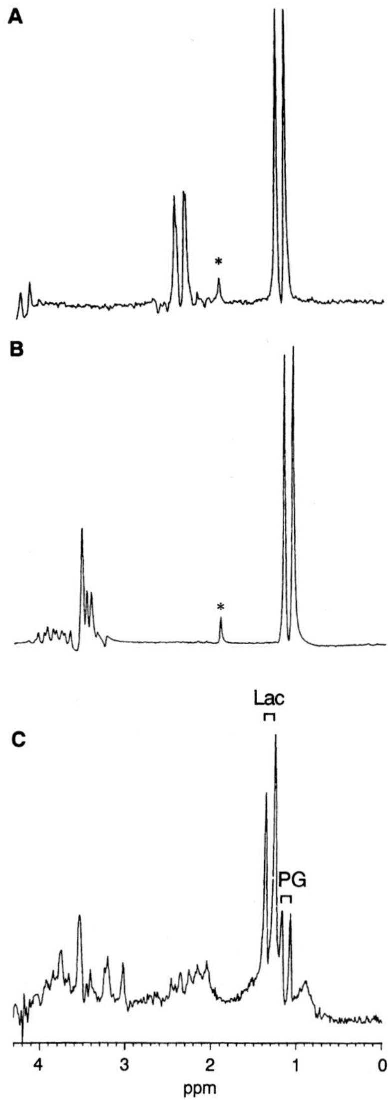

Fig. 5.230 Considered sources of the extra doublet at 1.14 ppm in the patient were β-OH buterate (A) and propylene glycol (also called 1,2 propanediol) (B). Propylene glycol is clearly identified as the source of the resonance in the infant brain (C) at 1.14 ppm and part of the resonance at 3.52 ppm, which overlaps with mI at 3.56 ppm. Part of the lactate is possibly that of D-lactate, a toxic metabolite of D,L propylene glycol. Propylene glycol is a commonly used solvent of pediatric drugs.

to be assigned. It helps to display a spectrum (from Location 2) with less line-broadening see Fig. 5.230C. The first peak is part of a doublet centered at 1.14 ppm, and the second peak of this doublet is hidden as a shoulder on the right hand side of the lactate in the line-broadened spectrum. The 1.14 ppm doublet is assigned to propylene glycol (also called 1,2 propanediol).

Continuing reading the spectrum from right to left, no NAA is found, but a broad hump, mainly from β,γ-Glx, is seen, followed by Cr, Cho, a spurious mI peak, and α-Glx. Part of the apparent mI peak is due to a secondary peak of propylene glycol see Fig. 5.230B.

2. Interpretation and Analysis

Absolute concentrations are not available, but comparing the Cr peak to the β,γ-Glx region we get an estimate of [Cr] being less than 30 or 40% of normal. Glx could in theory be increased, but not enough to fit with Cr being normal. A comparison of

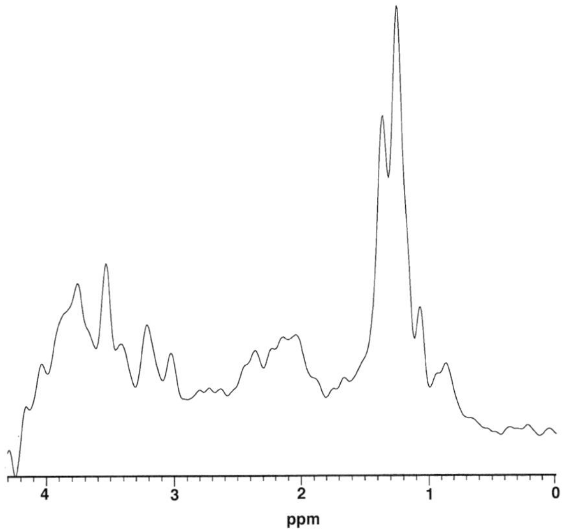

Fig. 5.231 MRS in the second location confirmed the presence of the unusual doublet at 1.14 ppm, as well as confirming the severity of the hypoxic damage.

the Cr peak to the noise also confirms that Cr is low. Since [Cr] is abnormal it becomes meaningless to compare the precise ratios to an age-matched healthy control. Instead, based on the above discussion, we can state that Cr, Cho and mI are all reduced, that NAA is absent, and that lactate and propylene glycol dominate the spectrum.

D. Location 2 (Grey Matter)

1. Peak Identification and Quality Assurance: Reading the Spectrum from Right to Left

The second spectrum figure, Fig. 5.231, is similar both in appearance and technical quality. Reading the spectrum from right to left, the following resonances are identified: propylene glycol, lactate, β,γ-Glx, Cr, Cho, mI overlapping with propylene glycol, and α-Glx.

2. Interpretation and Analysis

Similar to the first spectrum, there is a dominant propylene glycol peak, and lactate, NAA is absent, and Cr, Cho, and mI are low.

E. Conclusion

The huge lactate doublet, the absence of NAA, and the reduced Cr indicate that the MRS is consistent with severe hypoxic encephalopathy. Propylene glycol is a xenobiotic arising from a commonly used solvent of pediatric drugs, which sometimes is seen in the brain of pediatric patients.

F. Summary of Patient's History Post-MRS

The infant remained on a ventilator. She suffered seizures, cerebral edema, and hemorrhage, acute renal failure, disseminated intravascular coagulopathy, and she expired on the fifth day after delivery. Medications included ampicillin, gentamicin, cefotaxime, sodium bicarbonate, dopamine, dobutamine, furosemide, ethacrynic acid, insulin, kayexalate, phenobarbital, sedatives/tranquilizers (ativan), dilantin, clonazepam, and norcuron.

G. Other Comments

The solvents of the kayexalate (several grams administered), the phenobarbital, the sedatives/tranquilizers (ativan), and the dilantin, all contain DL-1,2-propanediol, and are likely to have contributed to the propylene glycol in the brain. More problematic is the source of lactate. Most probably it is intrinsic, and accumulated as L-lactate in hypoxic tissues. However, there is a strong possibility that propanediol is metabolized.

Fig. 5.230 illustrates how unusual resonances can be identified. Possible metabolite and a chemical shift reference (denoted by the *) are dissolved in water at the closest possible in vivo like conditions (saline, adjusted pH, etc.). In this case one candidate was β-hydroxy-butyrate, (β-OH-butyrate), but as demonstrated in Fig. 5.230A, β-OH-butyrate does not match what was seen in the infant—propylene glycol does.

The detection of propylene glycol in both locations first of all confirms that the unusual peaks are not an artifact. Secondly, it indicates that propylene glycol is found throughout the brain. The medications dissolved in D,L-propylene glycol may in part be metabolized into D-lactate, which is toxic. Since D- and L-lactate both resonate at 1.33 ppm, it cannot be excluded that part of the lactate was in fact D-lactate.

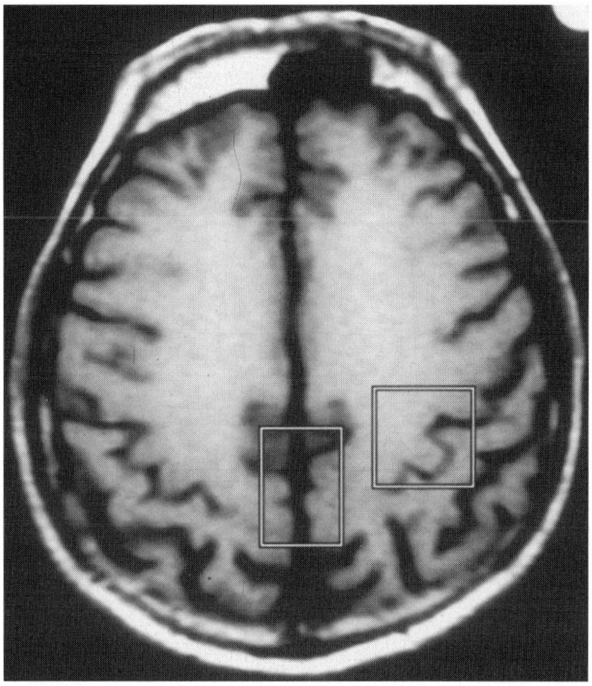

Fig. 5.232 T_1-weighted MRI showing the two MRS locations.

5.10.2 ETHANOL IN THE BRAIN

A. Purpose of MRS

MRS was requested to confirm presence of, and to evaluate the severity of hepatic encephalopathy in a patient awaiting liver transplantation. The MRS described here was a repeat examination. An MRS one week earlier (not shown) was probably consistent with hepatic encephalopathy, but the technical quality was not sufficient to make a definite diagnosis, and also not sufficient to evaluate the severity of the disease.

B. Summary of Patient's History

Male, age 46 years, with clinical diagnosis of liver disease and probable hepatic encephalopathy; hepatitis at age 12. He was awaiting liver transplantation, and denied any history of alcohol abuse.

C. MRS Method, Volume of Interest, Comments about Examination

STEAM (TE = 30 ms, TM = 13.7 ms, TR = 1500 ms) was performed in HMRI's parietal white matter VOI (Location 1), and HMRI's occipital grey matter VOI (Location 2) (see image in Fig. 5.232). Patient motion was avoided in this repeat examination, facilitating better acquisition of spectra than the week before.

D. Location 1 (White Matter)

1. Peak Identification and Quality Assurance: Reading the Spectrum from Right to Left

It is immediately obvious that the spectrum (Fig. 5.233) is quite unusual. The experienced MRS reader will see a peak missing, and a set of three new peaks, but the spectrum is of high quality, all peaks are narrow, the S/N is good, the baseline is flat and the lipid region is not troubled by broad humps or inverted signals. Altogether the peaks seem real, and are identifiable.

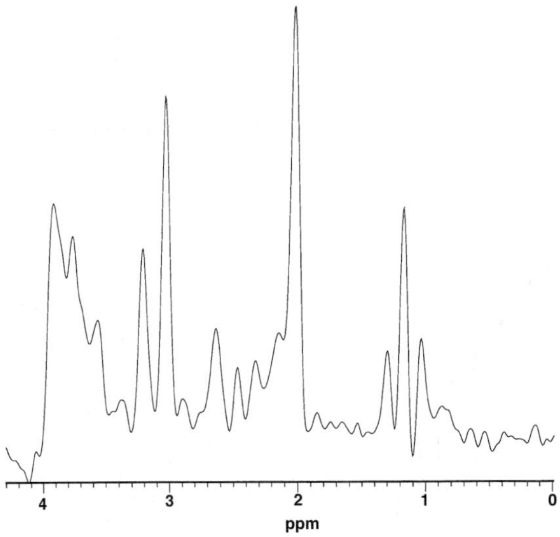

Fig. 5.233 The white matter spectrum shows an ethanol triplet, and changes consistent with chronic hepatic encephalopathy.

Reading from right to left, three new peaks are seen from 1.0–1.4 ppm. They are all from the same metabolite, and appear in a pattern where the middle peak is approximately double the height of the outer peaks. This is called a "triplet", just as the set of two peaks from lactate is called a "doublet". The triplet is centered at 1.16 ppm, and therefore assigned to ethanol.

After the ethanol triplet follows NAA, β,γ-Glx, the second NAA, Cr, and Cho at 3.22 ppm. But at 3.56 ppm, where a tall peak of mI is normally seen, only a very much reduced mI peak is seen, as a small shoulder on the right hand side of the α-Glx resonances. The second Cr is not very clearly separated from α-Glx, but it is seen. A second, but small set of peaks from ethanol (a quartet) is found in this region. This quartet is individually identifiable, but in this MRS its overlapping signal probably causes the somewhat unusual appearance of the second Cr and the α-Glx resonances. The second mI is not seen, in good agreement with the severe depletion of the first mI peak.

2. Interpretation and Analysis

[Cr] is not available, but can be assumed normal as there is no significant deviation from normal in liver disease patients. The ethanol triplet is huge; the center peak is about ⅔ the height of Cr. A more precise quantitation will be mentioned below, but very roughly estimated, a triplet of this size corresponds to a concentration of 4 to 5 mM.

The metabolite ratios are: NAA/Cr = 1.25 (11% decreased); Cho/Cr = 0.64 (23% decreased); mI is virtually absent, but a misfit by the computer resulted in the ratio mI/Cr = 0.49 (when a peak is absent, computer fitting routines often attempt fitting the peak anyway, usually finding erroneous numbers based on the best possible fit of the baseline and underlying broad resonances to the requested peak). Glx is not significantly increased.

E. Location 2 (Grey Matter)

1. Peak Identification and Quality Assurance: Reading the Spectrum from Right to Left

The spectrum (Fig. 5.234) is similarly of good quality. Reading from right to left, we see the same picture as in the white matter spectrum. The ethanol triplet at 1.16 ppm, NAA, β,γ-Glx, the second NAA, Cr, Cho, a small mI, α-Glx, and the second Cr.

2. Interpretation and Analysis

[Cr] is again not available, but can be assumed normal. The ethanol triplet is comparable to Cr. The metabolite ratios are: NAA/Cr = 1.2 (about 10% decreased); Cho/Cr = 0.6 (normal); mI/Cr = 0.4 (about 30% decreased). Glx is not significantly increased.

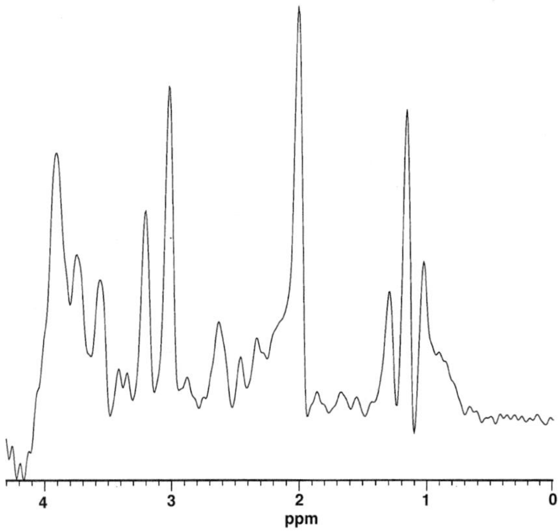

Fig. 5.234 The grey matter spectrum confirms the results found in the white matter. However, the signs of chronic hepatic encephalopathy are less marked in the grey matter location.

F. Conclusion

This MRS is clearly a case of chronic hepatic encephalopathy in view of the biochemical changes of mI and Cho, but there is no significant elevation of glutamine, indicating that the disease may be sub-clinical. The striking finding in these two spectra is that alcohol intake has occurred, despite vigorous denial on the part of the patient, and the absence of clinical signs.

G. Summary of the Patient's History Post-MRS

The possibility of alcohol abuse was reconsidered in relation to a possible future liver transplantation. The patient, after six months, recognized the need to abstain, and, remarkably, remains well after six years without a liver transplantation.

H. Other Comments

Quantitation of the ethanol triplet by comparison to internal [Cr] = 6.3 mM in white matter, permits an estimated concentration of 5 to 7 mM. This was obtained by comparing to MRS of known ethanol solutions measured under identical conditions.

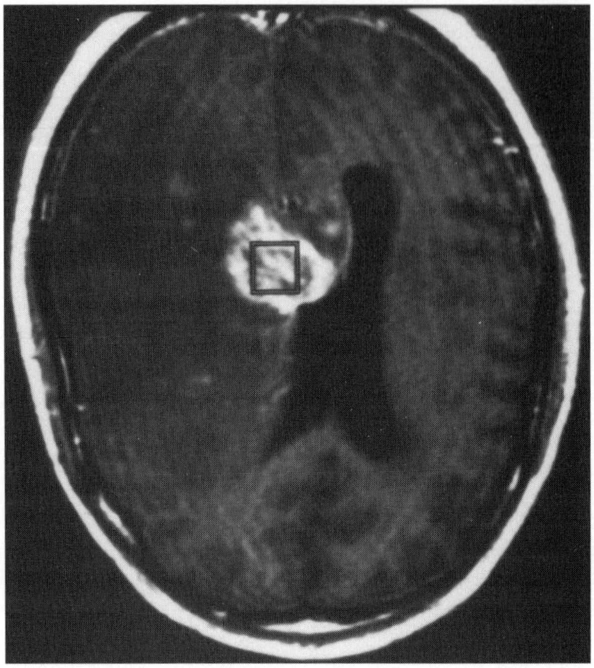

Fig. 5.235 The axial MRI showing the MRS (PROBE-S) location within a treated astrocytoma. (Courtesy of Dr. Joseph Norfray, Chicago Northside MRI Center, Chicago, IL.)

5.10.3 MANNITOL IN A TUMOR

A. Purpose of MRS

To evaluate a tumor post treatment.

B. Summary of Patient's History

The patient had an astrocytoma which was treated, and later recurred. The recurrence produced life-threatening intra-cranial hypertension which was alleviated only by administering dexamethasone and mannitol, intravenously. MRI and MRS were performed with a continuous mannitol infusion.

C. MRS Method, Volume of Interest, and Comments about Examination

Automated PROBE-STEAM (TE = 30 ms, TM = 13.7 ms, TR = 1500 ms). A small VOI was placed within the tumor as shown on Fig. 5.235. Note that this lesion enhances strongly with gadolinium; the blood-brain barrier is inoperative.

D. Location in Tumor

1. Peak Identification and Quality Assurance: Reading the Spectrum from Right to Left

The PROBE spectrum, Fig. 5.236, is well resolved and has no artifacts. Reading the spectrum from right to left lipid is seen at 0.9–1.0 and 1.2–1.4 ppm. The latter peak has a hint of splitting at the top, so a small fraction of the peak may be due to lactate. It cannot be a major fraction of the peak, because Cr and Cho are well separated, requiring lactate to appear as a clear doublet.

After the lipid, the absence of NAA is noticed, but β,γ-Glx is seen followed by Cr, Cho, and a very little mI. The large peak at 3.8 does not fit any of the standard metabolites found in the brain, and the spectral quality is excellent, so it is not an artifact.

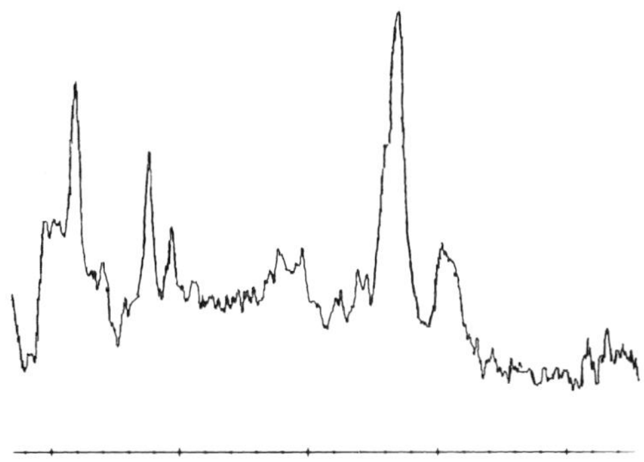

Fig. 5.236 Mannitol resonating at 3.8 ppm in a PROBE-S spectrum of an astrocytoma. (Courtesy of Dr. Joseph Norfray, Chicago Northside MRI Center, Chicago, IL.)

The peak cannot be due to α-Glx, because β,γ-Glx is not elevated, and also because the shape does not fit α-Glx. It can also not be mI which resonates at 3.56 ppm or the second Cr which resonates at 3.9 ppm. The peak at 3.8 ppm was assigned to mannitol by comparing to the spectrum to one of a mannitol solution, Cr was used as a chemical shift reference.

2. Interpretation and Analysis

The lipid and possible lactate peaks are the dominant peaks of the spectrum. [Cr] is not available, but judging from the appearance of the spectrum (compare to Glx and noise level), [Cr] can be estimated to be considerably decreased. The inversed Cho/Cr ratio can therefore to some extent be attributed to low Cr, rather than to increased Cho only. The mI is definitely decreased. Without absolute quantitation it is difficult to say anything precise about the mannitol concentration, but considering the size of the peak compared to the noise and VOI size, the concentration must be 4 mM or more.

E. Conclusion

The spectrum is typical for a tumor with prominent Cho, possibly decreased Cr, vanished NAA, and prominent lipids and possible lactate. The rare finding of mannitol in this spectrum must be related to two events. First, the large intake of mannitol, which was documented. Second, the intact blood-brain barrier has allowed good clinical benefit in reducing brain swelling, but as shown with Gd, the barrier is inoperative in the tumor itself, allowing mannitol to accumulate within the VOI.

F. Summary of Patient's History Post-MRS

Not available.

G. Other Comments

Patients commonly receive xenobiotics, most of which are not demonstrable in routine MRS. However, erroneous peak assignment (as mI) in this case should encourage systematic right to left reading of all spectra, and assignment only if frequency is appropriate.

Note that mannitol, a sugar, is clearly seen only in short TE spectra, due to strong coupling (see p. 28; Fig. 3.6).

6

MRS Patterns and Referral Guidelines

I. CLINICAL DECISION MAKING WITH ^{1}H MRS

A brief justification for MRS case-by-case simplifies the task of referring physicians and (neuro) radiologists, while providing at the same time a basis for claims to reimburse the costs of the examination through insurance companies. Table 1 lists syndromes and clinical presentations that are frequently investigated by MRS.

Following is a summary of MRS referral guidelines for a selection of these indications. The obvious standard contraindications for a patient to be in a MR scanner apply but are not mentioned for each case.

A. Dementia

Indications: Any patient with memory deficit, progressive cognitive difficulty, prolonged depression, "pseudodementia," agitation and/or confusion. Certain patients with unresolved cognitive change in presence of hypertension, TIA or stroke. N.B.: Normal CT, MRI, PET, or SPECT are *not contraindications* to ^{1}H MRS.

Contraindications: Frank psychiatric disease, e.g., schizophrenia; focal lesions; extradural hemorrhage; coexistent systemic diseases with neurological sequence (diabetes mellitus, uremia, cirrhosis, hyponatremia) may complicate interpretation of MRS.

Acquisition: STEAM TE = 30 ms of GM.

Major Findings on ^{1}H MRS: (1) Decreased NAA/Cr (and absolute [NAA]). (2) Increased mI/Cr (and absolute [mI]). Note: 1 without 2 is diagnosed as other dementia (not AD). 1 and 2 are diagnosed as AD, see Fig. 6.1A. (3) Note, normal MRS in an apparently demented patient is consistent with *treatable* depression.

1. Decreased NAA/Cr and absolute [NAA]. Fig. 6.1B.
2. Increased mI/Cr and [mI] (for Alzheimer's disease), Fig. 6.1C.

Confounding Entities: Stroke or copious deep white matter infarcts (DWMI) may increase Cho/Cr. Systemic diseases with increased cerebral mI/Cr = uremia, diabetes.

Sensitivity of MRS: Versus normal 83%; versus Other Dementia 82%, Fig. 6.1D.

Specificity of MRS: Versus normal 98%; and a negative predictive rate of 80% versus Other Dementia, Fig. 6.1E.

Value of MRS (PROBE) (rating): Secondary and screening.

Other Useful Investigations: Many, but less sensitive and less specific.

Key Cases: 5.2.1–5.2.5, 5.3.6, 5.4.9, and 5.7.5.

Table 6.1 Patients with the Listed Syndromes and Clinical Presentation Are Obvious Candidates for MRS

Elderly	Confusion; depression; **dementia**; hydrocephalus; **hypoxia**; stroke; TIA; **tumor**; **trauma**; metabolic (renal diabetes).
Adult	**Hypoxia**; stroke; **tumor**; **encephalopathy**; **liver disease**; liver transplant; **trauma**; hypertension; hydrocephalus; MS; **AIDS (mass and dementia)**; **HIV positive**; metabolic disease
Children	**Abnormal development** (inborn error); white matter disease, hydrocephalus; **trauma**; infection; cardiac transplant; **hypoxia** (arrest); **near-drowning**; diabetes; hypo-, hypernatremia
Newborn	**Hypoxia**; ischemic (stroke); ECMO; **shaken baby syndrome**; **abnormal development** (inborn error); white matter disease; hydrocephalus; **trauma**; infection; cardiac transplant

The MRS indications, contraindications, major findings, confounding entities, sensitivity, specificity, and key references are summarized in the section MRS referral guidelines for the syndromes types in **bold face**.

B. Hepatic Encephalopathy (HE)

Indications: MRS should be considered as screening examination in patients with abnormal liver function *plus* any abnormal neurology; dementia; liver transplant candidates; alcoholics; post-op monitoring of TIPS candidates; acute liver failure. N.B.: Normal CT, MRI, PET or SPECT are *not contraindications* to ^{1}H MRS.

Contraindications: None.

Acquisition: STEAM TE = 30 ms first of WM, but add or start with GM if dementia is also an issue.

Major Findings on ^{1}H MRS: A triad of changes (loss of mI, increased Glx, and reduced Cho) characterizes patients with chronic HE, and, more usefully, defines subclinical HE, Fig. 6.2A. Patients with TIPS develop these characteristic metabolite changes as evidence of HE after shunting, Fig. 6.2C. Transplant recipients undergo complete reversal of these biochemical abnormalities within weeks after successful transplantation, Fig. 6.2D. Exaggerated

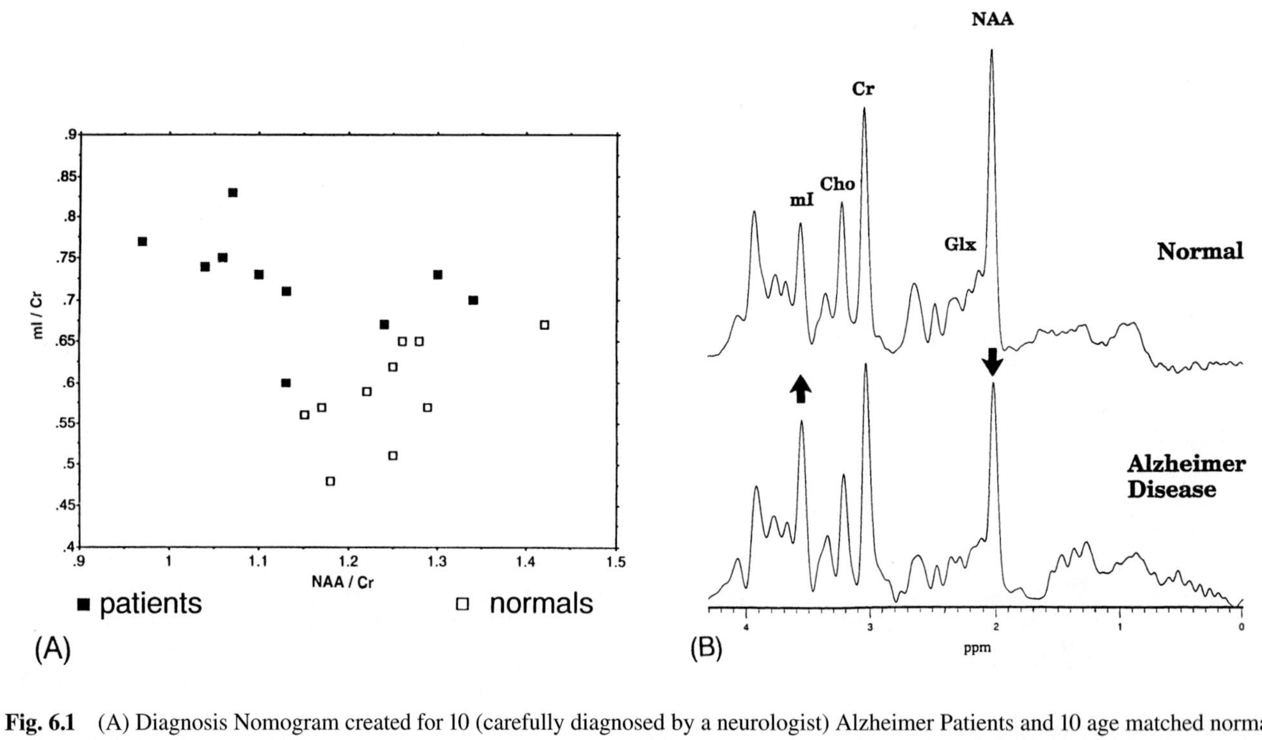

Fig. 6.1 (A) Diagnosis Nomogram created for 10 (carefully diagnosed by a neurologist) Alzheimer Patients and 10 age matched normal volunteers. The Alzheimer patients appear to be distinguished from normals by the presence of elevated mI/Cr and reduced NAA/Cr in grey matter. (B) Alzheimer's disease compared to normal. Quantitative MRS of occipital grey matter. (C) Alzheimer's disease: Automated PROBE: Effect of spectral processing. (D) Alzheimer's disease: diagnosis nomogram. Despite overlap, diagnosis in the individual case can be suggested by entering the NAA/Cr and mI/Cr ratios onto this graph. Left upper quadrant: AD likely; left lower quadrant: another dementia more likely; right upper quadrant; AD likely, but exclude diabetes or renal failure; right lower quadrant: "normal". (E) NAA/Cr versus mI/Cr should decide is it Alzheimer's disease (AD), Other dementia (OD), or Pick's (frontal lobe dementia; FLD).

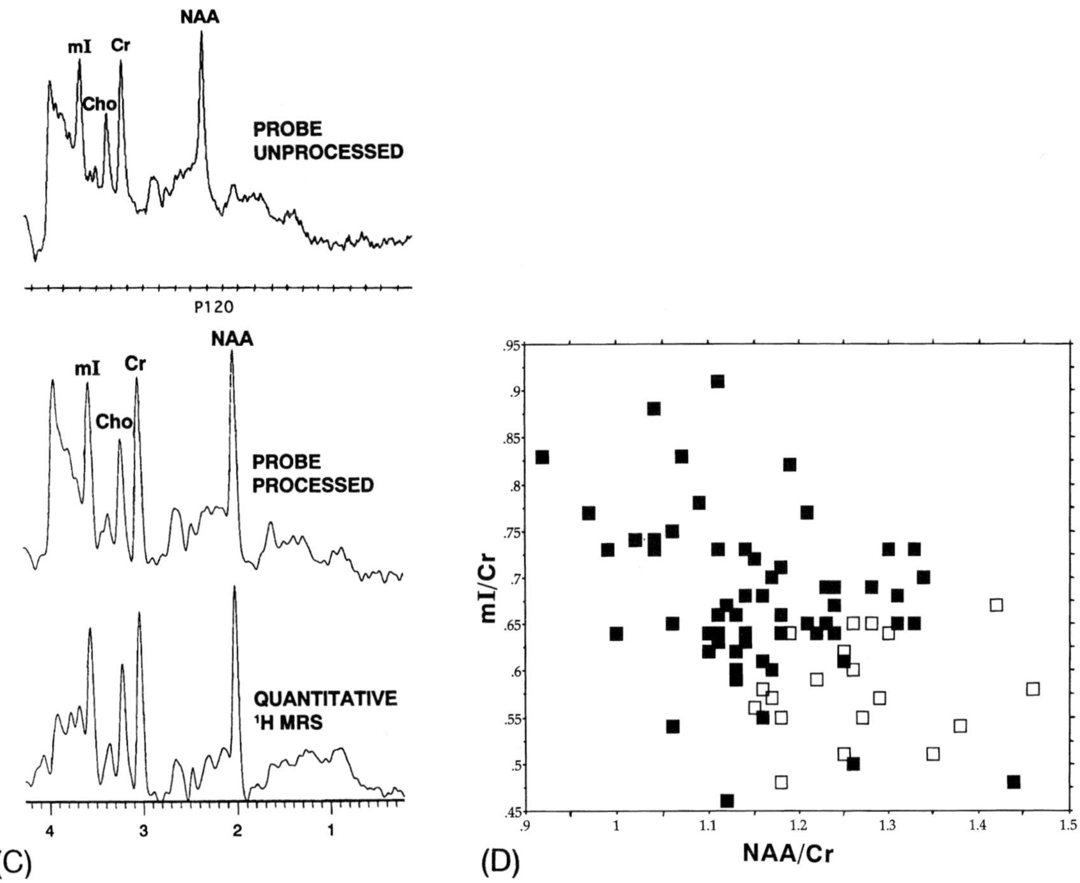

MRS changes are seen in acute fulminant HE, Fig. 6.2B, and the rare Reye's syndrome even when the MRI results are normal. MR spectra of HE are shown in Fig. 6.2A.

1. Increased α- and β,γ-Glx.
2. Decreased mI/Cr and [mI] (mI almost absent).
3. Decreased Cho/Cr (in white matter; but not invariably in grey matter).

Confounding Entities: Hyponatremia, even 2 weeks after normalizing plasma sodium.

Sensitivity of MRS: Close to 100% for subclinical HE; and twice as sensitive as clinical testing for overt HE.

Specificity of MRS: Almost 100%, together with clinical history and serum sodium or osmolality. More specific than "bright" basal ganglia on T_2 MRI or number-connection test and other current neuropsychiatric, and EEG tests.

Value of MRS (PROBE) (rating): Principal and secondary.

Other Useful Investigations: EEG; number connection test; abdominal ultrasound (shunt); blood ammonia; CSF glutamine (now obsolete).

Key Cases: 5.3.1–8, 5.4.4, and 5.10.2.

C. Hypoxic Encephalopathy and Stroke

Indications: Cardiac arrest; unidentified coma; asphyxia; near-drowning; stroke; TIA; congestive cardiac failure (CCF); respiratory failure; trauma. N.B.: Normal EEG, CT, MRI, PET or SPECT are *not contraindications* to ^{1}H MRS.

Contraindications: None (however technical problems due to susceptibility from hemorrhage may arise).

Acquisition: STEAM TE = 30 ms first of GM, add WM if time permits.

Major Findings on ^{1}H MRS: Because simple biochemical reactions underlie the universal ^{1}H MRS response to hypoxia: 1) lactate (and Glx) accumulate when O_2 is insufficient; 2) neurons bear the main impact and probably cannot recover: NAA is lost in proportion; 3) PCr (which comprises 50% of the Cr peak) needs ATP and hence O_2, and so falls markedly after hypoxia, see Fig. 6.3.

1. Increased lipid and overlying Lac doublet peak (at 1.3 ppm).
2. Decreased NAA.
3. Increased α- and β,γ-Glx.
4. Decreased [Cr].

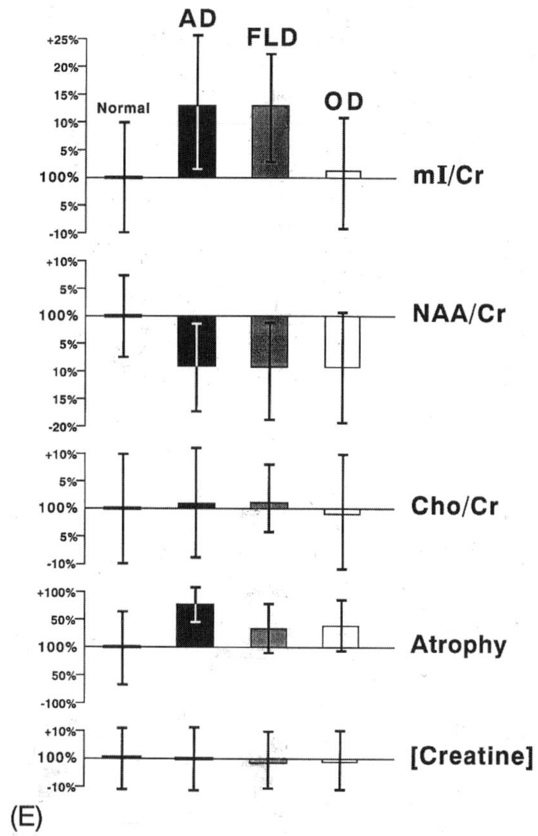

Fig 6.1 Continued.

Confounding Entities: Mitochondrial disease; lactic-acidosis any cause; cannot distinguish PVS from lethal injury. The Cho/Cr ratio appears increased because [Cr] is decreased.

Sensitivity of MRS: Appears highly sensitive to *severe* global hypoxic insult.

Specificity of MRS: Some overlap with post-traumatic spectrum. Prognosis: Best indicator of prognosis after near-drowning; lack of detailed analysis of other PVS.

Value of MRS (PROBE) (rating): Screening or secondary.

Other Useful Investigations: EEG.

Key Cases: 5.4.6, 5.4.7, 5.5.1–8, 5.6.1–4, and 5.10.1.

D. Abnormal Development

Indications: Any anatomical or metabolic disorder in the neonatal or pediatric period. N.B.: Normal US, CT, MRI, PET or SPECT are *not contraindications* to ^{1}H MRS.

Contraindications: None.

Acquisition: STEAM TE = 30 ms first of GM then WM, (the terms GM and WM here indicate the position of the location rather than the true composition of the tissue in the yet not differentiated infant brain).

Major Findings on ^{1}H MRS: Common, but non-specific findings are: 1) low NAA; 2) excess lactate. Unique metabolite profiles are to be expected in several relatively common disorders. An abundance of rare specific findings includes phenylalanine (phenylketonuria), glycine (non-ketotic hyperglycinemia, note not hyperglycemia), elevated NAA (Canavan's disease), and absent Cr (Frahm-Hanefeld Bianchi syndrome), see Fig. 6.4 and Fig. 3.15 in Chapter 3.

Confounding Entities: Gestational age, drugs, parenteral feeding solutions, xenobiotics which appear in the ^{1}H spectrum. Includes propanediol, mannitol, ethanol, glucose.

Sensitivity of MRS: Unknown. Leukodystrophies are unique; ^{1}H MRS appears more sensitive than MRI.

Specificity of MRS: High specificity for single enzyme (gene) defects.

Value of MRS (PROBE) (rating): Screening.

Fig. 6.2 (A) In parietal white matter, the majority of patients with cirrhosis show decreased Cho/Cr. Depletion of mI remains even after "recovery" from acute or chronic hepatic encephalopathy. It is considered a reliable marker of sub-clinical hepatic encephalopathy. Modest elevation of Glx is variable in sub-clinical HE and a useful marker of increasing severity of clinical hepatic encephalopathy and fulminant hepatic failure. (B) ^{1}H MRS in acute hepatic encephalopathy due to Reye's syndrome. *In vivo* ^{1}H MRS spectra acquired from a parietal white matter region in infant brain from a) 10 month old normal; b) Patient, day 2 after admission; c) Patient, day 8 after admission; d) Difference (b–c) between day 2 and day 8 in patient; e) Solution with 15 mM Glutamine. All spectra are scaled. Spectral assignments: Lactate (1.33 ppm); NAA (2.02 ppm); Gln = glutamine (2.10–2.50 ppm, and 3.65–3.90 ppm); Cr (including PCr) (3.03 ppm); Cho (3.23 ppm); mI (3.56 ppm). U = Unassigned (3.62 pm). Notable abnormalities concern a huge accumulation of cerebral glutamine, reduced Cho, and later reduced mI, all of which reflect liver failure. In addition there is a decrease in NAA and [Cr] and appearance of the unassigned peak. Reye's syndrome, a toxic-viral disease associated with aspirin intake is known to produce severe neuronal damage, and may not therefore completely correspond to the picture of acute liver failure. An occipital grey matter region gave almost identical results. (Reproduced with permission from T. Ernst, B. D. Ross and R. Flores, *Lancet* 1992, **340**, 486.) (C) Almost all TIPS patients show MRS changes of Hepatic Encephalopathy. (D) As expected, successful liver transplantation reverses all MRS abnormalities. MRS can detect portal vein thrombus or other sources of failure.

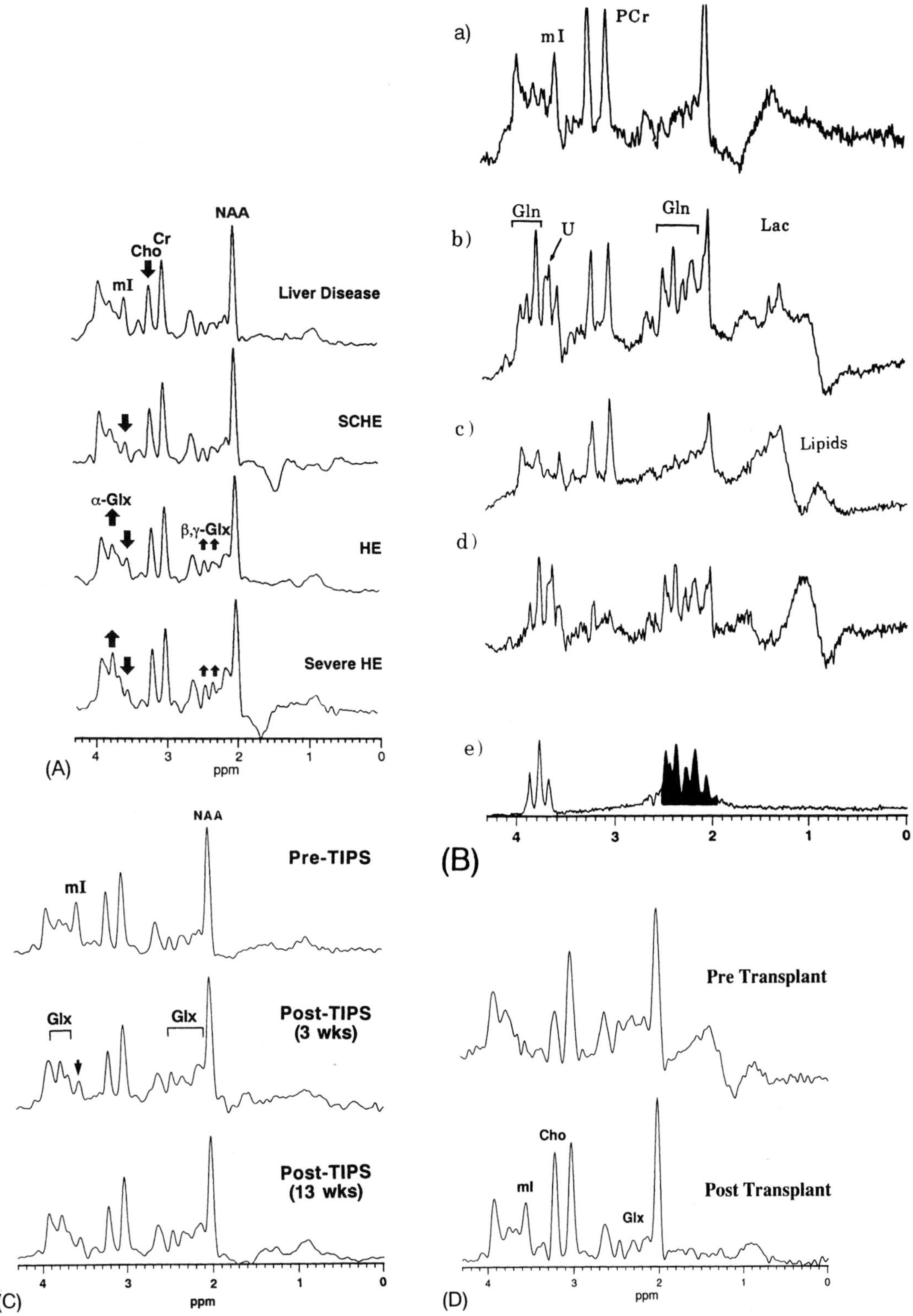
a)
mI
PCr
b)
Gln
U
Gln
Lac
c)
Lipids
d)
e)
4
3
2
1
0
(B)
NAA
Cho
Cr
mI
Liver Disease
SCHE
α-Glx
β,γ-Glx
HE
Severe HE
4
3
2
1
0
(A)
ppm
NAA
Pre-TIPS
mI
Glx
Glx
Post-TIPS
(3 wks)
Post-TIPS
(13 wks)
4
3
2
1
0
(C)
ppm
Pre Transplant
Cho
mI
Glx
Post Transplant
4
3
2
1
0
(D)
ppm

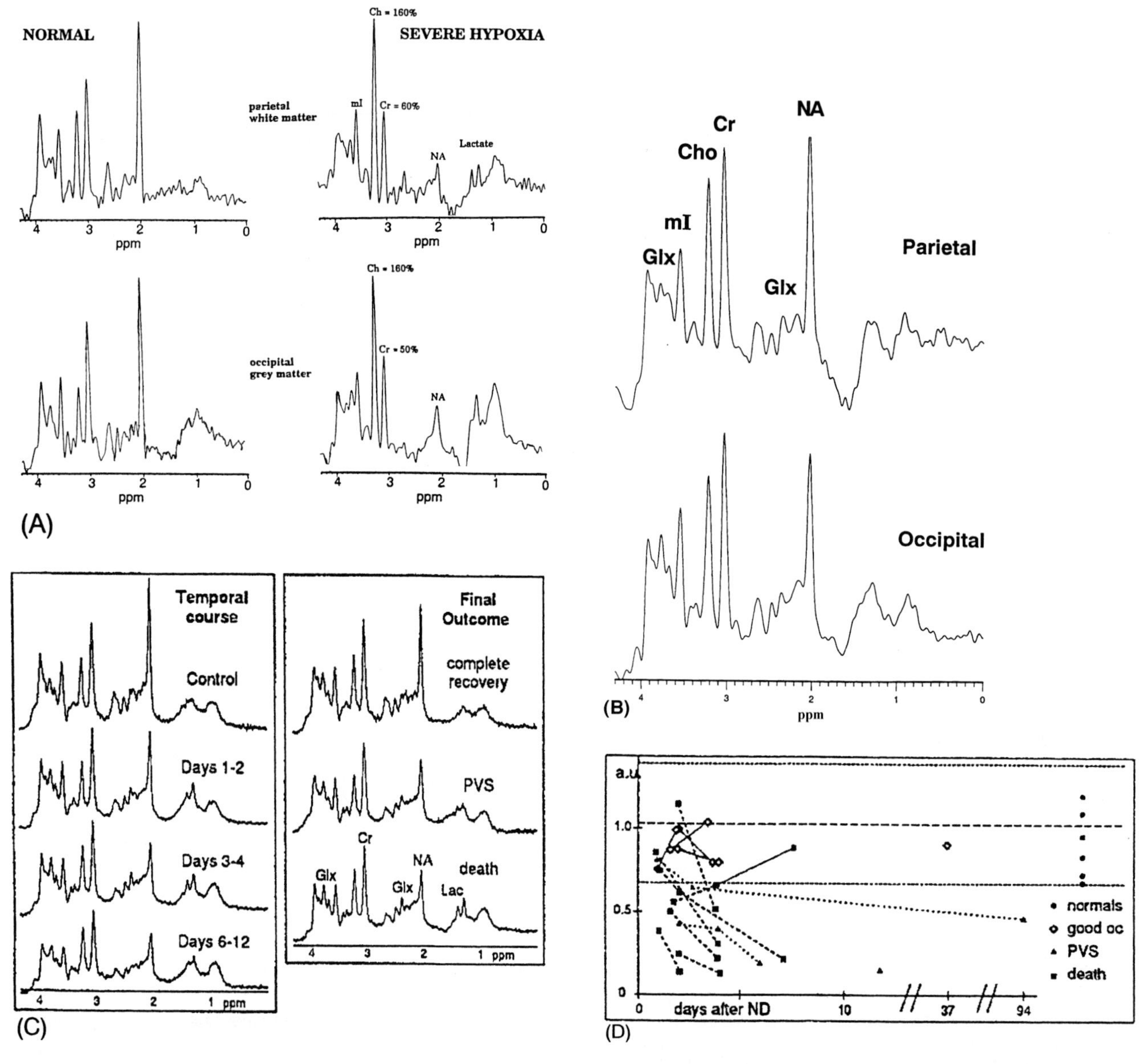

Fig. 6.3 (A) Severity and duration of initial hypoxic injury is reflected in MRS by degree of reduction of NAA ("loss of neurons"). MRS was performed 14 days after cardiac arrest. Gliosis (replacement of neurones by glia) is reflected in matched elevation of Cho, see also case 5.6.3. (B) Long-term effect of partial anoxia. Note that NAA/Cr is reduced, with the effect being most marked in the grey matter of occipital cortex. Cho/Cr is inappropriately elevated in the occipital VOI, possibly another consequence of hypoxia. (C) ^{1}H MR spectra as function of time after near drowning; sum of bad outcome group, (left panel). Cerebral ^{1}H MR spectra of occipital grey matter 3–4 days after ND as a function of eventual outcome. All spectra are scaled by unsuppressed water signal, (right panel). PVS: persistent vegetative state, and "Days" refer to days after resuscitation. (D) Predictive MRS index as function of time after near drowning. Individual data marked according to final outcome; including normal range (± 2 standard deviations). Details in Kreis et al. 1996 J. Clin. Invest., 97, 1142–1154.

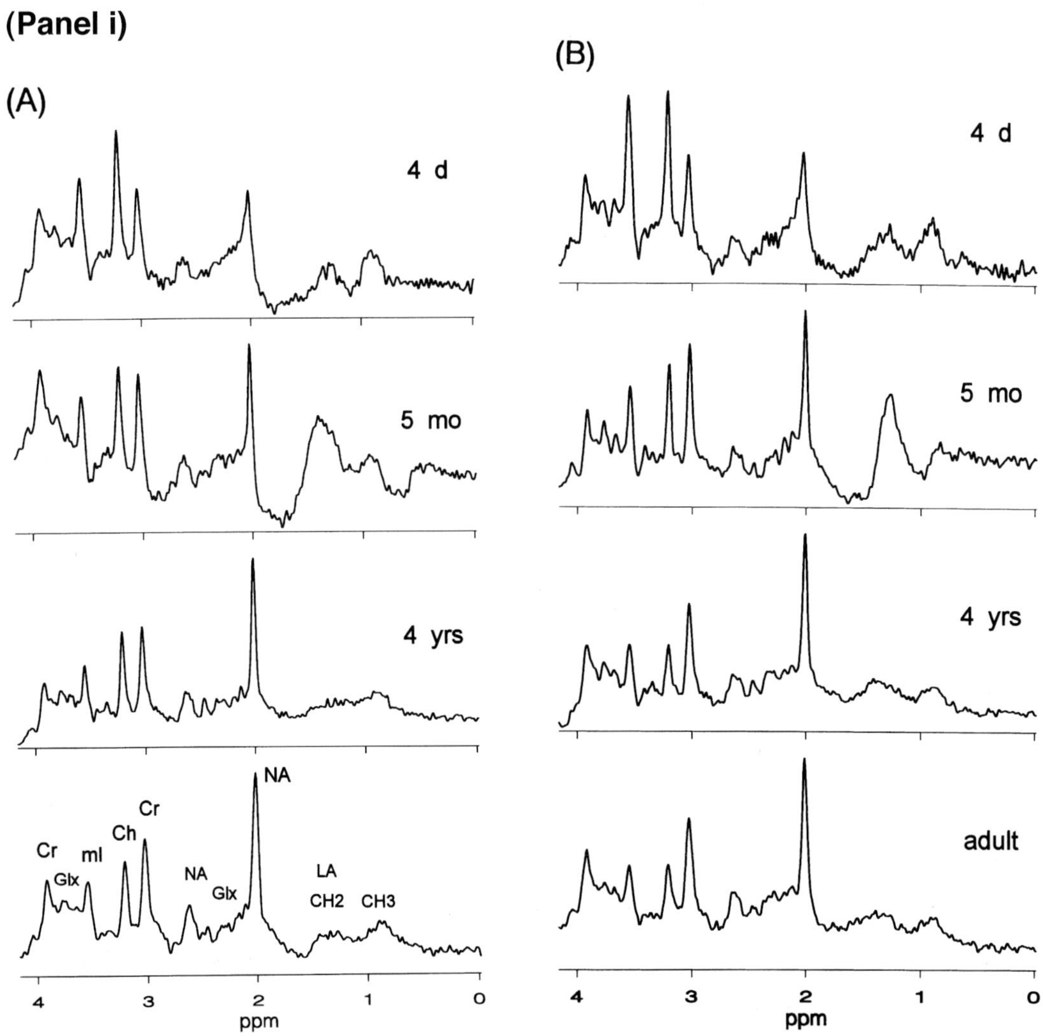

Fig. 6.4 (Panel i) Normal development: Typical cerebral ^{1}H MR spectra from subjects of different ages. Spectra were obtained from a VOI of primarily white matter in the parietal cortex (A) and from a VOI of primarily grey matter in the occipital cortex (B). (Panel ii) Inborn errors: Examples at 2 T, 20 ms TE (short echo). All spectra are courtesy of Dr. Jens Frahm. (A) Normal grey matter at 4 months. (B) Normal grey matter at 2 years. (C) Leigh's disease: white matter at 11 months. (D) Zellweger syndrome variant: grey matter at 15 months. (E) Guanidino-acetate methyl transferase deficiency (Frahm-Hanefeld-Bianchi syndrome) with total Cr deficiency; partially restored posttreatment. (G) Metachromatic leukodystrophy. (H) Severe adrenoleukodystrophy (ALD). (I) Adrenal insufficiency only (ALD variant). Spectra reproduced with permission from the following: A, B and D (JCAT 1992, 16, 335); C (Radiologie); E, F, H and I (SMRM 1993 and SMR 1994); G (J Neurol 1993, 241, 68–73). (Panel iii) Inborn errors: Examples at 1.5 T, 135/270 ms TE (long echo), PRESS. All spectra are courtesy of Dr. Wolfgang Grodd. (A) Normal white matter at 1 month. (B) Normal white matter at 4 years. (C) Leigh's disease at 5 years (135 upper/270 lower). (D) Canavan's disease at 7 months (135/270). (E) and (F) Alexander's disease at 14 months showing marked regional variation between white matter (E) and grey matter (F). (G) and (H) Marked inter-patient variability in presumed cases of Pelizaeus-Mersbacher disease, a disorder with debated diagnostic criteria; (G) = 17 month early onset; (H) = 6 years late onset. (Panel iv) Mitochondrial encephalopathies: Significant effects of field strength, sequence and echo-time. (I) Kern-Sayer syndrome (PRESS 272). (J) MERRF (PRESS 272). (K) Region variations in MELAS (CSI). (L) Ornithine transcarbamylase deficiency (PRESS 135). (M) Normal (PRESS 41 ms) Glx dramatically re-focused at 0.5 T; barely present at 1.5 T. (N) Glx phantom used to ensure accurate assignment. (O) Non-ketotic hyperglycinemia (PRESS 135), Gly or mI? (P) Age-matched normal STEAM 30–270 ms. The slow disappearance of mI increases certainty that the peak labelled Gly originates from glycine rather than mI. (Q) Phenylketonuria (PRESS 20). Difference spectroscopy ensures identity of resonance at 7.37 ppm.

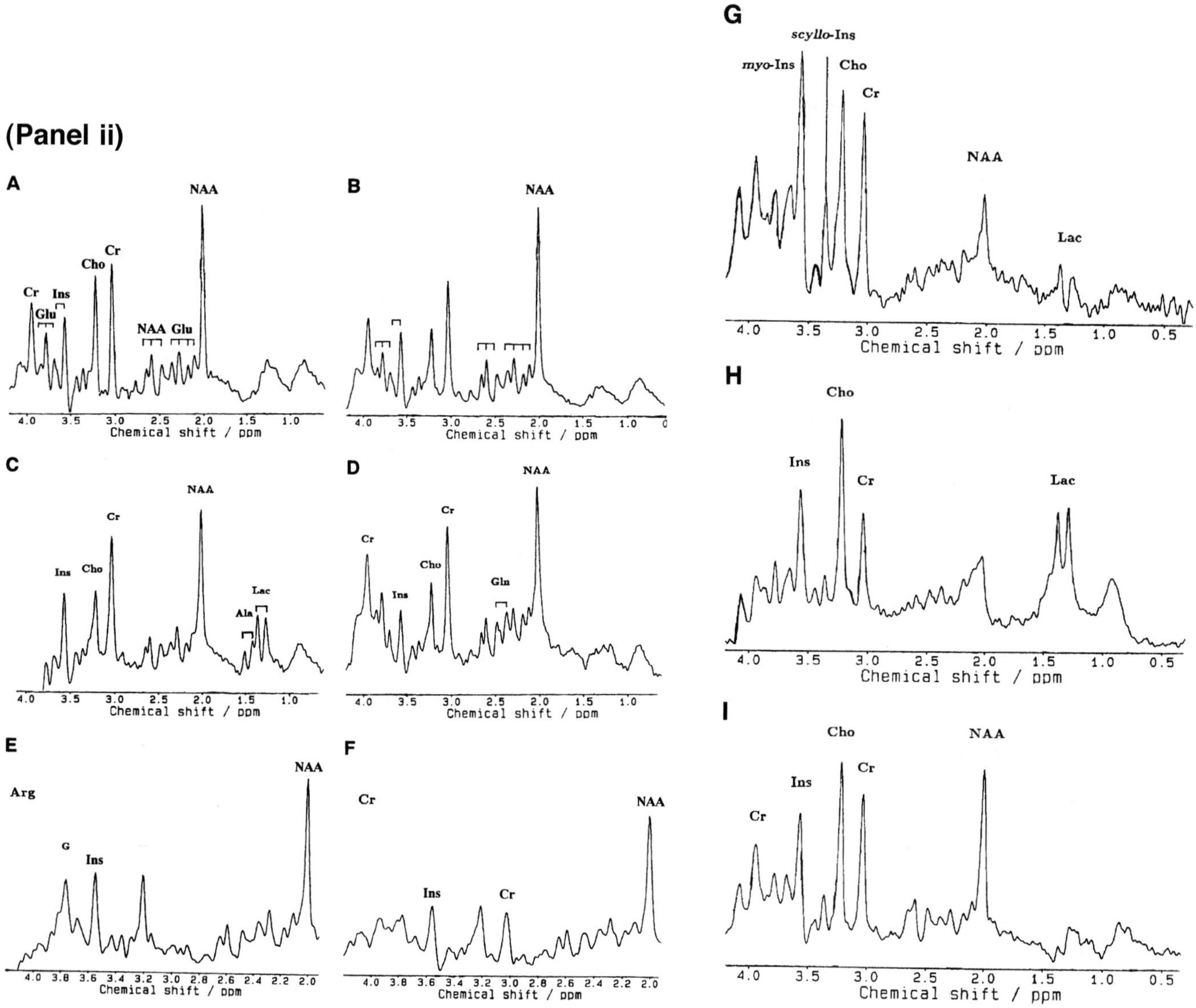

Fig. 6.4 Continued.

Other Useful Investigations: Many.

Key Cases: 5.9.1–8, and 5.10.1, (see also "normal development" 5.1.1–4).

E. Neonatal Hypoxia

Indications: N.B.: Normal US, CT, MRI, PET or SPECT are *not contraindications* to ^{1}H MRS.

Contraindications: None.

Acquisition: STEAM TE = 30 ms first of GM then WM, (the terms GM and WM here indicate the position of the location rather than the true composition of the tissue in the yet not differentiated infant brain).

Major Findings on ^1H MRS: The distinction between severe hypoxic encephalopathy with very poor prognosis and a prognosis of full recovery is made on the basis of: 1) excess lactate; 2) decreased NAA; 3) loss of Cr. Several grades of severity can be distinguished between these two extremes, see Fig. 6.5.

Confounding Entities: Gestational age, drugs, parenteral feeding solutions, xenobiotics which appear in the ^{1}H spectrum. Includes propanediol, mannitol, ethanol, glucose.

Sensitivity of MRS: Much higher than U.S. or CT.

Specificity of MRS: Not proven; overlap with shaken baby syndrome and inborn errors.

(Panel iii)

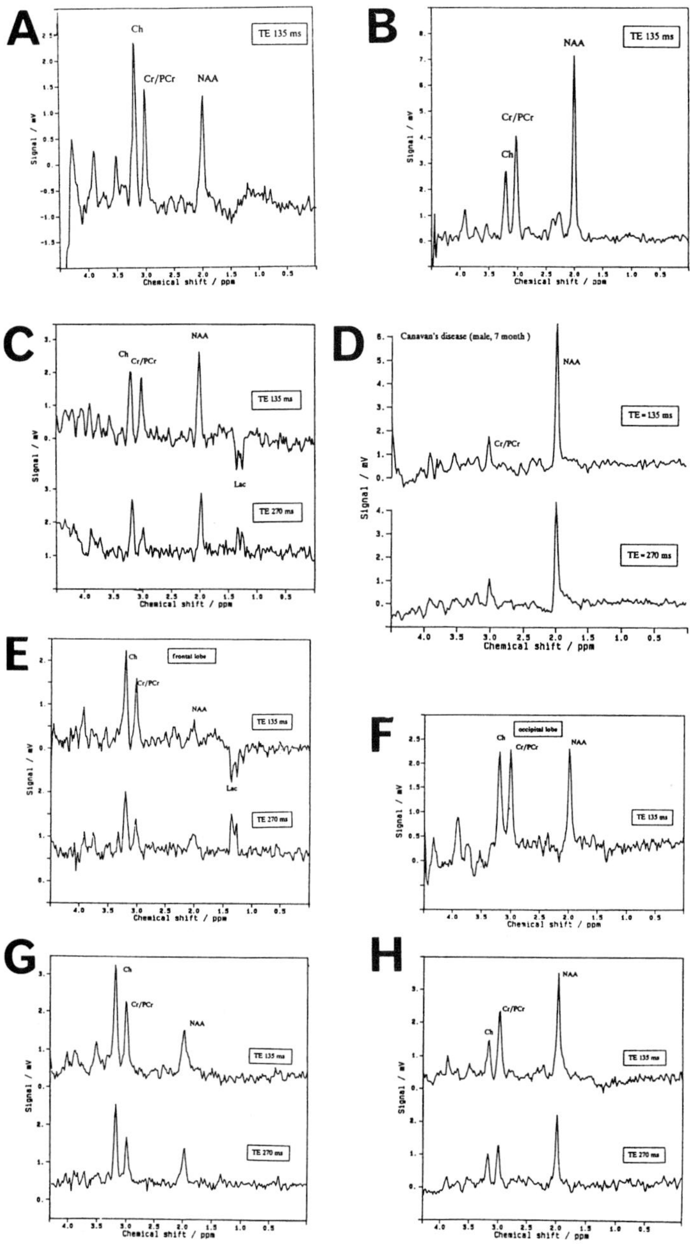

(Panel iv)

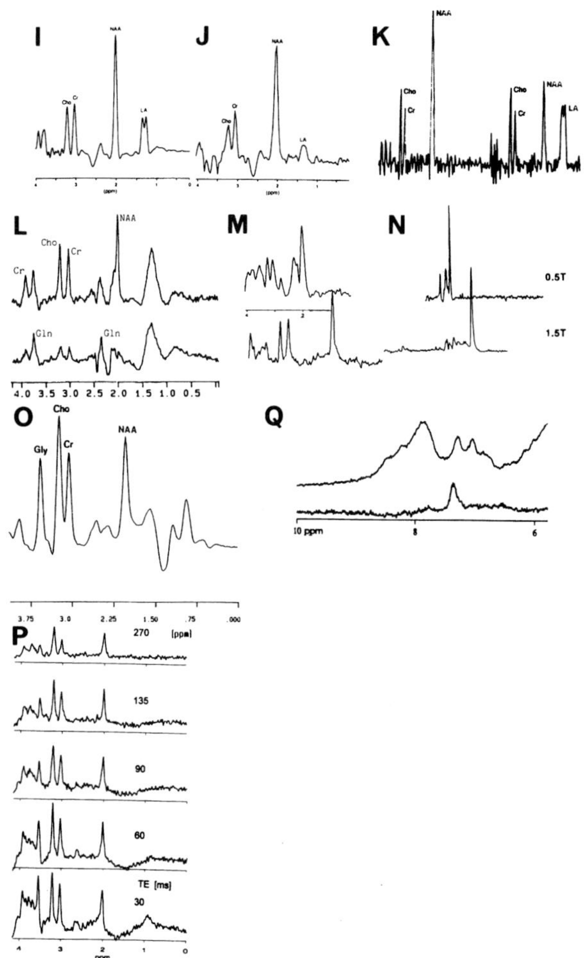

Value of MRS (PROBE) (rating): Primary or secondary.
Other Useful Investigations: U.S.; CT; MRI.
Key Cases: 5.5.5, 5.5.7, and 5.5.8.

F. Cerebral Tumor

Indications: Differential diagnosis of any focal or space-occupying lesions; presurgical planning: postradiation ("normal brain"); ? recurrence.

Contraindications: Postsurgical clips.

Acquisition: STEAM TE = 30 ms in the lesion, with minimal outer VOI contamination plus contralateral control voxel or unaffected region.

Major Findings on ^1H *MRS*: Key issues concern tumor versus stroke; tumor recurrence versus radiation necrosis; edema versus tumor; benign versus malignant. Research reports answer most of these questions, and individual experience indicate widespread enthusiasm for the (use) of

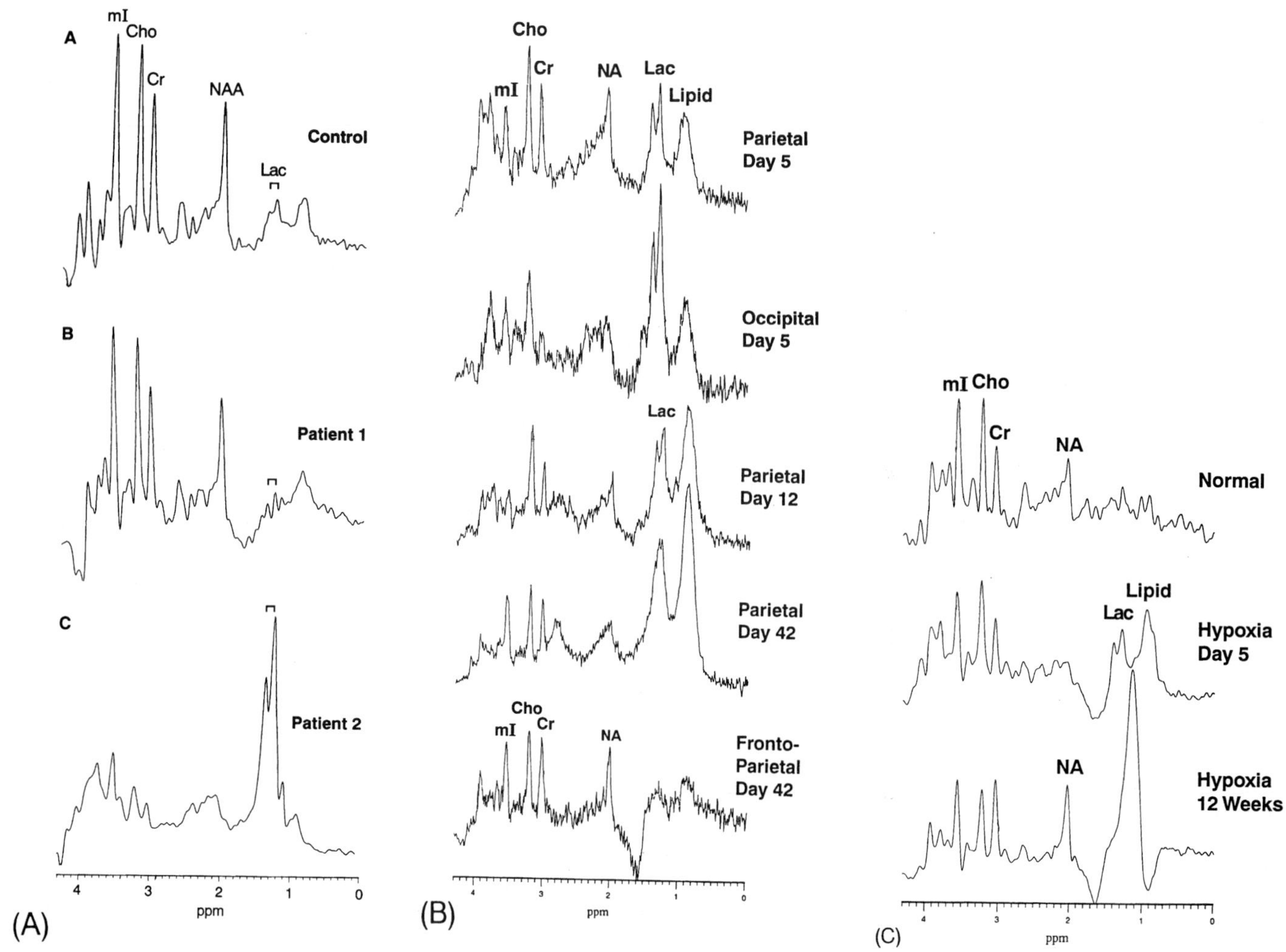

Fig. 6.5 (A) Good prognosis after hypoxia (Patient 1) compared to severe hypoxic encephalopathy with very poor prognosis (Patient 2 = case 5.10.1). (B) Anatomical heterogeneity of hypoxic damage and progression of biochemical damage over time. (C) Hypoxic encephalopathy in prematurity. ^{1}H MR spectra illustrating the paradoxical recovery of cerebral metabolites after severe hypoxic encephalopathy in an infant at 31 weeks gestational age (control = 34–35 weeks).

MRS for these purposes in Clinical Practice, see also Chapter 3 and the clinical cases in Chapter 5.

1. Low or absent NAA
2. Lowered Cr
3. Elevated Cho
4. Lac present or absent
5. Notable excess lipid

Confounding Entities: Heterogeneity, drugs, mannitol.

Sensitivity of MRS: Very high.

Specificity of MRS: The specificity for diagnosis of tumor is variable, due to overlap with other focal lesions. However, in the context of clinical tumor management specificity is high.

Value of MRS (PROBE) (rating): High.

Other Useful Investigations: Chemical Shift Imaging (CSI) for biopsy guidance.

Key Cases: 5.7.3, 5.7.6–12, and 5.10.3.

G. Closed Head Trauma

Indications: Closed head injury, trauma, coma, seizures, mental retardation, motor disorder posttrauma, suspicion of neuronal damage. N.B.: Normal US, CT, MRI, PET, or SPECT are *not contraindications* to ^{1}H MRS.

Contraindications: None (however technical problems due to susceptibility from hemorrhage may arise).

Acquisition: STEAM TE = 30 ms in WM and GM

starting with WM unless strong suspicion of hypoxic injury in which case GM is first priority.

Major Findings on ¹H MRS: 1) Low NAA/Cr and/or low [NAA]; 2) High Cho/Cr and/or high [Cho]; 3) Low [Cr]; 4) Hyperosmolar pattern: increased [mI], [Cho], and [Cr] in WM; 5) SIADH (hypo-osmolar pattern decreased [mI], [Cho] and [Cr]; 6) Accumulation of lactate and lipid. Low [NAA] and [Cr] are generally consistent with neurological deficit, and in combination with accumulation of lactate and lipid, hypoxic damage and poor outcome is indicated. High [Cho] is viewed as a sign of axonal shear injury. The hyperosmolar state (increased mI, Cho and Cr in white matter) persists long after the injury, but is generally a sign of moderate to good outcome. The mechanisms of the hyperosmolar pattern are not known, see Fig. 6.6.

Confounding Entities: Hypoxia.

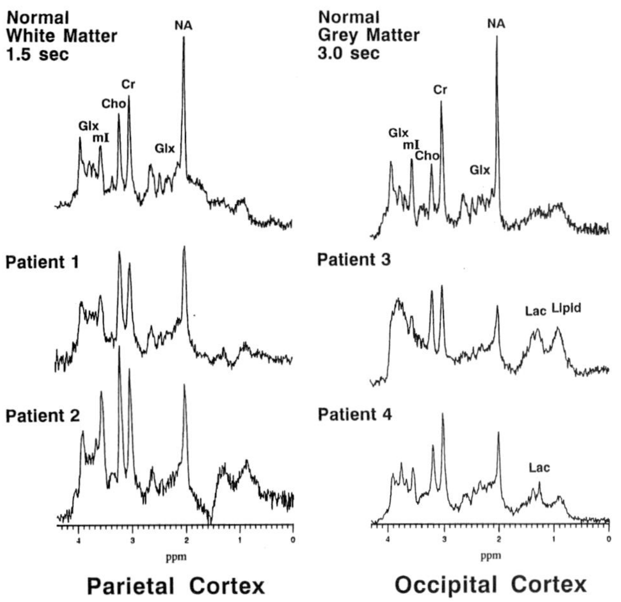

Fig. 6.6 Quantitative ¹H MRS for progression after closed head injury. The location of voxels was selected to represent tissue that is apparently normal on MRI. The exception is patient 3 who had suffered extensive and obvious tissue damage. Patient 1: Elevated cerebral [Cho] and reduced [NAA] in 32 year-old male, unconscious 7 days after head injury. Patient 2: Elevated cerebral [Cho], [Cr] and [mI] with nearly normal [NAA] in a female unconscious 2.5 months. Patient 2 = Case 5.4.2. Patient 3: Reduced [Cr], [NAA] and [mI] with intracerebral lactate and lipid accumulation in 2.5 year-old child unconscious 5 weeks after closed head injury. Patient 4: Reduced [Cr] and [NAA] with lactate (~2.5 mM) in a 7 year-old child on ventilator, 5 days after closed head injury. The top spectrum on each side is the appropriate age-related control spectrum.

Sensitivity of MRS: MRS appears highly sensitive to different patterns of disease in the heterogeneous disease of closed head trauma.

Specificity of MRS: Some overlap with hypoxic encephalopathy. Good indicator of outcome.

Value of MRS (PROBE) (rating): Secondary, Higher in prolonged posttraumatic coma.

Other Useful Investigations: Many.

Key Cases: 5.4.1–9.

H. Shaken Baby Syndrome

Indications: Suspicion of shaken baby syndrome (SBS), coma, seizures, mental retardation, motor disorders, suspicion of neuronal damage. N.B.: Normal US, CT, MRI, PET, or SPECT are *not contraindications* to ¹H MRS.

Contraindications: None (however technical problems due to susceptibility from hemorrhage may arise).

Acquisition: STEAM TE = 30 ms in occipital GM location and (secondary) in the parietal (developing) WM location.

Major Findings on ¹H MRS: Rapidly decreasing [NAA], [Cr], and [mI], increased lactate and "lipid and macromolecules" in cases with bad outcome, and near normal values in cases with good outcome, see Fig. 6.7.

Confounding Entities: Hypoxia, age-related corrections, prematurity.

Sensitivity of MRS: Very reliable for outcome prediction.

Specificity of MRS: Some overlap with hypoxic encephalopathy. Good indicator of outcome. Overlap with trauma caused by other factors than shaking, however the rapid decrease of the primary metabolites appears to be specific for SBS. Thus, most neurological conditions result in slower loss of NAA or Cr, and accumulation of mI and Cho. In SBS, a "catastrophic" metabolic insult, NAA may disappear completely in less than 7 days (see also Chapter 3).

Value of MRS (PROBE) (rating): High.

Other Useful Investigations: Many

Key Cases: SBS cases are not included in Chapter 5 but summarized in Fig. 6.7 and Fig. 3.24 in Chapter 3.

I. HIV CNS Lesions and AIDS Dementia*

Indications: Any HIV patient with changes in neurological function or altered mental status such as memory deficits, progressive cognitive difficulty, seizures, headaches, visual difficult or weakness. N.B. Normal CT, MRI, PET, or SPECT are *not contraindications* to ¹H MRS to evaluate

*Courtesy of Dr. J. Cousins, University Hospital, Albany, New York.

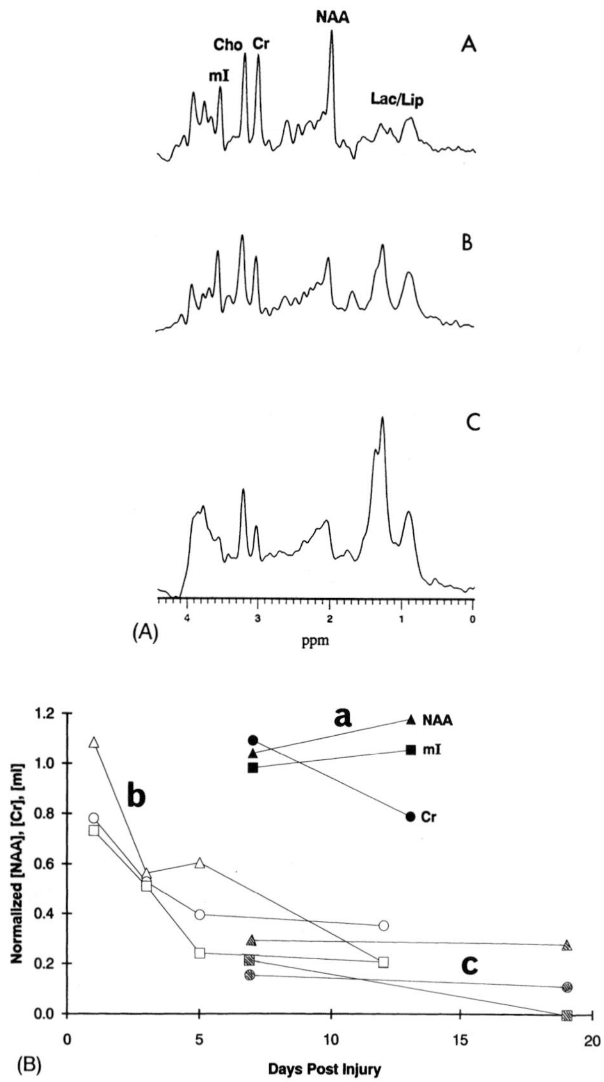

Fig. 6.7 (A) Proton MR spectra from the VOIs in parietal locations in three infants (labeled a, b, and c) with shaken baby syndrome. All three were in coma and on ventilator on the day of MRS. Infant a was 22.5 weeks old, b was 5 weeks old, and c was 31.5 weeks old. The spectrum of infant a at 7 days posttrauma is normal appearing. The spectra from infants b and c at 5 and 7 days posttrauma, respectively, show major abnormalities. Greatly reduced NAA, lower Cr, and the appearance of peaks in the lactate/lipid and lipid regions in infant b. Almost absent NAA, low Cr and mI, and dominant peaks in the lactate/lipid and lipid regions in infant c. (B) Time course of metabolite concentrations after trauma (NAA: Triangles, Cr: Circles, and mI: squares; closed symbols—infant a, open—infant b, and hatched—infant c). The concentrations were normalized to values obtained from aged-matched controls, for further details see Haseler et al. Pediatrics, 99, 1, 1997.

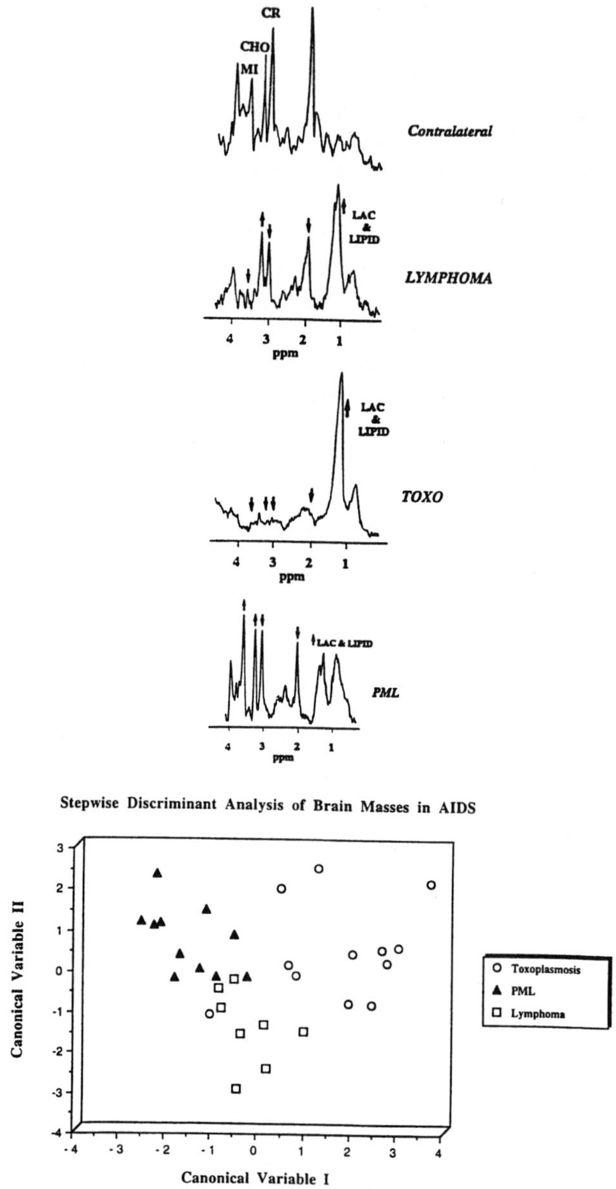

Fig. 6.8 Proton MRS readily distinguishes the differing metabolite profiles of commonly occurring AIDS brain masses. Top two spectra: MRS of a lymphoma in the lateral frontal region and from a contralateral VOI. Middle: MRS of a toxoplasmosis lesion in the parietal region. Second from bottom: MRS of a PML lesions in the subcortical posterior frontal region. Bottom: Scatter plot of discriminant analysis of proton MR spectra of metabolites in toxoplasmosis, lymphoma, and PML. Canonical variable I = 0.155 Cho − 3.08 Cr − 1.03 mI + 0.27 lactate; canonical variable II = −1.6 Cho + 3.15 Cr + 1.59 mI + 0.35 lactate. (Courtesy of Dr. B. L. Miller, Harbor UCLA Medical Center, Los Angeles).

AIDS dementia or HIV encephalopathy. Gadolinium injection for T_1 enhancement of lesions is not contraindicated for CNS lesion studies. Sedation may be required for patients with progressive disease status. Focal lesions smaller than $10 \times 10 \times 10$ mm^3 will be difficult to evaluate. MRI/MRS studies are recommended before stereotactic biopsy procedures.

Contraindications: Patients unable to tolerate an MRI scan for standard reasons (claustrophobia, metallic implants, pacemakers, etc.).

Acquisition: STEAM TE = 30; voxel sampling the CNS lesion (T_2 hyper-intense area or Gd enhancing region), with minimal outervoxel contamination; plus contralateral control voxel, or unaffected region. MRS acquisitions add about 15 minutes to a routine MRI examination.

Major Findings on ^{1}H MRS:

1. Toxoplasmosis (toxo): decreased mI, Cho, Cr and NAA with 15–20 fold increase in lactate/lipids near 1.3 ppm, there is a predominant lipid resonance near 0.9 ppm as well.
2. CNS lymphoma: increased Cho, lactate and lipids, but on the same order of magnitude, decreased mI and NAA and Cr.
3. PML (progressive multifocal leukoencephalopathy): increased mI and Cho; decreased Cr, NAA. Lactate and lipids are observed but at lower concentrations than toxo or lymphoma.
4. Cryptococcoma: decreased mI, Cho, Cr, NAA and increased lipids.
5. AIDS dementia: Decreased NAA/Cr and increased mI/Cr.

Confounding Entities: Drugs, or systemic disorders.

Diagnostic Accuracy: With lambda statistics overall diagnostic accuracy 94%: 100% Toxo, 75% CNS lymphoma, 83% PML, see Fig. 6.8.

Value of MRS (PROBE) (rating): High.

Key Cases: 5.4.9, and 5.7.2–5.

Appendix

List of Abbreviations

^{1}H	Proton
α-Glx	alpha-glutamine and glutamate
AcCoA	Acetyl Coenzyme A
AIDS	Acquired Immune Deficiency Syndrome
AD	Alzheimer's disease
Ala	Alanine
ALD	Adrenoleukodystrophy
ApoE	Apo-lipoprotein E
ATP	Adenosine triphosphate
β,γ-Glx	beta,gamma-glutamine and glutamate
ca.	circa
CHESS	CHEmical Shift Selective
Cho	Choline containing compounds
CNS	Central nervous system
CPM	Central pontine myelinolysis
Cr	Total creatine (creatine and phosphocreatine)
CSF	Cerebrospinal fluid
csi	chemical shift imaging (also seen as CSI or MRSI = magnetic resonance spectroscopic imaging)
CT	Computed Tomography
cyclic AMP	cyclic adenosine monophosphate
DS	Down's syndrome
DWMI	deep white matter infarcts
ECMO	Extracorporeal Membrane Oxygenation
EKG	Electrocardiogram
FDA	Federal Drug Administration
FLD	frontal lobe dementia
fMRI	functional Magnetic Resonance Imaging
GABA	Gamma amino butyric acid
Gd	Gadolinium
Gln	Glutamine
Glu	Glutamate
Glx	glutamine and glutamate
GM	grey matter
GPM	glial plasma membrane
HE	hepatic encephalopathy
HIV+	Human Immunodeficiency Virus Positive
HMRI	Huntington Medical Research Institutes
ICH	Intracranial hemorrhage
Lac	Lactate
Lip	Lipid
MELAS	myopathy, encephalopathy, lactic acidosis with stroke-like lesions
MERRF	Myopathy, encephalopathy with ragged-red fibers
mI	*myo*-Inositol
mM	milli Molar
mmol	millimoles
MR	Magnetic Resonance
MRA	Magnetic Resonance Angiography
MRI	Magnetic Resonance Imaging

MRS	Magnetic Resonance Spectroscopy
MS	multiple sclerosis
μs	microsecond
ms	milliseconds
N/A	not available
NAA	N-acetylaspartate
NAAG	N-acetyl-aspartylglutamate
ND	Near Drowning
NMR	Nuclear Magnetic Resonance
NPM	neuron plasma membrane
OTC	Ornithine transcarbamylase
PCr	Phosphocreatine
PET	Positron Emission Tomography
PML	progressive multifocal leukoencephalopathy
ppm	parts per million
PRESS	Point RESolved Spectroscopy
PROBE	PROton Brain Examination
PROBE-Q	Quantitative PROton Brain Examination
PVS	Persistent Vegetative State
Rf	Radio frequency
S/N	Signal to Noise ratio
SBS	Shaken Baby Syndrome
sI	*Scyllo*-Inositol
SIADH	Syndrome of inappropriate antidiuretic hormone secretion
SPECT	Single positron emission computed tomography
STEAM	STimulated Echo Acquisition Mode
T	Tesla
T_1	Longitudinal relaxation time
T_2	Transverse relaxation time
TCA	Tricarboxylicacid cycle
TE	Echo time
TIA	transient ischemic attack
TIPS	transjugular intrahepatic portasystemic shunt procedure
TM	The intermediate delay time in the STEAM pulse sequence (mixing time)
Toxo	Toxoplasmosis
TR	Repetition time
VOI	Volume of interest
WM	White matter

Index

Abcess, 241, 257
Ac CoA, 29 (*see also* 2-carbon fragments)
Aceto-acetate, 31
Acetone, 31
Acetylcholine, 24, 36
Acidosis, metabolic, 289–292
AD (*see* Alzheimer's disease)
Adaptations, metabolic, 29
Adjustments, prior to MRS, 10, 11, 20
Adolescent, 137
Adrenal insufficiency, 309, 310
Adrenoleukodystrophy, 257, 309–310
Adult
  common MRS indications, 304
  in NAA synthesis, 35
  normal, 46–52, 63–69
Age, corrections for, 12, 13, 14–19, 21, 50–51
Agitation, confusion and, 303
Agonal changes, 173–175
AIDS
  cryptococcoma, 197
  dementia complex, 39, 145, 202–205, 304
  differentiation from trauma, 145, 146
  encephalitis, 205
  lymphoma, 197, 198–199
  masses, 196, 304
  PML, 197, 200–203
  toxoplasmosis, 196–197
Alanine, 6, 24, 25, 207, 234, 248
[Alanine]
  in meningioma, 206–207, 210
  in multiple sclerosis, 243
Alcohol, 6, 25–26, 95, 98, 128 (*see also* Ethanol)
  liver disease and, 168, 297–299
  psychosis and, 113–114
Alexander's disease, 273, 309, 311
  lactate in, 30
Alzheimer's disease, 29, 74–91, 303–304 (*see also* Dementia; possible Alzheimer's disease; Down's syndrome)
  as example of neuronal loss, 38
  myo-Inositol in, 34
  NAA-reduction in, 38
  nomogram, diagnostic, 76, 79, 84, 91, 304, 305
  PROBE in, 87–91, 305
  versus hepatic encephalopathy, 108–110
Amines, 23
Amino acids, 23
  aromatic, 31
γ-Amino-butyric acid, 6, 50, 284 (*see also* GABA)
Ammonia, 37, 262 (*see also* NH_3)
Amyloid, link with *myo*-Inositol, 39
Anoxia, 30
Apgar score, 293
Apnoea, 53, 185, 289
ApoE, 88
Arginine, as source of creatine, 33, 282
Aromatic amino-acids, 31
Arrest, cardiac, 153, 161, 168–169, 182–184, 193

[Arrest, cardiac]
  referral guidelines, 305
Artefacts, 2, 46, 66, 225–226, 229 (*see also* Quality assurance)
  baseline, 230, 249
  eddy-current induced, 149, 259
  Glx, 165
  lipid, outer volume, 97, 188, 254, 265
  movement, 46, 59, 144, 185, 189
  phasing, 215
  spectral resolution, 46
  water suppression, 59–60, 97, 113, 215
Aspartate, 24, 182
Asphyxia, 153, 161 (*see also* Hypoxia)
  referral guidelines, 305
Aspirin, Reyes and, 306
Assignments, 24, 25, 26, 27, 46–50
  chemical shifts, 6
  glycine, 260
  lactate doublet, 136, 178, 182, 183, 197, 198, 210
  pitfalls in, 26
Asthma, 182–184 (*see* Hypoxic encephalopathy)
Astrocyte (*see also* Gliosis)
  marker, 32, 39
  Myo-Inositol, 39
  swelling, 112
ATP, 24, 32, 284
  link with PCr, ADP, hydrogen ion, 32
Atrophy, 71, 164, 166, 183, 191, 285 (*see also* Consolidation; Ventricular dilatation)
  versus hydrocephalus, 285–288
Automation (*see also* PROBE)
  failures, 180
  numerical printout, 217
  prescan, 10–11, 56

Baseline, 47
  "rolling," 62
Biopsy, 206, 232, 240
  avoidance of, 199, 203, 223
  stereotactic, 235
Biosynthesis, 27
Blood–brain barrier, 301
Brachytherapy, 212–223
Brain
  -cell volume, 27, 42
  disease, focal, 176–214
  fetal (*see* Neurotransplant)
  immature, 41
  injury, common MRS patterns of, 41
  metabolite, 5
  water, 54, 175, 281 (*see also under individual disease*; Water)

C_2H_5OH, 26 (*see also* Ethanol)
Canavan's (disease), 30, 175, 273–281, 309, 311
2-carbon fragments, 29 (*see also* Ac Coenzyme A)
Carboxyl, effect on chemical shift, 262
Cardiac failure, referral guidelines, 305
Cases (*see also* Localization)
  guide to reporting and reading clinical, 45–52
  history, 44, 46
  individual clinical, 53–301
  MRS-method, 46
    PRESS, 46
    STEAM, 46
  purpose of MRS, 45
Cell volume, 32
  regulators, 34
Central pontine myelinolysis, 80–82, 118, 119
  lipid droplets in, 31, 118
Cerebellum, metabolites of, 283, 284
Cerebral edema, 116
  palsy, 166
Cerebrospinal fluid (CSF), 38
  contribution of lactate and/or glucose from, 38, 71
  reduced S/N due to, 38
CH_3, 25, 26, 222 (*see also* Methyl)
CH_2-protons, 25, 26, 222
Chemical shift
  axis, 5
  imaging (CSI), 3
  table of, 6
    creatine, 6
    ethanol, 6
    GABA, 6
    glucose, 6, 38
    glutamine/glutamate, 6
    glycine, 6, 263
    lactate, 6, 38
    mannitol, 6, 28
    myo-Inositol, 6
    N-acetylaspartate, 6, 25
    propylene glycol, 6 (*see also* Propane diol)
    scyllo-Inositol, 6
Chemotherapy, 230–231
Children, common MRS indications, 304
Cholangitis, sclerosing, 99
Choline, 6, 15, 21, 25 (*see also* Phosphocholine; Glycerophosphorylcholine)
  anabolism/catabolism, 41
  constituents of, 36–37
  /creatine ratio vs gestational age, 15, 18
  differential diagnosis using, 30–31
  free, 36
  hepatic encephalopathy, 104
  misnomer, 36
  phosphatidyl-NMR-invisible in tumors, 41
  systemic versus local disorders, 38
  trimethylamine, 36–37
  variations, 50
Cirrhosis, 103
Cognition, difficulty with, 303
Coma, 185–193
  Glasgow score, 120, 140, 142, 143, 144, 147

[Coma]
MRS for differential diagnosis, 115, 193
Concentration
corrections for, 13, 20
corrections for, by volume or by mass, 20
normal, 20–21
reduced, in tumors, 212
Confusion, mental, 117, 304
Consolidation, 164, 166 (*see also* Hypoxia)
Contrast-agents vs MRS, 46, 198
enhancement, 300
Controls—healthy, 4
COO (*see* Carboxyl)
Corrections, 13, 20
for compartments, 20
for CSF, 20, 51
for glutamine/NAA, 268, 269
Cortex, 5
occipital, 5
thinning of in hydrocephalus, 43
Coupling, pattern, 25, 46, 301
CPM (*see* Central pontine myelinolysis)
Creatine, 4, 6, 21, 24, 25, 32 (*see also* Phosphocreatine; Total creatine)
as reference, Cr = 1.00, 4, 12, 13, 14–19
deficiency, 32, 282–284
interorgan pathways, 33
kinase, 29, 32
lacking from secondary tumor, 41
pool, 33
significance of, 32
Creatinine, 4, 50
Cryptococcoma, 30–31, 315
Cytomegalovirus, 264–266
Cytoplasm, redox of, 24

D-lactate, 294
D-L propylene glycol, 294
Deep-white matter, 303
Degenerative disease (*see* Neurodegenerative disease)
Dehydration, 173
Dementia, 29, 39, 304 (*see also* Specific disorders)
Alzheimer's, 39
complex of AIDS, 39
Down's, 39
MRS in, 74 passim
multi-infarct, 39
other, 83–85
Pick's disease, 208
plaques and tangles in, 39
specificity and sensitivity of MRS, 85
theory of prevention, 39
Density map, 34
Depression, 78, 86, 303, 304
Development
abnormal, 259 passim, 282, 304
delayed, NAA and, 30, 58, 284

[Development]
normal, 58, 309
Diabetes insipidus, 30, 43 (*see also* Syndrome of inappropriate antidiuretic hormone secretion (SIADH))
Diabetes mellitus, 30, 38, 42, 92, 115–116, 128
confounding entity, 303
effect on choline, 38
Dialysis (renal), effect on choline, 38
Difference-spectroscopy, 268
in phenylketonuria, 309, 311
Differential diagnosis, using MRS, 30–31 (*see also* Individual metabolites)
tumor recurrence from necrosis, 41
Diffuse axonal injury, 126–127 (*see also* Trauma)
DNA, 23
Donnan equilibrium, 27, 32 (*see also* Equilibrium; Gibbs Donnan)
Dopamine, 23
Doublet, 136, 137
β-hydroxybutyrate, 294
lactate, 174, 293 (*see also* Lactate; Assignment; Alanine)
propylene glycol, 293
Down's syndrome, 34, 39
Drowning (*see* Near-drowning)
Drugs, affecting MRS, 295–296

Echo-time (*see also* TE)
effect on MR signal, 8, 27
metabolite identification, and, 27
Ectopic grey matter, 194–195
Eddy-currents (*see* Artefacts)
Edema, 42, 174, 216, 246 (*see also* Brain water)
Elderly, common MRS indications, 304
Electroencephalogram, 305
Electrolytes, 32
Embolism, fat, 187, 193
Encephalitis
acute, 255
AIDS, 255
herpes, 256–258
NAA and, 30
Rasmussen's, 252 passim
Encephalopathy, global, 269 (*see also* Specific disorders)
Energy metabolism, 32
Epilepsy
NAA and, 30
status, 173
Equilibrium, 27, 32
osmotic, 42 (*see also* Gibbs, Gibbs-Donnan; Osmotic; Osmolytes)
Ethanol, 24–26, 297–299
quartet, 298
triplet, 6, 298
Ethanolamine, in phospholipids, 38, 41
Extra-corporeal membrane oxygenator (ECMO), 304

Fat-embolism, 187, 193
Fatty-acid, 24 (*see also* Triglyceride; Lipid)

FDA (Food and Drug Administration), 2
Fetal brain (*see also* Neurotransplant)
  distress, 58
"Floppy baby," 270, 271
Focal lesions, 194 passim
  MRS-strategy, 244
Frequency, rf, 5
  receiver, 5
  transmitter, 5

GA (*see* Gestational-age)
GABA, 6, 37, 50, 284 (*see also* γ amino-butyric acid)
  effect of vigabatrine, 37
Gadolinium (*see* Contrast agent)
Gain
  receiver, 10
  transmitter, 10
Gaussian, 13
Genetic disease, 24
Gestational-age, 14
  cerebral metabolite-ratios, 14–19
  correction for, 274
  metabolite-concentrations, 21
  MRS interpretation, 172
  prematurity, 53–56
  at term, 60
Gibbs, Gibbs Donnan, 32
Glasgow, coma scale (*see* Coma)
Glioblastoma (*see* Tumor)
"Gliosis," 42, 137, 191
Glucose, 6, 13, 24
Glucuronic acid, 28
Glutamate, 6, 24, 25, 35–36
  after near-drowning, 37
  as neurotransmitter, 35
  enzymatic conversion to glutamine, 35, 37
  in synthesis of NAAG, 35
Glutaminase (GS), 37 (*see also* PAG)
Glutamine, 6, 24, 25, 35–36 (*see also* Glx; β,γ-Glx; α-Glx)
  after valproate, 37
  astrocyte marker, 35
  clinical significance, 36–37
  elevation in hepatic encephalopathy, 37
  in hypoxic-ischemic disease, 35, 37, 138
  liver transplant and, 37
  in near-drowning, 35
  in OTC deficiency, 37
  in Reyes syndrome, 37
  synthetase, 37
Glx
  α, 37, 50, 268
  β,γ, 35, 37, 49, 268–269
  definition of, 24, 25, 105
  difference spectroscopy of, 269
  estimation of peak-intensity, 49–50
Glycerol, 24
Glycerophosphorylcholine, 24, 36, 50
Glycine, 6, 31, 262, 263 (*see also* "*myo*-Inositol"; "Choline")
  versus mI, 259–263
Glycolysis, 24
Guanidinoacetate, 33 (*see also* Creatine)
  methyl transferase deficiency, 33, 284, 309, 310

HE (*see* Hepatic encephalopathy)
Head-injury (*see* Trauma)
Heart disease, 147
  congenital, 170–172
Hemangioblastoma (*see* Tumor)
Hemiplegia, 231
Hemorrhage, 128, 176
  contraindication to MRS, 180
Hepatic encephalopathy, 29, 30, 38, 42, 92 passim (*see also* SCHE)
  acute, 30
  dementia and, 94, 110
  diagnosis, 304–307
  early detection, 98
  effect on choline, 38
  ethanol and, 297–299
  MRS and severity, 92, 307
  MRS-definition, 98
  psychosis, rule out, 113–114
  recovery, 101, 105
  referral guidelines, 304–305
  Reyes syndrome and, 306–307
  subclinical, 97, 169, 306–307
  transplantation and, 93, 307
  typical PROBE-spectrum in, 29
  versus Alzheimer's disease, 108–110
Herpes II, 30
HIV-infection, 194, 196–205, 304, 313–315
  AIDS dementia, 145–146, 204–205
  effect on cerebral-choline, 38
Homeostasis, 32
Hunter's angle, 47, 69 (*see* Reading a spectrum)
Hydrocephalus, 30, 285, 304
  arrested, 288
  shunt surgery for, 285
  versus metabolic disease, 285–288
Hyperammonemia, 267 (*see also* Reyes syndrome; Ammonia)
Hyperglycinemia, 259–263 (*see also* Glycine; Ketones)
Hypernatremia, 173, 304
Hyperosmolar, 30, 84
  coma, 115
  diabetes mellitus, 115–116
  posttrauma, 123–125, 145
Hypertension, 128, 303
  cerebral, 300
Hyperventilation, 30
Hypo-osmolar, 138
Hyponatremia, 42, 73, 80, 117–119, 304 (*see also* Central pontine myelinolysis)
  versus hepatic encephalopathy, 114, 305
Hypotonia, 283

Hypoventilation, 30
Hypoxia, 30, 39, 158, 304
  consolidation after, 164
  encephalopathy, 139, 152, 161–167, 171, 186, 266, 293–296
  energy failure in, 39
  focal, 176 passim, 191
  global, 147–175
  injury, 58
  ischemic cascade, 39
  MRS-visible lipids, 39
  perinatal, 264 passim
  recovered, Glx in, 30
  referral guidelines, 305–306, 308
  sensitivity of neurones to, 163
  time-course of MRS changes, 308
  timing of MRS in, 148

ICH, lactate and, 30
Inappropriate ADH secretion (SIADH), 30, 118, 121, 122, 138
Inborn-error, 30, 259 passim, 282–288, 304, 306, 309–311
Indications (medical), 2
Infants (*see* Newborn)
Infarct, 176 passim
Infection, indication for MRS, 304
Inflammatory disease, 242 passim, 246–247
  MRS-lipid in, 31
Injury, neuronal, 38
Inositol polyphosphate, 24, 34
Inositols, 24, 25 (*see also myo*-Inositol; *scyllo*-Inositol)
  NMR-invisible in phospholipids, 38
Institutional units, 52 (*see also* Quantitation)
Interpretation, of spectra, 46–52, 72–73 (*see also* Spectra, reading)
Intra-cerebral hemorrhage, 30 (*see* ICH)
Irradiation, effect on cholines, 38 (*see also* Radiation)

J-coupling, 37 (*see also* Coupling)
  of glutamate, 37

Kayexalate, propanediol in, 296
Kern-Sayer syndrome, 309, 311
Ketone bodies, 50, 115
  effect of β,γ-Glx region of spectrum, 50
  hyperglycinemia, nonketotic, 262
Korsakoff, 113 (*see also* Psychosis)
Krebs cycle, 24, 29 (*see also* Tricarboxylic acid cycle)
"Kumoscopy," 1

L-lactate, 294
Lactacidosis, 115, 292
Lactate, 6, 24, 25, 32, 174
  cerebral energy failure and, 174
  cerebrospinal fluid, 138
  D versus L, 296
  differential diagnosis with, 30, 286, 287
  doublet assignment, 6, 136, 178, 182, 183, 197, 198, 210
  effect of TR, 54
[Lactate]
  and lipid after trauma, 312, 313
  outcome with, 167
  quartet, 173
  sources of, 33, 138
  versus glutamine, 138
Leigh's syndrome, 270, 309, 310, 311
Linewidth, water, 211
Lipid (*see also* Triglyceride)
  clinical significance of, 36
  in shaken baby syndrome, 31, 37
  infantile pattern, 141, 151
  infection and, 36
  inflammation and, 36
  intracerebral, 164, 165
  juvenile (*see* infantile pattern)
  loss at long-TE, 277
  multiple sclerosis plaque, 243
  necrosis and, 36, 209
  nomenclature, 62
  outer volume (out of phase), 97, 123, 264–266
  resonances vs low molecular weight proteins, 47, 62
  tumor necrosis, 40
  use in differential diagnosis, 31
Lipolysis, hypoxia and, 39
Liver disease (*see also* Hepatic encephalopathy; Transplantation; TIPS)
  effect on choline, 38
  MRS in, 42, 92–114, 304
  TIPS in, 42
  urea synthesis, 42
Liver transplantation, 99 passim, 306 (*see also* Hepatic encephalopathy; Liver disease)
Localization, 3, 6
Localizer, 2 (*see also* Specific localization sequences STEAM; PRESS, CSI)
  lymphoma, lactate and lipid in, 30

Macromolecules, 24, 185, 284
Macrophages, 118
Magnetic resonance imaging (MRI), 1, 5
Mannitol, 6, 26, 31, 293
  in tumor, 300–301
Markers, 3
Mass, undiagnosed, 233
MELAS, 31, 309, 311
Meningioma, 206, 210
MERRF, 209, 311
  regional variations in, 311
Metabolic error, 282 (*see also* Inborn error)
Metabolite, 5
  brain function and, 29
  clinical significance, 23–44
  concentrations, 4, 12, 20, 21
  ratios, and location, 13
    and age, 13
Metachromatic leukodystrophy, 309, 310

Methyl-protons, 24, 25, 26 (*see also* CH_3)
Minimental score, 72, 74, 200
Mitochondria, 29
Mitochondrial encephalopathy, 270–272, 288, 292 (*see also* Leigh's syndrome, MERRF, MELAS)
Model-solutions, 27 (*see also* Individual metabolites)
Motion, artefacts due to, 46 (*see also* Artefacts)
MRS-markers, 3
  astrocyte, glial, 3, 24, 32
    myo-Inositol as, 34
  confounding entities, 303, 306, 310
  contraindications, 303
  energy, 3, 24
  neuronal, 3, 32, 34
MRS-patterns in diagnosis, 38–43
  edema and atrophy, 42
  hypoxic-ischemic disease, 39
  necrosis, gliosis and brain injury, 41
  neurodegenerative disease, 38
  neuronal injury, 38
  of referral, 303–315
  systemic disease, 42–43
  tumors, 39–41
MS (*see* Multiple sclerosis)
Multifocal leukoencephalopathy, 200, 201–202 (*see also* PML; AIDS)
Multiple sclerosis, 242–251
  myo-Inositol/Cr, 250
  NAA and, 30
Mycobacterium, 241
Myelin, myelination, 12, 24, 251
  choline-constituents of, 36
  delayed, 278
  MR-invisible, 36
  myo-Inositol in, 34
  precursors, 42
Myelinolysis, 118 (*see also* Central pontine)
myo-Inositol, 6, 16, 19, 21, 24, 25, 27, 34
  chemical structure, 34
  /creatine ratio vs age, 16, 19
  depletion, 80, 112
  diabetes mellitus, 303
  differential diagnosis using, 30–31
  fluctuations in, 34
  hepatic encephalopathy and, 81, 104, 110
  hyponatremia and, 81
  multiple-sclerosis and, 250
  reactions involving, 27
  second peak, 50
  uremia, 303
  versus Glycine, 259–263
Myopathy, 270–272, 282 (*see also* MELAS)

N-acetylaspartate (NAA), 6, 14, 21, 24, 25, 27 (*see also* NA)
  absent, 291
  adolescent, 137
  as osmolyte, 36, 116, 125
[N-acetylaspartate (NAA)]
  biosynthesis, pathways, 35
  concentration, 275
  correction for gestational age, 292
    examples, 274–280
    onset of, 56
  correction for glutamine, 269
  /creatine ratio vs age, 14, 17
  deacylase deficiency (Canavan's disease), 273–281
  diabetes and, 36
  functions of, 34
  identification of, 47
  in axon, 34
  recovery vs resynthesis of, 36, 166
  reduction in, 127 (*see also other individual cases for examples*)
  second-peak, 49
  third peak, 49–50
  transport of, 34, 35
N-acetylaspartylglutamate, 34, 35 (*see also* NAAG)
NA (alternative naming of the "NAA-peak" at 2.02 ppm), 308, 309, 312, 313
NAA, 27
NAD, NADH, 24
NAAG, 34, 35
ND (near-drowning), 29, 30, 147 passim, 304
  increased Glx in, 37
  glutamine versus glutamate anomaly, 37
  nomogram, outcome after, 308
  PROBE spectrum in, 29
Near-drowning (*see* ND)
Necrosis, 265 (*see also* Lipid; Tumor)
  lipid in, 209
  versus tumor recurrence, 220
Neonate
  encephalopathy, 273
  myo-Inositol in, 30
Neoplasm
  low grade and *myo*-Inositol, 30
  NAA and, 30
Neurochemicals, neurochemistry, 5, 23, 24
Neurodegenerative disease, 74–91 (*see also* Alzheimer's disease; Dementia)
Neuronal marker, 24 (*see also* Markers)
Neuropsychological tests, 108
Neurospectroscopy, 23–24
Neurotransplant, 36, 41
Newborn, indications for MRS, 304
NH_3, 37, 262 (*see also* Ammonia)
Noise (from the switched magnetic fields), 2
Nomogram (*see also* Alzheimer's disease)
  AIDS masses, 199, 314
  Alzheimer's disease (*see* Nomogram; Dementia)
  dementia, 76, 79, 84, 91
  drowning, 308
  hypoxic encephalopathy, 308
  shaken-baby syndrome, 314

Nonketotic hyperglycinemia, 259–263, 309, 311 (*see also* Glycine; Ketone bodies)
Norepinephrine, 24
Normal
  adult, 13, 20, 46–52, 63–69, 309
  baby, 14–19, 21, 53–60
  child, 61–62
  grey matter, 51–52, 63–69, 309
  newborn, 53–60
  premature, 53–57
  white matter, 46–51, 67–69, 309
Nucleotides, 24

Oligodendrocyte, -glia 36, 39
  NAA in precursor cells, 36
Ornithine transcarbamylase deficiency (*see* OTC)
Osmolyte, 27, 32
  abnormalities in MRS, 92, 115–120
  brain-cell volume and, 27
  choline as, 38
  creatine as, 32
  depletion in HE, 96
  hyperosmolar response, head-injury and, 30
  idiogenic, 42
  myo-Inositol as, 34
OTC, 30, 31, 309, 311 (*see also* Ornithine transcarbamylase deficiency)
Outcome, 140, 141, 142, 143 (*see also* Prognosis)
  after arrest, 184
  after near-drowning, 151, 160, 308
  encephalopathy, 269
  hypoxia and, 308, 312
  lactate and, 167, 290
  late-, 264
  NAA and, 184, 263, 290
  neonatal, 312
  premature infant, 312
  shaken baby syndrome, 313, 314
  trauma, 313
Outer-volume lipid, 97, 123, 264–266
Oxidative phosphorylation, 29

PAG, 37
Parkinson's disease, 23
Partial volume, 9, 234, 238
Patient
  positioning, 49
  referral guidelines, 303–315
    abnormal development, 306
    cerebral tumor, 307
    closed head trauma, 311
    dementia, 303
    hepatic encephalopathy, 304
    HIV, AIDS, 314
    hypoxic encephalopathy, 305
    neonatal hypoxia, 307
    shaken baby syndrome, 313
Peak
  difference between STEAM and PRESS, 46
  height vs area, 5–6
  -identification, 46
  -ratios, 4, 46, 50–52
Pediatrics, MRS in, 259–292
Pelizaeus-Mersbacher disease, 309, 311
Perinatal hypoxia, 53
Periventricular leukomalacia, hypoxic encephalopathy and, 39
PET, 3, 24 (*see also* Positron emission tomography)
pH, 262
Phenolic, 241
Phenylketonuria, 309, 311
  phenylalanine in, 31
Phosphocholine, phosphorylcholine, 24, 36, 50
Phospholipids, 24, 38
  NMR-visible vs invisible, 38
Phosphocreatine, 24, 25 (*see also* Total creatine)
  deficiency, 282
Physics, 5
Pick's disease, 208, 304 (*see also* Dementia)
Pitfalls, 2, 26
  elevated NAA, due to Glx, 269
  outer-volume lipid, 36
Plaque
  active, 243
  multiple sclerosis, 242, 245
  quiescent, 243
  vs tumor, 245–251
PML (*see* Progressive multifocal leukoencephalopathy)
Pontine myelinolysis, 80 (*see also* Central pontine myelinolysis (CPM))
  hyponatremia and, 80
Portal hypertension, 111
Positron-emission tomography, 24, 108, 304, 305, 306, 310, 312, 313 (*see also* PET)
Possible Alzheimer's disease, 30 (*see also* AD; Alzheimer's disease)
Premature
  developmental delay, 58
  infant, 57
  reversed "Hunter's angle" in, 57–58
PRESS, 4, 7
PROBE, 1, 2, 46 (*see also* Automation; PROBE-Q)
  normal adult, 63–64, 65–66, 67–69
  normal baby, 59–60
  normal child, 61–62
  normal elderly, 70–71
  proton brain exam (GE™), 1
PROBE-Q, 46, 86, 180–181, 194, 195, 217–218, 233–234
  in Alzheimer's diagnosis, 86
  failure in, 222
  P (PRESS), 46
  S (STEAM), 46, 60, 64
Processing
  post-in PROBE Q, 68
Prognosis, 52 (*see also* individual cases)

Progressive multifocal leukoencephalopathy, 197, 200–203, 314, 315
lactate in, 30
Propane diol, 6, 31 (*see also* Propylene glycol)
Propylene glycol, 6, 31, 293–296
Proton spectra (*see* Spectra)
Protons, magnetically equivalent, 5
Pseudodementia, 303
Psychosis, Korsakoff-Wernicke, 113 (*see also* Alcohol)
Putamen, 12
Pyridine nucleotide, 24 (*see also* NAD, NADH)
Pyruvate, 24, 32
Pyruvate dehydrogenase, 32, 33

Quality assurance, 46–47, 211, 226, 238
patient positioning, 47, 49
peak identification and, 46–52
resolution, 178, 211
resolution sufficient for lactate assignment, 136, 178, 182, 183, 197
Quantitation, 4, 11–22, 50–51 (*see also* individual cases)
by peak-ratio, 4
creatine, 225
Quartet, 298

Radiation, MRS-changes, 227
Rasmussen's encephalitis, 252–255
Reading a spectrum, 46–50 (*see also* Hunter's angle; Spectra)
Receptor, targeted-ligands, 24
Reciprocity, 13
Redox-state, 3, 33 (*see also* NAD, NADH)
Reference, external (*see* Calibration)
Referral (*see* Patient-referral)
Rehydration, 116
Relaxation times, 4, 278 (*see also* T_1; T_2)
Renal failure, 88, 128, 304
myo-Inositol in, 30
Resolution
spatial, 238
sufficient for lactate assignment, 136, 178, 182, 183, 197
Reyes syndrome, 42, 259, 267–269
fulminant hepatic failure in, 305

Saturation, 8
Scaling
methods, 12
peak intensity, 222, 266
SCHE, 97, 169, 306–307
scyllo-Inositol, 6, 25, 50
increased, 276
scyllo-inosose, 28 (*see also* Inositols)
Sedatives, 55
barbiturate, effect on MRS, 55
Sepsis, 173
Serine-phospholipids, 38
Serotonin, 24
Shaken baby syndrome, 304 (*see also* Trauma)
Shim, shimming, 5, 10
Shunt surgery (*see* Hydrocephalus)
SIADH, 30, 118 (*see also* Inappropriate ADH secretion)
trauma and, 121, 122, 138
Signal-to-noise (S/N), 9, 10, 38
impact of CSF in VOI on, 38, 71
reduced, 218
Signal averaging, 8
Single-volume, 3 (*see also* Volume; Voxel)
localization, 3
philosophy of, 3
Single photon, 3, 304, 305, 306, 312, 313 (*see* SPECT)
Slice-selection, 6
SPECT, 3, 304, 305, 306, 310, 312, 313
Spectra
adult, normal, 63–64
Alzheimer's, negative, 78
central pontine myelinolysis, 81
child, normal, 62
elderly, normal, 70–73
gestational age, 53
grey matter, 68
hypoxic encephalopathy, 152
infant, normal at term, 60
interpretation and analysis, 50–51
normal, 53, 54
premature infant, 53, 55
reading from right to left, 46–50, 72–73
visual cortex, 192, 193
visual inspection, 63
white-matter, 68
Spectral resolution, blood and, 178
Spectroscopy, difference, 26
Spin-coupling, 7, 8
Splitting (*see* Doublet)
Sodium, 122
STEAM, 4, 7
Stroke, 30, 176, 303, 304
referral guidelines for, 305
Sugars, 23
Syndrome of inappropriate ADH secretion (*see* SIADH)
Systemic disease, 92–119
differential diagnosis and, 73
impact on MRS, 82

T_1
correction for metabolite T_1, 278
correction in infants, 55
quantitation correction, 13
weighting, in MRS, 55
T_2
correction for metabolite T_2, 278
correction for residual short T_2 resonances, 284
quantitation correction, 13
Tau-protein, 88
Taurine, 25, 27
TE, 28

[TE]
effect on peak ratios, 51, 106, 260–261
long, 261, 277
Thallamus, 285, 286
Therapeutic monitoring, with MRS, 212
radiation, 220
TIA (*see* Transient ischemic attack)
Time
echo, 7 (*see also* TE)
relaxation, 8 (*see also* T_1, T_2)
repetition, 8 (*see also* TR)
TIPS, 42, 96
TM, 51
Total creatine, 6, 25, 50
Toxoplasmosis, 30, 314, 315
TR, 51
Transient ischemic attack (TIA), 303, 304
Transjugular intrahepatic portasystemic shunt, 42 (*see* TIPS)
MRS, pre- and post-, 103
Transplantation
cardiac, 170, 304
liver, 99–102, 103–107, 299, 304
"choline-overshoot," 102
patient-selection, 107
neural, 41
Trauma
closed head, 30, 120–146, 304
diffuse axonal injury in, 120, 126, 127
Glasgow coma score, 120
hyperosmolar state after, 120, 123–125, 127
hypoxia and, 120
long-term effects of, 139
outcome, 126, 140, 142
patterns, MRS, 124–125
SIADH and, 120
Tricarboxylic acid cycle, 24 (*see also* Krebs cycle)
Triglyceride, MR-visible, 36, 118, 119, 187, 188
clinical significance, 36
Tryptophan, 24
Tuberculosis, 239–241, 256
Tumor, 194, 206–238
biopsy, specific lipids in, 40, 41
craniopharyngioma, 40
cysts, 40
glioblastoma, 212–223, 231
brachytherapy in, 212 passim
glioma, 40, 233–236
hemangioblastoma, 237–238
heterogeneous anatomy, 39
heterogeneous lactate distribution, 40
meningioma, 40, 206–207
[Tumor]
alanine in, 40, 206
necrosis, 227
neurochemical changes, 39–41
postsurgical, 224
progression, 221
quantitative MRS of, 39, 40
recurrence vs necrosis, 221, 224, 227
secondary, 40
tuberculoma, 194
tuberous sclerosis, 41
versus infarct, 178, 180–181

Uremia, 303

Valproate, toxicity of, 42
Vegetative state, 184
Ventricle, 164
Ventricular dilatation (*see also* Consolidation)
atrophy, 71, 164, 166, 183, 191, 285
atrophy versus hydrocephalus, 285–288
Viral encephalitis, 267–269 (*see also* Reye's syndrome)
Volume, cell volume markers, 3
Volume of interest, 46 (*see also* Artefacts; Outer volume lipid)
dementia diagnosis, 77
imperfections in prescribing, 65
landmarks, 65
occipital-grey matter, 46, 77
parietal white matter, 46
prescribing, 46, 65
Voxel, 3 (*see also* Volume-of-interest)
-size, effect on signal to noise, 9

Water (*see* Edema)
concentration, 51, 150
increased, 275
pure, as reference, 279, 280, 281
suppression, 2, 5, 10–11
artefact, 215
excessive, 289
ideal, 87
reference, error from, 281
residual, 10, 129
White matter disease, 273–281 (*see also* Adrenoleukodystrophy; Canavan's disease)

Xenobiotics, 23, 31, 293–301 (*see individual metabolites*)

Zellweger's syndrome, 309, 310
Zeugmatography, 1